PENGUIN REFERENCE BOOKS

THE PENGUIN DICTIONARY OF BRITISH NATURAL HISTORY

Richard Fitter was born in London in 1913 and educated at Eastbourne College and the London School of Economics. He was assistant editor of the *Countryman* from 1946 to 1959 and subsequently Director of the Intelligence Unit in the Council for Nature until 1963. He has been Hon. Secretary of the Fauna Preservation Society since 1964 and in 1977 became a Trustee of the British National Appeal of the World Wildlife Fund.

Richard Fitter is the author of twenty-one books on various aspects of natural history, especially birds and wild flowers, including *London's Natural History* in the New Naturalist Series, and *The Birds of Britain and Europe*. He is a member of many societies for the conservation of wildlife, and has travelled widely in connection with saving the world's wildlife. His wife edits the Fauna Preservation Society's journal, *Oryx*, and they have both played an active part in the work of the Berkshire, Buckinghamshire and Oxfordshire Naturalists' Trust, of which she is currently Chairman and he Past President. They live in the middle of a beech wood in the Chilterns and have three children and seven grandchildren.

THE PENGUIN DICTIONARY OF

British Natural History

RICHARD FITTER
assisted by
MAISIE FITTER

ALLEN LANE

ALLEN LANE
Penguin Books Ltd
17 Grosvenor Gardens
London SW1W 0BD

First published by Penguin Books 1967
Reprinted with revisions 1978
Published in Allen Lane 1978

ISBN 0 7139 1135 2

Set in Monotype Times
Printed in Great Britain by
Richard Clay (The Chaucer Press), Ltd,
Bungay, Suffolk

ACKNOWLEDGEMENTS

We are grateful to John Clegg, Alastair Fitter and John Sankey, who have read all or part of the proofs and eliminated many of the minor errors that inevitably creep into a compilation of this nature, to Professor W. H. Thorpe, F.R.S., for allowing us to quote his definitions of animal behaviour terms, and to Brian Armitage who ably prepared the index to scientific names. We shall also be grateful if any readers who notice other errors will write to tell us about them.

R.F.

Chinnor, August 1966 M.F.

THE SCOPE OF THE BOOK

IN this dictionary of the natural history of the British Isles the term 'natural history' is interpreted in the broadest sense to include all living things and natural phenomena of the earth and its atmosphere. Because it is restricted to the British Isles, including the whole of Ireland, the Isle of Man and the Channel Isles, the adjective 'British' is usually to be understood as qualifying all statements; it is only occasionally inserted where a definition without it would be manifest nonsense.

The dictionary is designed for the use of field naturalists and country lovers and for young people and students in the field. The bulk of the entries therefore consist of the names of animals and plants. It is not primarily concerned with biology (for which, see the excellent *Penguin Dictionary of Biology*), but a number of ecological, ethological and genetical terms in general use by field naturalists have been included. As far as possible, words relating to the internal organs of animals and plants have been omitted, but many words relating to their external organs are included. In some cases this means that only one meaning is given for a word that has several meanings in biology as a whole, e.g. Follicle.

A selection of entries deal with various physical, meteorological, geological and geographical phenomena of the natural environment, but a line has been drawn between geology, pedology and mineralogy. This means that the names of some rocks commonly used by field naturalists are given, but none of minerals, e.g. quartzite but not quartz. Another line is drawn between the earth's atmosphere and outer space, astronomical phenomena being omitted.

So far as possible all animals and plants with a vernacular name in general use have been included, together with a few local or regional vernacular names known to be in current use, but most English names that are purely book names are omitted. There are no alphabetical entries of scientific names below the rank of family, with a very few exceptions where the scientific name is also the accepted vernacular term. A name ending in -id refers to a species from a similar family ending in -idae, e.g. Muscid, Muscidae. In the higher reaches of both botanical and zoological classification, all phyla and classes are included, with an arbitrary selection of the more important orders and families. In brackets after each entry of an organism or grouping of organisms are two names leading the reader back to the phylum to which it belongs.

If the entry is an English name these two names will normally be the scientific name of the organism or group preceded by the Order (in animals and lower plants) or the Family (in flowering plants); e.g. Adder (Serpentes: *Vipera berus*), Adderstongue (Ophioglossaceae: *Ophioglossum*). If, on the other hand, the entry is a Latin name, the brackets will normally contain the two preceding steps in the classification; e.g. Acari (Arthropoda:

Arachnida), Acari being an Order in the Class Arachnida in the Phylum Arthropoda. A class name in Latin or a phylum name in English may have only one bracketed name following.

Zoological classification follows Rothschild's *A Classification of Living Animals* (1961). Botanical classification, on which there appears to be no agreement, follows that in use by the Botany Department of the University of Oxford. We are grateful to the late Dr E. F. Warburg for making this available.

If an organism is mentioned in the text unaccompanied by its scientific name it will have a separate entry of its own. Any name or word in the body of the text not followed by a scientific name is given a capital initial if it has a separate entry of its own which is important to the understanding of the entry in question.

The word 'related' is used to indicate another species belonging to the same family as the main species mentioned in the entry. With a few exceptions, entries are given under the main name if there are only a few species with that name, e.g. Bladder, Moss, Red, Sea and White Campions are all included under Campion. Where a name covers many species, however, as with Beetle, Moth, or Grass, they are entered separately under their adjectival name, e.g. Cabbage Moth, Cabinet Beetle, Canary Grass.

LIST OF ABBREVIATIONS

adj. = adjective	sp. = species (singular)
agg. = aggregate	spp. = species (plural)
pl. = plural	var. = variety
⇨ = see	⇨⇨ = see also

A

Aaron's Rod. ⇨ Common MULLEIN.

Abdomen. The rear section of an arthropod's body, with one, or in insects two (head and thorax) sections in front; also the part of a vertebrate's body containing the digestive organs.

Abele. ⇨ White POPLAR.

Aberdevine. Bird fanciers' name for the siskin.

Acacia, False. Locust Tree (Leguminosae: *Robinia pseudacacia*). A tall deciduous tree from North America, widely planted in the south and sometimes more or less naturalized on light soils.

Acanthocephala. A small phylum of endoparasitic worms, known as Spiny-headed or Thorny-headed Worms, chiefly parasitic on aquatic vertebrates and using crustaceans as intermediate hosts, though some species which have adapted to terrestrial vertebrates use insects as intermediates. They are named from the hooked spines which cover the proboscis with which they fasten on to their host. Six of the nine British avian acanthocephalids parasitize water birds.

Acari (Arthropoda: Arachnida). The sub-class or order containing the Mites and Ticks, small or very small arachnids, many of them parasitic, some practising Phoresy, with a life history containing several larval or nymphal instars. There are at least 1,700 British species, grouped in five orders or sub-orders, of which four consist of mites and one of ticks.

Acaricide. A pesticide that kills mites and ticks.

Acarine. Relating to mites or ticks.

Acarology. The scientific study of mites and ticks.

Accentor (Passeriformes: *Prunella*). Small song birds. Alpine Accentor *P. collaris*, a rare straggler. Hedge Accentor, ⇨ HEDGESPARROW.

Acclimatization. The adaptation of alien animals and plants to living in the open in another country, though not yet to reproducing themselves or feeding naturally throughout the year, especially in winter, e.g. in Britain the Australian Budgerigar *Melopsittacus undulatus* will breed in the open, but must be fed and sheltered in winter.

Accrescent. In botanical description, of an organ, especially a calyx, that becomes larger after flowering.

Achene. A dry fruit containing one seed and not splitting open. ⇨⇨ NUT, SAMARA.

Acicle. The slenderest prickle of a bramble, often no more than a stiff bristle.

Acid Soil. ⇨ SOIL.

Aconite, Winter (Ranunculaceae: *Eranthis hyemalis*). A low-growing garden plant with yellow flowers appearing before the leaves in January and February; sometimes well naturalized in plantations.

Acorn. The fruit of the oak tree.

Acorn Worm (Hemichordata: Enteropneusta). A very primitive class of worm-like chordates, of which several species occur in sand and gravel on our shores.

Actaeon Shell (Tectibranchiata: *Actaeon tornatilis*). An aberrant tectibranch sea slug with a reddish winkle-like shell, frequent on sandy shores.

Actinomorphic. Radially symmetrical, e.g. the shape of sea-anemones and many flowers.

Aculeata (Hymenoptera: Apocrita). The group of hymenopterous families containing 531 species of bees (Apidae), ants (Formicidae) and wasps (numerous families), characterized by their stinging adults.

Acuminate. ⇨ LEAF-SHAPES.

Adaptation. The process of selection which enables an individual animal or plant to survive more easily in its particular environment, and so gives it a better chance of leaving offspring than an individual not so adapted. ⇨ NATURAL SELECTION.

Adder or Viper (Serpentes: *Vipera berus*). The only British poisonous snake, with venom capable of killing a man; viviparous and characterized usually by a dark zigzag stripe down the back; widespread and frequent in Great Britain.

Adderstongue (Ophioglossaceae: *Ophioglossum*). Two species of curious small fern, with the spores in a narrow plantain-like spike rising from the middle of a leaf, *O. vulgatum* being much the commoner, widespread but local in grassy places.

Adephaga (Insecta: Coleoptera). The sub-order of beetles containing the carnivorous beetles, in five families: Carabidae (tiger and ground beetles), Dytiscidae (water beetles), Haliplidae (small water beetles) and Hygrobiidae (screech beetle).

Adnate. In botanical description, of an organ joined to a different kind of organ, e.g. a leaf to a stem.

Adventive or Casual. An alien or introduced plant growing unaided by direct human intervention, i.e. not planted, but not yet permanently established, when it becomes a denizen or colonist.

Aerobe. Organisms, especially bacteria, which need oxygen.

Agar. A culture medium for bacteria prepared from seaweeds.

Agaric (Basidiomycetes: Agaricaceae). Typical toadstools, including Fly Agaric or Bug Agaric *Amanita muscaria*, one of the best known poisonous fungi, with a bright red, white-spotted cap, common in birch and pine woods and formerly used as an insecticide; and Verdigris Agaric *Stropharia aeruginosa*, a slimy greenish toadstool common in grassy places. Velvet-stemmed Agaric. ⇨ WINTER FUNGUS.

Agonistic. Term used in animal behaviour studies to describe aggressive behaviour.

Agrimony (Rosaceae: *Agrimonia*). Two yellow-flowered grassland plants, *A. eupatoria* being the commoner. The unrelated Hemp Agrimony (Compositae: *Eupatorium cannabinum*), with pinkish-purple flowers, is common by ditches and fresh water and in damp woods, especially in the South.

Alate. Winged; used of aphids and other insects with both winged and wingless forms.

Albatross (Procellariiformes: Diomedeidae). A family of long-winged sea-birds, mainly in the southern hemisphere. On about half a dozen occasions one has been seen in British waters, or storm-blown inland; when identifiable it has proved to be the Black-browed Albatross *Diomedea melanophrys*.

Albinism (Adj. albino, albinistic). The total or partial absence or suppression of the normal coloured pigments in an animal, due to the absence of the enzyme tyrosinase. In a pure albino there are no coloured pigments in any external part of the body, the legs and eye of a bird, for instance, appearing pinkish or yellowish-white. Most bird albinos, however, are partial, showing white patches on various parts of the plumage. ⇨ LEUCISM.

Alcyonarian (Anthozoa; Alcyonaceae). The order of soft corals to which Dead Men's Fingers belongs.

Alder (Betulaceae: *Alnus glutinosa*). Medium to tall deciduous catkin-bearing, waterside tree; widespread and sometimes forming small woods known as carr in swamps and fens.

Alder Fly (Megaloptera: *Sialis*). Two species of insect with smoky-brown wings, found near fresh water; the larvae are aquatic.

Ale-Hoof. Old name for ground-ivy.

Alevin. The young salmon until it absorbs its yolk-sac, when it becomes a parr.

Alexanders (Umbelliferae: *Smyrnium olusatrum*). An early-flowering yellow umbellifer with glossy leaves, formerly used as a pot-herb, and widely naturalized on banks and waste ground near the sea.

Alfalfa. American name for lucerne.

Algae. A large assemblage of lower plants, formerly regarded as a single group, but now usually classified in eight separate divisions, or phyla, including the blue-green algae (Cyanophyta), green algae (Chlorophyta), brown algae (Phaeophyta), red algae (Rhodophyta), and diatoms and their allies (Chrysophyta). Marine algae are commonly known as seaweeds. At one time the algae were classed together with the Flagellata as Protophyta, and indeed two groups of flagellates are today classed by botanists as phyla, or major divisions, of the lower plants, Euglenophyta and Pyrrophyta, while zoologists rank them as the sub-class Phytomastigina of the protozoan class Mastigophora. ⇨ BLANKETWEED.

Algal. Relating to algae.

Algivorous. Feeding on algae.

Algology. The study of algae.

Alien. An animal or plant whose native range lies outside the British Isles. ⇨ INTRODUCED SPECIES.

Alison. Two cruciferous plants: the now very rare arable weed Small Alison *Alyssum alyssoides*, and the not infrequent garden escape Sweet Alison *Lobularia maritima*.

Alkaline Soil. ⇨ SOIL.

Alkanet (Boraginaceae). Several blue-flowered plants of the Borage Family, mainly garden escapes, but Green Alkanet *Pentaglottis sempervirens* may be native in the West.

Alleles or Allelomorphs. Pairs or sets of genes on homologous chromosomes (i.e. the corresponding chromosomes derived from each parent), which control particular characters; if the pairs are similar in relation to these characters they are said to be homozygous; if dissimilar heterozygous.

Allen's Rule. The tendency for the relative size of exposed portions of the body of a warm-blooded species, such as limbs, tail and ears, to decrease with a fall in mean temperature, i.e. going northwards in the northern hemisphere.

All-Heal. ⇨ Common VALERIAN.

Allometric. Adjective used of organisms which grow at different rates.

Allopatric. Adjective used of different species or subspecies with areas of distribution that do not overlap. ⇨ SYMPATRIC.

Allopolyploid. An interspecific hybrid, which has doubled its chromosomes and so is able to breed; most hybrids are sterile. ⇨ POLYPLOIDY.

All-Seed. Name for several tiny plants, including Flax-seed (Linaceae) *Radiola linoides*, and Four-leaved All-seed (Caryophyllaceae) *Polycarpon tetraphyllum.*

Alluvium. Fine particles of soil or rock washed down by rain or river water and deposited in a valley or estuary. Alluvial soils are highly fertile.

Alternation of Generations. The alternation in the life cycles of certain animals and plants, e.g. gall wasps, jellyfish, ferns, of a generation reproducing sexually with a generation reproducing asexually. The two forms, and in the case of gall-wasps their galls, are often very different from one another; some have even originally been classified as different species or genera. In Spermatophytes the alternate generation is suppressed.

Altocumulus, Altostratus. ⇨ Cloud.

Amaryllidaceae (Angiospermae: Monocotyledones). The family of bulbous plants including snowdrops, snowflakes, daffodils, and, in some classifications, garlics.

Amber. Fossilized yellow or orange resin from prehistoric coniferous forests in the Baltic Sea area, washed across the North Sea to the beaches of eastern England, from Yorkshire to Essex and especially between Cromer and Felixstowe, where diligent search may still turn up nuggets up to the size of a hazel nut. Amber is soft, and burns with a yellow flame. When rubbed in cloth it can generate enough electricity to attract small scraps of paper. Electricity is named from the Greek for amber *electron.*

Ambergris. An aromatic waxy substance, produced in the intestines of the sperm whale, and sometimes found floating in the sea or cast up on shore. It is prized for fixing and intensifying perfumes in making cosmetics.

Amber Snail (Pulmonata: *Succinea*). Four, mainly yellowish, land snails frequenting damp places, the commonest being *S. putris.*

Ambrosia Beetle (Rhynchophora: *Trypodendron*). Beetles closely related to the bark beetles, whose larvae burrow in the wood instead of in or just under the bark.

American Element in the British Fauna and Flora. A good many bird species exist in both Europe and North America, often in differing subspecies, e.g. peregrine, pintail, moorhen. A few non-European birds are blown across the Atlantic on migration every year, mostly waders, and especially the Pectoral Sandpiper *Calidris melanotos*. Most of the North American plant species growing in Britain have been introduced by man, e.g. Canadian fleabane, but a few exist naturally on both sides of the Atlantic, e.g. pipewort which has a relict European distribution in Scotland and Ireland.

Ammonite. Fossil animals which lived in spirally curled shells reminiscent of the ramshorned Egyptian god Ammon, or of coiled snakes, whence their popular name of snakestone; not infrequent in Jurassic deposits, especially in the cliffs at Whitby, Yorks.

Amoeba (Protozoa: Rhizopoda). The best known of all protozoans, a minute irregularly shaped blob of protoplasm, just visible to the naked eye, reproducing itself by dividing in two; several species occur on the mud of ponds and ditches.

Amphibian (Vertebrata: Amphibia). A class of 'cold-blooded' vertebrates, whose temperature varies according to that of their surroundings; they are distinguished from the reptiles by their softer skin and need to resort to water to breed, because the young pass through a gill-breathing tadpole stage. Includes frogs and toads (Salientia or Anura) and newts (Caudata or Urodela). There are seven native British species.

Amphineura, Loricata or Polyplacophora. Chitons or coat-of-mail shells.

Amphipod (Malacostraca: Amphipoda). An order of small shrimp-like crustaceans, mainly marine, but including some fresh and brackish water species; typical amphipods are flattened laterally, carry their bodies somewhat curved, and progress by jumping. ⇨ FRESHWATER SHRIMP, SANDHOPPER, SAND SHRIMP, SKELETON SHRIMP.

Anacanthini (Vertebrata: Pisces). The order of bony fishes containing the Cod Family (Gadidae) and the Hake.

Anadromous. Adjective used of fishes that spend part of their lives at sea, but ascend rivers to spawn.

Anaerobe. Organisms, mainly bacteria, which can grow in the absence of free oxygen.

Anastomosing. In botanical description, of veins on a leaf when each loops round to join the next one and so form a network.

Anatidae (Aves: Anseriformes). A family of aquatic birds, consisting of the ducks, geese and swans, and characterized by their webbed feet, usually broad bills, and long or longish necks.

Anchovy (Clupeidae: *Engraulis encrasicholus*). A smallish marine fish, with snout longer than lower jaw; it spawns off the Dutch coast and migrates annually through the English Channel to the Atlantic.

Anemone. For Beadlet, Dahlia, Daisy, Gem, Opelet, Plumose, Snakelocks, and Wartlet Anemones, ⇨ SEA-ANEMONE.

Anemone, Wood, or Windflower (Ranunculaceae: *Anemone nemorosa*). A widespread and common small white flower carpeting woods in spring.

Anemophily. The pollination of plants by the wind.

Angiosperm (Spermatophyta: Angiospermae). The flowering plants, the larger and more advanced sub-division of the seed-plants reproducing by fruits and divided into Monocotyledons and Dicotyledons.

Angel Fish. ⇨ MONKFISH.

Angelica, Wild (Umbelliferae: *Angelica sylvestris*). A tall, stout white or pink-flowered umbellifer of damp woods and meadows. The sweetmeat is made from the stems of the alien *A. archangelica*, which has yellow-green flowers, and is well established along the banks of the Thames near London, and elsewhere.

Angler (Pisces: *Lophius piscatorius*). A singular fish with large head and small body; the front ray of the dorsal fin is extended into a long 'fishing-rod', which can be held in front of the capacious mouth to lure smaller fishes inside; widespread on our coasts. Its many other names include angel fish, monkfish, frogfish, and fishing frog.

Angler's Curse. Anglers' name for both dun and spinner of mayflies of the genus *Caenis*.

Animalcule. A general term for tiny, usually microscopic, animals. Bell and Slipper Animalcules ⇨ INFUSORIAN; Wheel Animalcule ⇨ ROTIFER.

Animal Kingdom. One of the two great traditional divisions of the living world, the other being the Vegetable Kingdom. The higher forms of animals and plants are easy enough to separate, animals being generally (though by no means always throughout their life history) mobile and feeding either on plants or on other animals, and plants (except for the slime fungi) immobile and feeding mainly on nutriment extracted from the soil, though a few groups are parasitic on other plants, and a very few even carnivorous. Some lowly groups, notably the flagellate protozoa, are claimed by both zoologists and botanists, and perhaps should really be allocated to a third kingdom. Though 'wild animal' is restricted in common usage to mammals, the term 'animal' strictly speaking applies to the whole range of the animal kingdom, from elephants and ostriches to fleas and protozoa, and includes man.

Annelida. The phylum containing the segmented, ringed or bristle worms, with bodies divided into rings or segments, each self-contained, with its own set of internal organs and protected by a tough cuticle, often armed with bristles. They are divided into five classes; the marine Polychaeta, Myzostomaria and Archannelida, the freshwater Hirudinea (leeches) and the Oligochaeta, which include the terrestrial earthworms as well as many freshwater forms. The Polychaeta, Myzostomaria and Oligochaeta were formerly included in one large class, Chaetopoda. Most annelids are hermaphrodites, and have the capacity for regenerating lost organs, such as tails, or even heads. There are about 670 known British species.

Annual. A plant which completes its life cycle in one year: i.e. grows from seed, flowers, fruits, and dies.

Anoplura (Insecta: Phthiraptera). The sub-order containing the sucking lice.

Anseriformes (Vertebrata: Aves). An order of aquatic birds consisting of the family Anatidae, containing the ducks, geese, and swans.

Ant (Aculeata: Formicidae). A family of highly organized social insects with thirty British species living in mostly underground nests, with fertile females (queens), sterile females (workers) and winged males. The queens are winged also at first, but after the mass emergence of winged males and females, usually in late summer, and consequent pairing, they tear off their wings prior to founding a new colony. The so-called ants' eggs fed to fish are not eggs but cocoons. The common ants of gardens and grassland are the Common Red Ant *Myrmica rubra*, the Yellow Lawn Ant *Acanthomyops mixtus*, and the Black Lawn Ant *A. niger*; these and other species of ants are known as Dairying Ants because they farm and milk aphides for their honeydew, both above and below ground. Several species, including the Common Guest Ant *Leptothorax acervorum*, and the Workerless Ant *Anergates atratulus* live in the nests of other ants. Other notable species include: the Fossil Ant *Ponera coarctata*, so named because it is hardly distinguishable from ants found in Baltic amber thirty million years old; Pharaoh's Ant *Monomorium pharaonis*, a tiny reddish-yellow alien ant which infests buildings; the Slave-making Ant *Formica sanguinea*, which captures cocoons of the large Black Ant *F. fusca* in 'slave raids' on their nests, and uses the resultant workers as 'slaves'; the Thief Ant *Solenopsis fugax*, the smallest British ant, so named because it lives in other ants' nests and steals their food; the large Wood Ant, Horse Ant or Hill Ant *Formica rufa*, which makes large anthills up to 3–4 ft high in woods; and the Yellow Hill Ant *Acanthomyops flavus*, which makes the anthills common in many grassy places. Velvet-ants are wasps.

Antenna (pl. Antennae). The feeler of which every insect has two on its head, forming its organs of smell and touch.

Anther. The pollen-producing part of a flower's male organ, borne at the tip of the stamen.

Anthocyanin. Red, blue, and purple pigments in the sap of plants, which produce some of the more spectacular autumn colour of dying leaves and also tinge some young shoots and some whole plants red or purple.

Anthozoa (Cnidaria). A class of coelenterate marine animals containing the sea-anemones and corals, all of which spend much the greater part of their lives in the sedentary zooid stage, attached to rocks, stones, seaweeds and other firm surfaces.

Anthropogenic Factor. ⇨ ECOSYSTEM.

Antibiotic. A substance secreted or generated by a living organism, especially bacteria and fungi, which inhibits or kills other species. Penicillin, produced by the mould *Penicillium notatum*, is one of the best known.

Anticyclone. A high-pressure weather system, in which the air flows in a clockwise direction, producing fine warm sunny weather in summer, and often very cold sunny weather in winter. ⇨ DEPRESSION.

Antlers. Bony growths from the skull in deer (males only except in reindeer), and covered with a thin skin called velvet which dies when the antlers are full grown and is rubbed off by the deer on trees and bushes (fraying).

Topography of an Antler

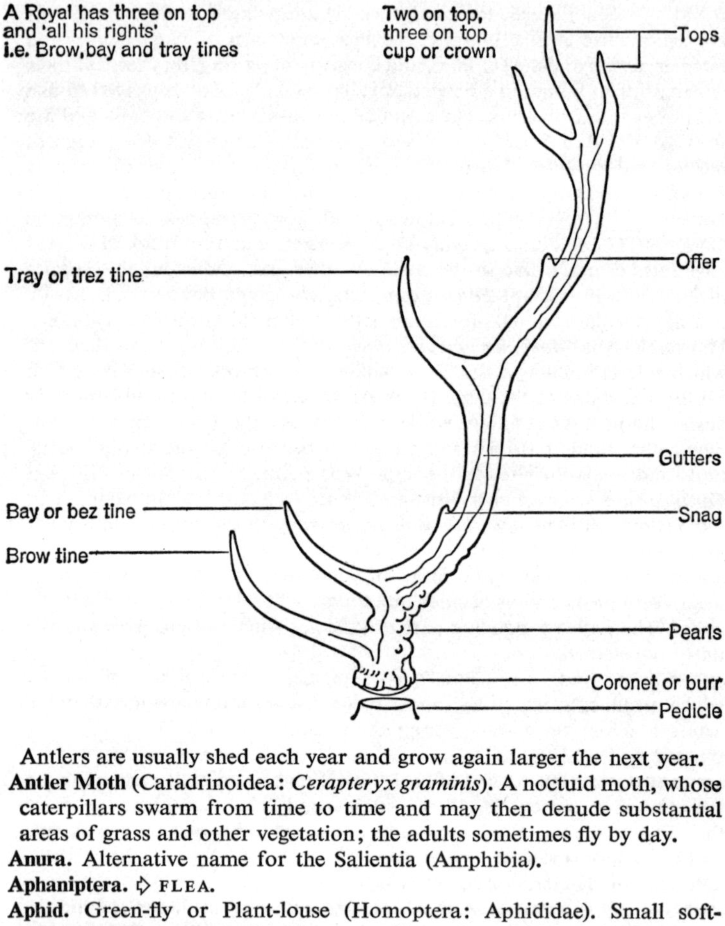

Antlers are usually shed each year and grow again larger the next year.

Antler Moth (Caradrinoidea: *Cerapteryx graminis*). A noctuid moth, whose caterpillars swarm from time to time and may then denude substantial areas of grass and other vegetation; the adults sometimes fly by day.

Anura. Alternative name for the Salientia (Amphibia).

Aphaniptera. ⇨ FLEA.

Aphid. Green-fly or Plant-louse (Homoptera: Aphididae). Small soft-skinned, often green, plant-bugs, sucking plant juices and often congregating in large numbers on both wild and garden plants; some species are milked by ants for their Honey-dew. They have a remarkable life history: winged and wingless generations alternate through the summer, reproducing by virgin birth, till the autumn, when males are at last produced; these mate with the females and produce fertilized eggs that overwinter to start the asexual reproduction again in the spring. Many aphids are regarded as pests, including the Green-fly *Siphocoryne rosarum* of garden roses, the Black-fly *Aphis fabae* of broad beans, which overwinters in the egg on the spindle-tree, and the alien Woolly Aphis or American Blight *Eriosoma lanigerum*, common in neglected apple orchards, which produces a waxy substance looking like cotton-wool.

Apiculate. Of a leaf with a small broad point at the apex.

Apocrita (Insecta: Hymenoptera). Much the larger of the two sub-orders of the Hymenoptera, with adults narrowed at the waist and ovipositors not serrated; subdivided into the Parasitica (ichneumons and other parasitic wasps), and Aculeata (bees, wasps, and ants).

Apodes (Vertebrata: Pisces). An order of elongated bony fishes including the eels.

Apodiformes (Vertebrata: Aves). The order of birds which includes the swifts (Apodidae).

Apomixis (adj. Apomictic). Self-fertilization or reproduction by a seed not fertilized from outside, as in the dandelions and hawkweeds, resulting in the production of a Clone.

Aposematic Coloration. ⇨ WARNING COLORATION.

Appetitive Behaviour. In animal behaviour, the variable introductory phase of an instinctive behaviour pattern or sequence. (W. H. Thorpe).

Apple. ⇨ CRAB APPLE, OAK-APPLE.

Appressed. In botanical description, an organ pressed close to another but not joined to it, e.g. sepals to petals.

Arachnida. A class of arthropods distinguished from the insects by the possession of four pairs of walking legs, and comprising the spiders (Araneae), harvestmen (Opiliones), false-scorpions (Pseudoscorpiones), and mites and ticks (Acari). Arachnida have six pairs of limbs, of which the first pair form the jaws (chelicerae), the second the palps; the remaining four are used for locomotion. Only the spiders have their body obviously divided into two parts. There are over 1,500 known British arachnids.

Arachnology. The scientific study of Arachnida, and more especially of spiders and harvestmen.

Araneae. Spiders.

Araneology. The scientific study of spiders.

Archaean. ⇨ PRE-CAMBRIAN.

Arctic-Alpine Plants. Are common to both alpine and arctic regions, both normally snow-covered in winter.

Arcuate. In botanical description markedly curved.

Ardeiformes (Vertebrata: Aves). The order of mostly large, long-legged marsh birds that includes the herons, egrets, bitterns (Ardeidae), storks (Ciconiidae), spoonbills and ibises (Plataleidae or Threskiornithidae). Only two species, the grey heron and the bittern, breed in Britain, but several others are more or less frequent visitors.

Arenaceous. Sandy.

Argentine (Pisces: *Argentina*). Two species of deep water fish, allied to the Salmonidae: sometimes caught by trawlers; the smaller *A. sphyraena* is the more frequent. Sheppy Argentine ⇨ PEARL-SIDE.

Aril. Fleshy cup surrounding the seed in a few plants, notably the yew, often brightly coloured to attract birds and other potential distributors of the seed.

Aristate. Awned.

Ark Shell. ⇨ NOAH'S ARK SHELL.

Arrow-Grass (Juncaginaceae: *Triglochin*). Two widespread and frequent, somewhat plantain-like monocotyledons, named from the shape of their fruits when they burst: Sea Arrow-grass *T. maritima* in saltmarshes and grassy places near the sea; Marsh Arrow-grass *T. palustris* in damp grassy places inland.

Arrowhead (Alismataceae: *Sagittaria sagittifolia*). Tall aquatic monocotyledon with arrow-shaped leaves and purple-spotted white flowers; frequent in England.

Arrow Worm (Chaetognatha). A small group of carnivorous, transparent, marine worm-like animals, isolated in a phylum of their own. They are among the commonest members of the oceanic plankton; two species of *Sagitta* are occasionally, and *Spadella cephaloptera* commonly found in rock pools on our coasts.

Artemis Shell (Lamellibranchia: *Dosinia*). Two species of widespread and common marine bivalves resembling rounder smooth cockles: Rayed Artemis *D. exoleta*, and Smooth Artemis *D. lupinus*.

Arthropod. The Arthropoda form the most numerous phylum of the animal kingdom, containing more than 85 per cent of all animal species in the world. They have an external skeleton of hardened cuticle, divided into more or less distinct rings or segments, some of which bear jointed limbs. There are twelve classes, of which the most important are the Diplopoda (millipedes), Chilopoda (centipedes), Insecta (insects), Crustacea (crustaceans), and Arachnida (spiders and mites).

Artiodactyla (Vertebrata: Mammalia). One of the two orders of herbivorous hoofed mammals or ungulates, whose native wild representatives in Britain are the deer. Among introduced ungulates are the wild goat and several more species of deer. There are no truly wild cattle, sheep or horses in the British Isles.

Asarabacca (Aristolochiaceae: *Asarum europaeum*). A creeping plant with dull purple, bell-shaped flowers, naturalized in a few widely scattered plantations and hedge-banks.

Aschelminthes. The invertebrate phylum which includes the rotifers or wheel animalcules (Rotifera), the bristle-backs (Gastrotricha), the Priapulida, formerly included in the Gephyrea, the horsehair worms (Nematomorpha), and the roundworms (Nematoda). The two latter classes are sometimes also grouped together as Nemathelminthes.

Ascidian. ⇨ SEA-SQUIRT.

Ascomycetes. A large class of Fungi, sometimes known as Sac Fungi because they carry their spores in a sac or ascus, and ranging in size from very small to quite large; the fruiting bodies of the larger species are often tuberous or rounded, not toadstool-shaped. Many species are the cause of plant diseases; others 'ripen' cheeses.

Ash (Oleaceae: *Fraxinus excelsior*). A tall deciduous tree with pinnate leaves, especially common on calcareous soils, where it may form woods; also a frequent constituent of oakwoods; fruits are known as keys. Mountain Ash, ⇨ ROWAN.

Asilid. A family of robber flies.

Asparagus, Wild (Liliaceae: *Asparagus officinalis*). The native plant is a prostrate maritime form in the south, west, and south Wales, but the upright cultivated form often escapes on to roadsides and waste places. Bath Asparagus is a local name for spiked star of Bethlehem; the young spikes used to be sold as asparagus in Bath market.

Aspen (Salicaceae: *Populus tremula*). A widespread and common small tree, catkin-bearing and deciduous like other poplars, with almost circular leaves borne on stalks so thin that they tremble in every breeze.

Asphodel. Neither of the two British plants so named is closely related to the true asphodel, though both also belong to the Lily Family. Bog Asphodel *Narthecium ossifragum* with golden-yellow flowers is common on wet heaths and moors and in bogs, mainly in the north and west. Scottish Asphodel *Tofieldia pusilla*, with greenish-yellow flowers, is almost confined to the Highlands.

Assassin Bug (Heteroptera: Reduviidae). Rather elongated predatory bugs; Heath Assassin Bug *Coranus subapterus* is the commonest species, found on heaths and dunes. ⇨⇨ FLYBUG, THREAD-LEGGED BUG.

Assassin Fly. ⇨ ROBBER FLY.

Aster, Sea (Compositae: *Aster tripolium*). A common saltmarsh plant, whose daisy-like flowers sometimes lack their purple ray-florets and look more like feeble dandelions. The garden aster, which occasionally escapes on to waste ground but is never really naturalized, is the Chinese Aster *Callistephus chinensis*.

Asteroidea. ⇨ STARFISH.

Astigmata (Arachnida: Acari). An order or sub-order of mites, including feeders on fungi, vegetable detritus and stored products, and vertebrate parasites. ⇨⇨ BULB, CHEESE, FEATHER, FLOUR, FUR, ITCH and SUGAR MITES, and MANGE.

Attire. The antlers of an adult male deer.

Auk (Charadriiformes: Alcidae). Medium to small diving sea-birds with black and white plumage. Four British breeding species: razorbill, puffin and the two guillemots. The Great Auk *Alca impennis*, a former breeder, became completely extinct in the 1840's. The Little Auk *Plautus alle* is an occasional winter visitor; large numbers are sometimes storm-blown inland in 'wrecks'.

Auricle. Small ear-like downward extension at the base of a leaf, especially in grasses.

Aurora Borealis or Northern Lights. Spectacular display of coloured lights in the northern sky at night, caused by streams of charged particles from the sun approaching the earth, and producing an electric current, which causes the rarefied gases in the upper atmosphere to glow in the same way that an electric current passing through a gas-filled neon tube causes the neon gas to glow; often associated with flares on the surface of the sun.

Autecology. The ecology of individual species of animals or plants, as distinct from Synecology.

Autotomy. The capacity to lose a portion of the body, such as a limb or the tip of the tail, and grow it again, as with lizards' tails and lobsters' claws.

Avens, Mountain (Rosaceae: *Dryas octopetala*). A small white-flowered alpine plant, local on mountains north from Snowdonia, and on lower ground in Scotland and Ireland.

Avens, Water (Rosaceae: *Geum rivale*). Widespread orange-pink flower of damp shady places, especially in the north and west.

Aves. ⇨ BIRD.

Avifauna. The bird fauna of a region.

Avocet (Charadriiformes: *Recurvirostra avosetta*). A large black and white wader, with a strikingly upcurved bill. Protected since 1947 on Havergate Island, Suffolk, for many years its only British breeding colony, by the Royal Society for the Protection of Birds; now also breeds at Minsmere, another R.S.P.B. sanctuary in Suffolk; some winter in south Devon; occasional passage migrant elsewhere.

Awlwort (Cruciferae: *Subularia aquatica*). An uncommon white-flowered plant of lake and tarn bottoms in the north.

Awn. A small bristle at the tip of a flower or fruit, especially in the grasses, or of a leaf. ⇨ p. 118.

Axil. The angle where a leaf or leaf-stalk joins the stem. The adjective axillary also applies to the arm-pits of vertebrates, especially birds.

Azotobacter. One of the nitrogen-fixing bacteria in the soil.

B

Bacillariophyta. ⇨ DIATOM.

Bacillus. A general term for rod-shaped bacteria; also restricted to a particular genus.

Back-Cross. The cross of a hybrid with one of its parents, or between a Heterozygote and a Homozygote.

Backswimmer. ⇨ WATER-BOATMAN.

Bacon Beetle (Diversicornia: *Dermestes lardarius*). A pest of bacon, hides and fatty foodstuffs, also known as the Larder or Lard Beetle.

Bacteria (Bacteriophyta: Schizomycetes). The most primitive phylum of plants, sometimes regarded as belonging to a separate kingdom; minute one-celled organisms which are among the smallest known organisms, mostly shaped as rods from 1/50,000 to 1/5000 of an inch long. It has been calculated that 250,000 bacteria could congregate on the full stop at the end of this sentence. Countless millions of them inhabit the soil, the water, air and the bodies of living things. Some, commonly known as 'germs' or 'microbes', are responsible for human diseases, such as typhoid fever. Others perform essential functions in maintaining the fertility of the soil and facilitating the decomposition of organic matter. Soil bacteria may be aerobes needing the oxygen in the air, or anaerobes, able to complete their life cycle without oxygen. Among the most important bacteria are those which fix nitrogen from the atmosphere and so introduce it into the soil and the food chain. These are often to be found on the roots of peaflowers and other plants, where they form tiny nodules which may be seen by the naked eye. ⇨⇨ AZOTOBACTER, BACILLUS, IRON BACTERIA, SEWAGE FUNGUS.

Bacteriology. The study of bacteria.

Bacteriophage. ⇨ PHAGE.

Bacteriophyta. The phylum or major division of the vegetable kingdom containing the bacteria.

Badger (Carnivora: *Meles meles*). The largest British mammalian carnivore, with striking black and white striped face, nocturnal and living in holes in the ground (sets); widespread and quite common.

Baker's Brat. ⇨ FIRE-BRAT.

Balance of Nature. A state of equilibrium or normal ecological succession, which is liable to be disturbed by human interference, either direct, by killing organisms, or indirect, by creation or destruction of Habitat or pollution of the environment.

Baleen or Whalebone. A fibrous horny material in the upper jaw of the baleen whales (Mystacoceti). ⇨ WHALE.

Balm (Labiatae: *Melissa officinalis*). A white-flowered cultivated herb, which sometimes escapes. Bastard Balm (L: *Melittis melissophyllum*), our showiest labiate, has large pink flowers; very local, chiefly in the south-west.

Balsam (Balsaminaceae: *Impatiens*). Plants with spurred, two-lipped flowers, of which our one native species, Yellow Balsam or Touch-me-not *I. noli-tangere* is much less common (very local in damp woods, mainly in the Lake District) than our three introduced species: Himalayan Balsam or Policeman's Helmet *I. glandulifera*, with large pink flowers, now widespread by fresh water; Orange Balsam or Jewel-weed *I. capensis*, spreading by fresh water in the south; and Small Balsam *I. parviflora*, with pale yellow flowers, frequent in waste places.

Banding. The American term for bird ringing.

Baneberry or Herb Christopher (Ranunculaceae: *Actaea spicata*). A somewhat foetid white-flowered, black-berried plant of limestone in northern England.

Barbastelle (Chiroptera: *Barbastella barbastellus*). An uncommon species of bat, confined to England and Wales.

Barbed-Wire Plant. ⇨ Spiny COCKLEBUR.

Barbel. (1) A fleshy filament hanging from a fish's mouth. (2) (Cyprinidae: *Barbus barbus*). A freshwater fish named from its four barbels, although the carp also shares this characteristic; local in rivers in England, especially the Thames and Trent.

Barberry (Berberidaceae: *Berberis vulgaris*). Deciduous thorny bush or shrub, with yellow flowers and edible red berries, widespread but now rather scarce in hedgerows and scrub; often rooted out because it is the alternate host of the black stem rust of wheat.

Bark Beetle (Rhynchophora: Scolytidae). A family of beetles related to the weevils, and noted for the curious hieroglyphics made in bark or on wood just under the bark by their burrowing larvae and often exposed when the bark breaks away, whence their vernacular names of Engraver Beetle, Shothole or Pinhole Borer and Typographer. Best known is the destructive Elm Bark Beetle *Scolytus scolytus*. ⇨⇨ AMBROSIA BEETLE.

Barkbug (Heteroptera: Aneuridae). Small plant bugs similar to the flatbugs; two British species, the commoner being the Common Barkbug *Aneurus laevis*.

Bark-Louse. ⇨ Psocid.

Barley, Wild (Gramineae: *Hordeum*). Four species of long-awned grass, resembling the cultivated barleys: the common Wall Barley *H. murinum* of waste places and roadsides, the taller Meadow Barley *H. secalinum* of grassland, the short-headed Sea Barley *H. marinum* of grassy places near the sea, and the local Wood Barley *Hordelymus europaeus* of calcareous woodlands. The cultivated Barleys *Hordeum vulgare* and *H. distichon* are frequent relics of cultivation.

Barnacle (Crustacea: Cirripedia). A sub-class of marine crustaceans with so remarkable a life history that until Darwin studied them in 1833 the adults were thought to be aberrant molluscs. The larvae start as free-swimming nauplii in the plankton, and only after passing through a stage when they resemble ostracods do they hang upside down on a hard surface and protect themselves with a series of hard plates, feeding with the aid of their feathery legs projecting through an aperture. Acorn Barnacles are the most familiar, covering the tidal rocks with a sharp encrustation most painful to unwary bathers: the commonest species are *Balanus balanoides* and *Chthalamus stellatus*, respectively mainly northern and southern in distribution, and the fast spreading invader from Australia *Elminius modestus*. Stalked or Ship's Barnacles attach themselves to floating objects, such as ships or wooden spars; best known is the Goose Barnacle *Lepas anatifera*, believed in the Middle Ages, on account of its superficially goose-like appearance, to be an early stage of the barnacle goose, whose Arctic breeding grounds were not known. There are also a number of parasitic barnacles, of which the best known are *Sacculina carcini* and *Peltogaster paguri*, which appear as brownish or yellowish shapeless masses under the tails of the shore and hermit crabs respectively.

Barograph. Self-recording barometer.

Barometer. Instrument for measuring atmospheric pressure, either with a column of mercury or, in an aneroid barometer, with a near vacuum.

Barrel-Fish (Percomorphi: *Leiurus*). Two species of surface-living oceanic fish, with the singular habit of following barrels and other flotsam for barnacles, on which they feed; also known as Log-fish and Rudder-fish; rare, and mainly off the Irish coast.

Bartsia (Scrophulariaceae). Three hemiparasites with two-lipped flowers, the commonest being Red Bartsia *Odontites verna* growing in disturbed ground. Yellow Bartsia *Parentucellia viscosa* grows in damp grassland in the south and west, and Alpine Bartsia *Bartsia alpina*, with purple flowers, on mountains.

Basalt. A basic rock composed of solidified lava, the result of ancient volcanic action; most frequent in Co. Antrim, where it covers 1600 sq. miles and has produced the famous Giant's Causeway, an array of natural hexagonal basalt columns formed as the lava cooled; similar columns are at Fingal's Cave, Staffa, in the Inner Hebrides.

Basic. Alkaline: used of soils which include ions of calcium, magnesium, and other elements having alkaline hydroxides, with which hydroxyl ions (bases) are usually associated.

Basidiomycetes. A large class of Fungi, containing the great majority of the most familiar fungi of the countryside, such as mushrooms, toadstools, bracket and jelly fungi and puffballs, as well as the smaller parasitic forms known as rusts and smuts. They carry their spores externally on club-shaped cells called basidia.

Basil, Wild (Labiatae: *Clinopodium vulgare*). A slightly aromatic labiate with pinkish-purple flowers, frequent in chalk and limestone grassland. ⇨ THYME, BASIL.

Basket Shell (Lamellibranchia: *Corbula gibba*). A small marine bivalve related to the gapers, widespread on muddy gravels below low water mark; its two shells are unequal, the larger being ribbed like basketwork, the smaller having a horny margin.

Bass (Percomorphi: *Morone labrax*). A marine fish related to the perch, common inshore and in estuaries of our southern and western coasts in summer and early autumn. The rare Stone Bass is a sea perch, ⇨ PERCH, SEA.

Bat (Mammalia: Chiroptera). A highly specialized group of insectivorous mammals, with wing membranes that make them the only mammals capable of flying like birds. Though they are not blind, as popularly supposed, they use a form of echo-location to guide them at night. They roost in hollow trees and caves as well as old buildings. There are twelve resident British species, of six genera, of which the most widespread are the Greater and Lesser Horseshoe Bats (*Rhinolophus ferrumequinum* and *R. hipposideros*) with remarkable horseshoe-shaped nose-leaves; the Pipistrelle, the Noctule, and the distinctive Long-eared Bat *Plecotus auritus*. Daubenton's Bat *Myotis daubentoni* is often seen flying over water. In 1954 a thirteenth species, the Continental Mouse-eared Bat *M. myotis* was found to be a local visitor of uncertain status in Dorset, and in 1964 *Plecotus austriacus*, closely related to the long-eared bat, was found to occur in the same area. ⇨ BARBASTELLE.

Bat Fowling. Catching birds roosting in trees, shrubs or hedges at night with a special U-shaped net attached to two light poles and capable of folding over to entrap any bird that flies into it. A light is used to dazzle the birds once they have been disturbed by a beater on the other side of the hedge or bush.

Batology. The scientific study of brambles, a largely apomictic group producing a large number of microspecies.

Batrachian. Outmoded term for amphibian.

Beak-Sedge (Cyperaceae: *Rhynchospora*). Two species occur: the local White Beak-Sedge *R. alba* in bog pools, mainly in the west, and the very local Brown Beak-Sedge *R. fusca*, also on wet heaths.

Bear, Brown (Carnivora: *Ursus arctos*). Extinct in Britain since the Dark Ages.

Bearberry (Ericaceae). Two trailing evergreen undershrubs of hill country, the commoner *Arctostaphylos uva-ursi*, with red berries, extending south to Derbyshire; the Black Bearberry *Arctous alpina*, with black berries, confined to north Scotland.

Beard Grass (Gramineae: *Polypogon monspeliensis*). A scarce grass of marshy places near the sea.

Beaufort Scale. The scale by which the speed of the wind is measured, devised by the nineteenth century British Admiral Beaufort, ⇨ p. 27.

Beaver (Rodentia: *Castor fiber*). Extinct in Britain since the sixteenth century.

Beck. A small stream in the north of England, equivalent to a burn in Scotland.

Bedbug (Heteroptera: *Cimex lectularius*). The formerly well known blood-sucking pest of ill-kept houses, once common, now scarce, thanks to DDT; subspecies *columbarius* preys on pigeons in dovecotes.

Bedeguar Gall. ⇨ ROBIN'S PINCUSHION.

Bedstraw (Rubiaceae: *Galium*). Plants with whorls of narrow leaves up their rather weak stems, and small, usually white flowers, formerly used

Beaufort Number	*Limits of Velocity in m.p.h. at 33 ft above level ground*	*Description of wind in Forecasts*	*Noticeable Effect of Wind* *On Land*	*At Sea*
0	Less than 1	Calm	Smoke rises vertically.	Sea is mirror-smooth.
1	1–3		Direction shown by smoke drift, but not by vanes.	Small wavelets like scales, but no foam crests.
2	4–7	Light	Wind felt on face; leaves rustle; wind-vanes moved.	Waves are short and more pronounced.
3	8–12	Light	Leaves and twigs in motion. Wind extends a light flag.	Crests begin to break. Foam has glassy appearance, not as yet white.
4	13–18	Moderate	Raises dust and loose pages and moves small branches.	Waves are longer. Many white horses.
5	19–24	Fresh	Small trees in leaf begin to sway.	Waves are more pronounced. White foaming crests seen everywhere.
6	25–31	Strong	Large branches begin to move. Telephone wires whistle.	Larger waves form. Foaming crests more extensive.
	32–38	Strong	Whole trees in motion.	Sea heaps up. Foam begins to blow in streaks.
8	39–46	Gale	Twigs break off. Progress generally impeded.	Waves increase visibly. Foam is blown in dense streaks.
9	47–54	Gale	Slight structural damage occurs. Chimney-pots fall.	Ditto.
10	55–63	Strong Gale	Trees uprooted. Considerable structural damage.	High waves with long overhanging crests. Great foam patches.
11	64–75	Strong Gale	Damage is widespread. Seldom experienced in Britain.	Waves so high that ships are hidden in the troughs. Sea covered with streaking foam. Air filled with spray.
12	above 75		Countryside is devastated. Winds of this force are only encountered in tropical revolving storms, or in Britain in short gusts on the coast.	

for stuffing pillows and mattresses; fourteen species, the commonest being Lady's Bedstraw *G. verum* with yellow flowers, Heath Bedstraw *G. saxatile*, Hedge Bedstraw *G. mollugo*, Marsh Bedstraw *G. palustre*, Goosegrass, Pawnbroker's Plant and Sweet Woodruff.

Bee (Aculeata: Apidae). Mostly furry hymenoptera, though some of the smaller species are almost hairless and wasplike, divided broadly into Social and Solitary Bees. The Social Bees are true Social Insects, attached to the colony after they become adult; this leads to a greater or less degree of social organization, and the failure to develop reproductive capacity in the great majority of the females, who are called workers. The Honeybee is the most highly organized of the social bees, Bumblebees having much smaller colonies; a few Mining Bees have developed subsocial habits. In the Solitary Bees the young fend for themselves on food provided for them and leave the colony on attaining maturity; the individual family thus ceases to exist, though customary colonial nesting grounds may continue to be used for many years. Solitary bees include the Carpenter Bees, the Leaf-cutter Bees, the parasitic Homeless Bees, the Mining Bees and the Yellow-faced Bees *Prosopis*.

Beech (Fagaceae: *Fagus sylvatica*). Tall deciduous catkin-bearing tree, making woods on the chalk and limestone of the south, where it represents the vegetation climax, and often planted elsewhere. Noted for autumn colour. Fruit a nut, known as beech-mast.

Beech Fern (Thelypteridaceae: *Thelypteris phegopteris*). A common fern of moist shady places in the north and west, not normally associated with beech trees.

Beech Tuft (Basidiomycetes: *Armillaria mucida*). A white toadstool, related to honey fungus, growing in clumps on beech trees.

Bee-Eater (Coraciiformes: *Merops*). Brilliantly coloured long-tailed birds that feed on bees and other flying insects and nest in holes in banks. European Bee-eater *M. apiaster* is an irregular visitor that bred in Sussex in 1955. Blue-cheeked Bee-eater *M. superciliosus* is a rare straggler.

Bee Fly (Brachycera: Bombylidae). A family of a dozen species of two-winged flies, some of which bear a striking resemblance to a small bumblebee; the commonest is *Bombylius major* which flies in April. The larvae of some species are parasitic on solitary bees.

Beefsteak Fungus. ⇨ POOR MAN'S BEEFSTEAK.

Bee-Louse (Cyclorrhapha: *Braula coeca*). A wingless species of two-winged fly, looking somewhat like a mite; highly adapted to cling to the hairs of honeybees, which it sometimes infests in some numbers.

Bee Moth (Pyraloidea: *Aphomia sociella*). A common pyralid moth; the larva lives in the nests of bumblebees and wasps.

Bee Orchid (Orchidaceae: *Ophrys apifera*). A frequent orchid of calcareous turf in the south, named from the resemblance of the lower lip of the flowers to the rear view of a small bumblebee. Wasp Orchid is a variety, var. *trollii*.

Beeswax. A waxy substance secreted by honeybees as a result of excess intake of nectar, which is used to construct the walls of the cells of the honeycomb.

Beet, Wild. ⇨ SEA BEET.

Beetbug (Heteroptera: Piesmidae). Elongated plant-bugs; half a dozen British species, mostly aliens; the Beet Leaf Bug *Piesma quadratum*, rare away from saltmarshes in Britain, is an important pest in central Europe as vector for the virus causing the beet disease crinkle.

Beet Fly. ⇨ ONION FLY.

Beetle (Insecta: Coleoptera). One of the largest orders of insects, including some 3,700 species, found in an exceptionally wide range of habitats, and characterized by having their forewings hardened into elytra. There are two sub-orders, one small, the Adephaga or Carnivorous Beetles, and the other much larger, the Polyphaga or Omnivorous Beetles.

Belemnite. Cigar-shaped fossils found in Jurassic and Lower Cretaceous rocks, being the internal shells of extinct cephalopods allied to the cuttlefishes.

Belladonna or Deadly Nightshade (Solanaceae: *Atropa bella-donna*). A highly poisonous plant, of medicinal value, with large dull purple bell-shaped flowers and black cherry-like berries; widespread in scrub and disturbed ground on limy soils, but local and mainly in the south.

Bellbine. ⇨ Great BINDWEED.

Bellflower (Campanulaceae: *Campanula*). Nine native or introduced species with handsome blue or purple bell-shaped flowers, including Harebell, Giant Bellflower *C. latifolia* of northern England, Nettle-leaved Bellflower *C. trachelium* of the south, and Clustered Bellflower *C. glomerata* of calcareous turf. The allied Ivy-leaved Bellflower *Wahlenbergia hederacea* is a small creeping plant of damp places in the south and west.

Beluga or White Whale (Cetacea: *Delphinapterus leucas*). An Arctic dolphin, white all over when adult, which has very occasionally been stranded on our shores.

Bennet. ⇨ BENT.

Bent or Bennet. A general term for short flowering stems of grasses projecting above the turf. Also applied especially to the genus *Agrostis*, of which the commonest species are the abundant White Bent *A. stolonifera* and Fine Bent *A. tenuis*, the latter especially on acid soils, and the common Black Bent *A. gigantea* and Brown Bent *A. canina;* and to our two members of the genus *Apera*, both of light soils, mainly in the Breckland: *A. spica-venti* and *A. interrupta*.

Benthos. The fauna and flora of the sea bottom.

Bergmann's Rule. The tendency for the body-size of a warm-blooded species or subspecies to increase with the decreasing mean temperature of its habitat, i.e. going north in the northern hemisphere.

Bergylt. Alternative name for both the ballan wrasse and the Norway haddock.

Bermuda Grass (Gramineae: *Cynodon dactylon*). A cosmopolitan weed of warmer countries, occasional in sandy places in the south, usually near the sea.

Berry. Fruit with a fleshy surround to the seeds. ⇨ DRUPE.

Betony (Labiatae: *Betonica officinalis*). A non-aromatic relative of the woundworts. Water Betony ⇨ WATER FIGWORT.

Bevy. A small flock or family party of quail.

Bib. ⇨ POUT.

Biennial. A plant that takes two years to complete its life cycle: the first year producing a shoot from the seed, the second year flowering, fruiting and dying.

Big Bud Mite (Prostigmata: *Cecidophyopsis ribis*). A gall-producing eriophyid mite, which causes big bud disease in black currants. A related species, *Phytoptus avellanae*, produces a similar condition in hazel.

Bifid. In botanical description, deeply divided into two.

Bilberry (Ericaceae: *Vaccinium*). Deciduous undershrubs, growing on moors and heaths and in woods on acid soils, whose sharp-tasting edible fruits are much esteemed in jams and tarts. The common species *V. myrtillus* is known as bilberry in the north of England, whortleberry (and the fruits as hurts) in the south, and blaeberry in Scotland. Northern Bilberry *V. uliginosum* grows only on the higher moors in the north and in Scotland.

Bindweed. Name of two groups of herbaceous climbing plants belonging to the Convolvulaceae (*Calystegia*, *Convolvulus*), twining anti-clockwise, and the Polygonaceae (*Polygonum*) twining clockwise. Great Bindweed or Bellbine *Calystegia sepium*, a persistent weed, has large 2-in. white trumpet-shaped flowers; *C. pulchra*, an increasing garden escape, has similar pink flowers; Sea Bindweed *C. soldanella*, found on sandy shores, also with large pink trumpets, is the only bindweed that does not climb, but is prostrate; it has kidney-shaped instead of arrow-shaped leaves like all the others; Field Bindweed *Convolvulus arvensis*, another persistent weed, has smaller pink or white flowers. Black Bindweed *P. convolvulus* is another weed, and the much less common Copse Bindweed *P. dumetorum* grows in hedges and scrub; both have small greenish flowers.

Binominal System (now preferred to Binomial System). ⇨ NOMENCLATURE.

Biocenosis. A community of organisms inhabiting a biotope.

Biology. The scientific study of living things, i.e. animals (zoology) and plants (botany).

Bioluminescence. The production of light by living animals, by a chemical reaction that gives rise to negligible amounts of heat, popularly but incorrectly called Phosphorescence. Many marine animals are bioluminescent, including the jellyfish *Pelagia noctiluca*, the bristle-worm *Chaetopterus variopedatus* and several opossum shrimps, and also the terrestrial glow-worm and various fungi and bacteria.

Biomass. The total weight of living organisms in any given area.

Biome. The living part of the Ecosystem.

Biosphere. That part of the Earth which includes living organisms.

Biota. The fauna and flora of a particular area or district.

Biotic Factor. ⇨ ECOSYSTEM.

Biotope. A habitat in which the main climatic, edaphic and biotic conditions are uniform.

Biotype. A group of animals or plants of equal genetic make-up.

Birch (Betulaceae: *Betula*). Graceful, deciduous, catkin-bearing small trees or shrubs. Three native species comprise two trees, the Silver Birch *B. verrucosa*, noted for its white bark, and the Downy Birch *B. pubescens*, and one small shrub, the Dwarf Birch *B. nana*, almost confined to the Highlands.

Bird (Vertebrata: Aves). A highly specialized group of 'warm-blooded' vertebrates maintaining a constant temperature and characterized by their feathers and the development of their forearms into wings, which enables them to fly. About 200 species breed in the British Isles, of which about fifty are summer visitors only; about 100 are regular winter visitors or passage migrants, and there are also some 180 stragglers of varying degrees of rarity. The 200 breeding birds cover eighteen orders, of which much the largest is the Passeriformes (song birds), with twenty families.

Topography of a Bird

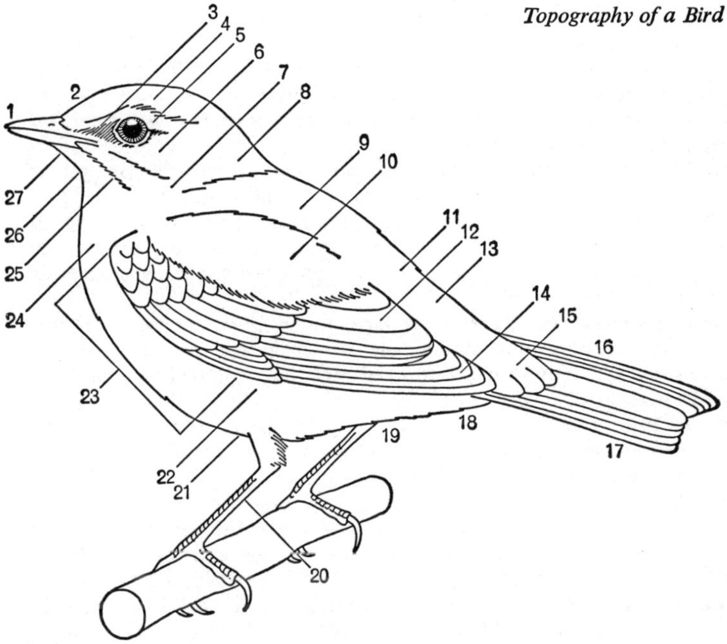

1. Bill	10. Scapulars	19. Vent
2. Forehead	11. Back	20. Tarsus
3. Lores	12. Secondaries	21. Belly
4. Crown	13. Rump	22. Flanks
5. Superciliary eyestripe	14. Primaries	23. Wing coverts
6. Ear coverts	15. Upper tail coverts	24. Breast
7. Neck	16. Tail	25. Moustachial stripe
8. Nape	17. Outer tail feathers	26. Throat
9. Mantle	18. Under tail coverts	27. Chin

Bird Banding. American term for Bird Ringing.

Bird Navigation. It is now established that birds, when on their migratory flights, navigate to their known nesting sites, and in some cases also to their known wintering grounds, by the sun and in the case of some night migrants at least by the stars, i.e., by the same means as human navigators. What is not yet known is how they do it.

Bird Ringing (America: Bird Banding). The largest and best known scheme for marking animals in Britain is the Bird Ringing Scheme run by the British Trust for Ornithology with a large grant from the Nature Conservancy. Some 600 amateur ringers, trained and licensed by the B.T.O., ring upwards of a quarter of a million birds in Britain every year, producing increasingly valuable information on migration and longevity of birds. Ringing consists in encircling a bird's leg with a light metal ring inscribed: Inform British Museum (Natural History), London, S.W.7.

Birdseye. ⇨ PRIMROSE.

Birdsfoot (Leguminosae: *Ornithopus*). Two small yellow or orange pea-flowers, named from the claw-like appearance of their seed pods (⇨ BIRDSFOOT TREFOIL); much the commoner is *O. perpusillus*, mainly on sand and gravel in the south.

Birdsnest, Yellow (Monotropaceae; *Monotropa hypopitys*). A yellow saprophytic plant locally frequent in woods in the south.

Birdsnest Fungi (Ascomycetes: Nidulariaceae). A family of sac fungi in which the spores are carried in a small cup, like eggs in a bird's nest; *Cyathus olla* sometimes grows in gardens, and several other species grow on dead sticks in woods.

Birdsnest Orchid (Orchidaceae: *Neottia nidus-avis*). Our commonest saprophytic orchid, widespread and locally frequent in open woods, especially beechwoods, mainly in the south; the roots are supposed to resemble a bird's nest.

Bird of Prey. Carnivorous birds with hooked beaks; divided into the diurnal – hawks, falcons, eagles, buzzards, and their allies (Falconiformes), and the nocturnal – the owls (Strigiformes).

Bird Song. A form of language among birds, mainly found among the passerines, that is still not fully understood but is obviously related to the holding of territories. It is impossible to draw a hard and fast line between song and call-notes; to some extent a song is a specialized call-note that sounds pleasant to humans. By this definition the finest British songsters are the nightingale, blackbird, song thrush, skylark, woodlark, blackcap, and garden warbler.

Bird Table. A flat surface, either on a pole or suspended from above on which food can be put for garden birds.

Birks. Yorkshire vernacular name for a birchwood.

Birthwort (Aristolochiaceae: *Aristolochia clematitis*). A medicinal plant used by medieval midwives and still surviving near a few old buildings in England and Wales; flowers yellow, tubular, with a long lip.

Bishop's Mitre (Heteroptera: *Aelia acuminata*). A grass-feeding brown shield-bug, frequent in the south; narrowed at both ends and so resembling a mitre.

Bishop's Weed or Bishopweed. ⇨ GROUND-ELDER.

Bistort (Polygonaceae: *Polygonum*). Plants with spikes of small pink or white flowers and whitish papery sheaths at the base of the leaves; eight native species, including Bistort *P. bistorta*, with many local names

(Easter-ledges, snakeweed, snakeroot), in damp meadows in the north; Amphibious Bistort *P. amphibium*, widespread in and by fresh water; the Alpine Bistort *P. viviparum*, frequent on mountains; ⇨ PERSICARIA, WATER-PEPPER.

Bittercress (Cruciferae: *Cardamine*). White-flowered crucifers allied to Lady's Smock; three of the four species are widespread: Hairy Bittercress *C. hirsuta*, a common garden weed; Wavy or Wood Bittercress *C. flexuosa*, in moist shady places; and the actually bitter Large Bittercress *C. amara*, of streamsides and fens.

Bitterling (Pisces: *Rhodus amarus*). A small European freshwater fish, breeding ferally in a number of ponds in north-west England.

Bittern (Ardeiformes: Ardeidae). Shy, marsh-dwelling members of the Heron Family; the European Bittern *Botaurus stellaris*, noted for its booming cry, is a rare breeding species, and the American Bittern *B. lentiginosus* and European Little Bittern *Ixobrychus minutus*, which has bred, are irregular visitors.

Bittersweet or Woody Nightshade (Solanaceae: *Solanum dulcamara*). A shrubby clamberer, with purple and yellow flowers and poisonous red berries; widespread and common.

Bivalve (Mollusca: Lamellibranchia). A large class of aquatic molluscs characterized by having twin valves to their shell, and including the mussels, cockles and oysters.

Black Beetle. A popular but highly inaccurate name for cockroaches, which are brown and not beetles.

Blackberry. The fruit of the bramble.

Blackbird (Passeriformes: *Turdus merula*). A large thrush, the cocks being all black in plumage, the hens and juveniles dark brown with speckled breast. One of the two commonest British birds (the other is the chaffinch), widespread and abundant, penetrating right into the centre of large towns. Lays brown-spotted greenish-blue eggs in a bulky and conspicuous nest in a bush, hedge or tree.

Black Bulgar (Ascomycetes: *Bulgaria inquinans*). A fairly common blackish-brown rubbery fungus of oak and beech trunks.

Blackcap (Passeriformes: *Sylvia atricapilla*). Warbler, named from the cock's black crown, with high quality song and somewhat mimetic; summer visitor, frequent in woods and scrub in England and Wales, rather local elsewhere.

Blackcock. ⇨ Male Black GROUSE.

Blackfish, Pilot Whale or Caa'ing Whale (Cetacea: *Globicephala melaena*). A large gregarious dolphin, not uncommon off our northern shores and formerly hunted in the Orkneys and Shetlands.

Black-Fish (Percomorphi: *Centrolophus niger*). A black or blackish surface-living oceanic fish, probably not uncommon off our western and south-western coasts, but rare inshore. Only four specimens of the Cornish Blackfish *C. britannicus* have ever been found; the first was washed up at Polperro, Cornwall, in 1859.

Black Flies (Nematocera: Simuliidae). Small two-winged flies, related to the gnats and midges, which cause great annoyance by swarming round people's heads and sometimes biting; their larvae are aquatic and swarms are also found near water.

Black-Fly. The aphid which infests broad beans, also known as Bob.

Blackgame. ⇨ Black GROUSE.

Blackthorn (Rosaceae: *Prunus spinosa*). A very thorny deciduous shrub, forming dense thickets; widespread and common; flowers white, appearing before the leaves in spring. Fruit a sloe ⇨⇨ PRUNUS.

Black Wasp (Aculeata: Sphecidae). A group of small black solitary wasps, nesting in rotten wood or plant stems; they include the Mournful Wasp *Pemphedron lugubris*, and the Mimic Wasps *Mimesa*.

Bladder Fern (Athyriaceae: *Cystopteris*). Two graceful small ferns, much the commoner being *C. fragilis*, on limestone rocks and walls.

Bladder-Seed (Umbelliferae: *Physospermum cornubiense*). A white-flowered umbellifer with swollen fruits, scarce in woods in the south-west.

Bladder Senna (Papilionaceae: *Colutea arborescens*). A small shrub with red-marked yellow peaflowers and inflated papery seed-pods; naturalized on waste ground and railway banks, especially near London.

Bladder Snail (Pulmonata: *Physa fontinalis*). A freshwater snail with a fragile oval shell, common in clear water and on waterweeds. The less common Moss Bladder Snail *Aplecta hypnorum* has a narrower shell and prefers still water, even occurring in temporary ponds and dried-up ditches.

Bladderwort (Lentibulariaceae: *Utricularia*). Four species of free-floating, insectivorous, aquatic plants, with submerged feathery leaves furnished with air bladders that capture small water animals and later fill with water and sink the plant to the bottom for the winter; Flowers spurred, two-lipped, yellow, in leafless spikes. Greater Bladderwort *U. vulgaris* is the commonest species.

Blaeberry. The Scottish name for the bilberry.

Blanketweed. A name for various filamentous green algae that grow abundantly in fresh or brackish water, e.g. *Enteromorpha*.

Bleak (Cyprinidae: *Alburnus alburnus*). A small gregarious fish of slow streams, lakes and ponds in England and Wales; sometimes used as bait.

Blenny (Percomorphi: Blenniidae). Five superficially goby-like small inshore marine fishes, the commonest being the Common Blenny or Shanny *Blennius pholis*, and the largest the Tompot Blenny or Gattorugine *B. gattorugine*. The southern Butterfly Blenny *B. ocellaris* has a large eye-like spot on one of its back fins. Viviparous Blenny, ⇨ EEL-POUT.

Blewits (Agaricaceae: *Tricholoma saevum*). An edible grassland toadstool, sometimes forming rings. Wood Blewits *T. nudum* and the related St George's Mushroom are also edible.

Blight. A general term for plant diseases. American Blight, ⇨ APHID.

Blind. American term for a Hide.

Blind-Worm. ⇨ SLOW-WORM.

Blinks or Water Blinks (Portulacaceae: *Montia fontana*). A small plant with tiny white flowers, often forming patches in springs and on wet mud, but also growing as a weed in drier soils.

Blister Beetle or Spanish Fly (Heteromera: *Lytta vesicatoria*). A rather uncommon bright golden-green beetle; the larvae have habits like those of the related oil beetles.

Blizzard. A snowstorm driven by a high wind.

Bloodworm. (1) The larva of chironomid midges. (2) ⇨ RIVER WORM.

Bloody-nosed Beetle (Phytophaga: *Timarcha*). Two species of leaf beetle named from their habit of exuding a bright red fluid from their mouths and 'knee' joints when alarmed.

Blow Fly. ⇨ BLUEBOTTLE.

Bluebell (Liliaceae: *Endymion non-scriptus*). An abundant, wood-carpeting, spring wild flower, known as Wild Hyacinth in Scotland, where bluebell means the harebell; also growing on sea-cliffs and mountains. White Bluebell is a name for the Three-cornered Leek *Allium triquetrum*, a well-established alien in the south-west.

Bluebottle. ⇨ CORNFLOWER.

Bluebottle or Blow Fly (Cyclorrhapha: *Calliphora*). Two large, metallic blue, two-winged flies (Calliphoridae), whose larvae feed on decaying animal matter and dung – and indoors on the family joint in the larder if they are given the chance: *Calliphora erythrocephala* and *C. vomitoria*. The related but smaller *Protocalliphora azurea* breeds in birds' nests.

Blue Bug (Heteroptera: *Zicrona caerulea*). A handsome blue shieldbug common in many grassy and heathy places.

Blue Butterflies (Papilionina: Lycaenidae). A group of eleven rather small butterflies, of which the males (except the Brown Argus *Aricia agestis*) are blue and the females brown. The commonest are the widespread Common Blue *Polyommatus icarus*, and the Holly Blue *Celastrina argiolus*, whose larvae feed on the buds and fruits of holly and dogwood. The Chalkhill Blue *Lysandra coridon* is locally frequent on the chalk downs of the south. The rare Large Blue *Maculinea arion* has a remarkable life history during which its larva spends most of its life in an anthill, feeding on ant larvae and supplying the ants with 'honey' from a special gland.

Blue-eyed Grass (Iridaceae: *Sisyrinchium bermudianum*). Blue flower native in Ireland but only a garden escape in Britain; part of the American Element in the British flora.

Blue-Green Algae (Cyanophyta: Cyanophyceae). A phylum of primitive, one-celled, colonial or filamentous lower plants, named from their colour due to the presence of the pigment phycocyanin. They are contained in a single class, and are often abundant in the freshwater plankton or in damp soil. Some form scums either, like *Coelosphaerium* and *Microcystis*, on the surface of still water, or, like *Oscillatoria*, on underwater objects; others, such as *Anabaena* and *Aphanizemon*, produce the phenomenon known as the Breaking of the Meres. *Nostoc* forms gelatinous masses up to the size of a pea, and may be free-floating in water.

Blue Mountain Grass. ⇨ MOOR GRASS.

Bluetail, Red-flanked (Passeriformes: *Tarsiger cyanurus*). Small song bird allied to the thrushes; a very rare straggler from eastern Europe.

Bluethroat (Passeriformes: *Cyanosylvia svecica*). Small song bird, allied to the thrushes; not infrequent as a passage migrant on the east coast;

named from the cock's brilliant blue throat, with either a red or a white spot on the breast.

Blusher, The (Agaricaceae: *Amanita rubescens*). An edible (when cooked) woodland toadstool, closely related to the death cap, and unfortunately rather too similar to its also highly poisonous relative, the panther, but distinguished by flesh that turns pink on being touched.

Boar, Wild (Artiodactyla: *Sus scrofa*). Extinct in Britain since the early seventeenth century.

Boar Fish or Cuckoo Fish (Pisces: *Capros aper*). Marine fish related to and slightly resembling the John Dory; not uncommon off our south-west coasts.

Boat-Fly. ⇨ WATER-BOATMAN.

Bob. A vernacular name for black-fly.

Bog. An area of wet acid peat with a characteristic vegetation of sphagnum mosses, forming especially in areas with very heavy rainfall, e.g. central Ireland, where the atmosphere is constantly moist and drainage very poor. Blanket Bog covers the whole surface of the ground, even low hills, like a blanket, and is not confined to the wet hollows like valley bog. Raised Bog may develop on top of valley bog, or more often on Fen. Valley Bog occurs where water from fairly acid rocks stagnates in the bottom of a valley or hollow, keeping the soil wet so that sphagnum moss and other bog plants can grow. Because it is fed by drainage water, bringing down some soluble salts, the water is less acid than in blanket bog, which depends on rain water. ⇨⇨ BOG-FLOW, SCRAWBOG.

Bogbean or Buckbean (Menyanthaceae: *Menyanthes trifoliata*). Handsome aquatic, with conspicuous spikes of white flowers and trifoliate leaves projecting from many ponds and tarns in districts with acid soils.

Bog Cotton. ⇨ COTTON-GRASS.

Bog-Flow or Bog-Slide. A phenomenon, more frequent in Ireland than elsewhere, whereby a bog becomes so full of water that it breaks its peaty crust and discharges its viscous contents, which spread over the surrounding countryside. In 1896 a whole family in Kerry were overwhelmed in a bog-flow.

Bog Moss (Sphagnales: *Sphagnum*). A family and genus of mosses, mostly growing in wet places, some species forming characteristic spongy cushions in bogs, especially *S. plumulosum* and *S. cuspidatum*.

Bog Myrtle. ⇨ SWEET GALE.

Bog Orchid (Orchidaceae: *Hammarbya paludosa*). Our smallest orchid, with yellow-green flowers, widespread but decreasing in wet sphagnum moss in bogs and on heaths and moors.

Bog Rosemary (Ericaceae: *Andromeda polifolia*). Pink-flowered undershrub, decreasing in bogs in the north and west.

Bog-Rush (Cyperaceae: *Schoenus nigricans*). A dark-flowered sedge of calcareous fens and the less acid bogs.

Bog-Slide. ⇨ BOG-FLOW.

Bohemian Chatterer. ⇨ WAXWING.

Bolder. Norfolk name for the bulrush.

Boletus. (Agaricaceae: *Boletus*). A group of largely edible, mainly woodland toadstools, of which the best known is the Cep *B. edulis*, while *B. satanas* is poisonous though not lethal.

Bombardier Beetle (Adephaga: *Brachinus crepitans*). A locally common ground beetle, noted for its habit of ejecting a highly volatile liquid, apparently containing oxides of nitrogen, from its anus, making a small jet of vapour and a tiny bang.

Bombycoidea (Lepidoptera: Heteroneura). A superfamily of large moths, comprising the lackeys and eggars (Lasiocampidae); it also includes the non-British Silkworm Moth *Bombyx mori*.

Bonito. ⇨ TUNNY.

Bonxie. Shetland name for the great skua.

Book-Louse (Psocoptera: *Liposcelis divinatorius*). The best known British psocid and the commonest of the wingless group that inhabits buildings, often occurring among old books and neglected entomological collections.

Bootlace Fungus. ⇨ HONEY FUNGUS.

Bootlace Worm (Nemertina: *Lineus longissimus*). A common ribbon worm on shores containing muddy shingle, frequently attaining 15 or 20 ft and sometimes much more; one instance of well over 30 yds is on record.

Borage. (Boraginaceae: *Borago officinalis*) A blue-flowered garden plant, sometimes escaping.

Boraginaceae or Borage Family. (Angiospermae: Dicotyledones). A family of mostly blue-flowered plants, with usually roughly hairy leaves and stems, and the flower spikes tightly curled at first, like a scorpion's tail; including the forget-me-nots, comfreys, buglosses, gromwells and houndstongues.

Bore. A tidal wave proceeding up a river and caused by the tide being funnelled into a narrow estuary. The most famous British bore is on the Severn, but there are others on the Somerset Parrett, the Great and Yorkshire Ouses, the Trent (known as the eagre) and the Solway Firth.

Borer or Stem Sawfly (Symphyta: Dephidae). A small family of sawflies specializing in boring in the stems of plants, including the Wheat-stem Borer *Cephus pygmaeus*, a notorious pest.

Botanical Society of the British Isles. The senior botanical society in Britain, founded 1836, covering flowering plants, vascular cryptogams and stoneworts. It has two journals, *Watsonia*, named after the Victorian botanist, H. C. Watson, and *Proceedings*. The Society was responsible for the Maps Scheme, an important scientific enterprise in which hundreds of amateur and professional botanists took part, which resulted in the publication in 1962 of the *Atlas of the British Flora*, containing 1,604 dot-maps on a 10-kilometre-square grid, of the distribution of flowering plants and vascular cryptogams in the British Isles.

Botanical Society of Edinburgh. The Scottish counterpart of the Botanical Society of the British Isles, founded in 1836; publishes *Transactions and Proceedings*.

Botany. The scientific study of plants.

Bot Fly (Diptera: Cyclorrhapha). Several species of two-winged flies, infesting ungulates, and with a similar life history to the warble flies. *Cephenomya auribarbis* (Calliphoridae) infests deer; *Oestrus ovis* (Larvaevoridae) infests sheep; and *Gasterophilus* spp. (Muscidae) infest horses, donkeys and mules. Bots are strictly speaking the larvae.

Bothidae (Pisces: Heterosomata). The family of flatfish with both eyes on the left side of the body, containing the brill, megrim, scald-fish, turbot and topknots.

Boulder Clay. ⇨ CLAY.

Bouncing Bett. ⇨ SOAPWORT.

Bourne. A stream in chalk and limestone districts which only flows above ground after heavy rain in a wet season.

Box (Buxaceae: *Buxus sempervirens*). An evergreen shrub or small tree, often planted in ornamental woodlands, and regarded as native in three places, notably Box Hill in Surrey.

Brachycera or Short Horns (Insecta: Diptera). One of the three sub-orders of the two-winged flies, containing the bee flies (Bombylidae), horse and gad flies (Tabanidae), long-headed flies (Dolichopodidae), pointed-wing flies (Lonchopteridae), robber and assassin flies (Asilidae), snipe flies (Rhagionidae), spider-parasite flies (Cyrtidae), soldier flies (Stratiomyidae), stiletto flies (Therevidae), and window flies (Scenopidae).

Bracken or Brake Fern (Dennstaedtiaceae: *Pteridium aquilinum*). A tall, widespread and abundant fern with far-creeping rhizomes, often a pestilential weed of woods, moors and heaths, where excessive grazing or burning often enables it to become established over extensive areas, to the exclusion of heather or grasses.

Bracken Bug (1) (Heteroptera: *Monalocoris filicis*). A widespread and common capsid bug feeding on the spores of bracken and other ferns. (2) ⇨ HARVEST MITE.

Bracken-Clock. ⇨ Garden CHAFER.

Bracket Fungus or Polypore (Basidiomycetes: Polyporaceae). A family of usually fan-shaped fungi named from their habit of growing bracket-like on live or dead wood. *Grifola gigantea* may grow up to 3 ft across. ⇨⇨ DRYAD'S SADDLE, POOR MAN'S BEEFSTEAK, RAZOR-STROP FUNGUS.

Braconid Wasp. ⇨ ICHNEUMON.

Bract. A small leaf immediately beneath a flower or cluster of flowers; bracts beneath the individual flowers in a cluster are known as bracteoles.

Bradyodonti (Chordata: Vertebrata). A curious group of scale-less fishes linking the primitive cartilaginous sharks and rays with the sturgeon and other bony fishes, and containing the chimaeras.

Brae. A bank or hillside above a valley in Scotland.

Brake Fern. ⇨ BRACKEN.

Bramble (Rosaceae: *Rubus fruticosus*). A vigorous half-evergreen thorny clamberer or shrub, with well known edible fruits called blackberries; widespread and abundant, often making impenetrable thickets; apomictic and so including many microspecies. Stone Bramble (*R. saxatilis*) is a herbaceous northern plant with red fruits.

Brambling or Bramble Finch. (Passeriformes: *Fringilla montifringilla*). A finch closely related to the chaffinch, but a winter visitor, from northern Europe; found especially in the neighbourhood of beeches; has bred in Scotland.

Branchiopoda (Arthropoda: Crustacea). A sub-class of primitive crustaceans of fresh or inland salt or brackish water, including the Fairy Shrimp and the Cladocerans.

Branchiura. ⇨ FISH-LOUSE.

Brandling (Oligochaeta: *Eisenia foetida*). A handsome earthworm common in soil rich in organic material, such as manure or compost heaps, easily recognized by its red, purple or brown segments alternating with yellow ones. Much used for bait, and said to be relished by trout. Also known as Gilt-tail.

Brandybottle. ⇨ Yellow WATER-LILY.

Breaking of the Meres, also known as a Water Bloom. A phenomenon of eutrophic lakes, i.e. lakes rich in salts and organic matter, especially the meres of Cheshire and Shropshire. It is caused by algae, especially blue-green algae (Myxophyceae), which on very calm days, usually in autumn, form a thick film on the surface of the water, which becomes opaque. A breeze will break the film.

Bream, Freshwater (Pisces: Cyprinidae). Both the Bream, Bronze Bream, or Carp Bream *Abramis brama* and the White or Silver Bream *Blicca bjoernka* are gregarious fish, occurring mainly in sluggish streams or lakes in east and south England.

Bream, Sea (Percomorphi: Sparidae). A family commoner in the Mediterranean than in British waters, though ten species have occurred from time to time. The commonest is the Common Sea Bream *Pagellus centrodontus,* often taken by fishermen off our south and east coasts. The Black Bream or Old Wife *Spondyliosoma cantharus* is regular off our south and west coasts, but the Spanish Bream *P. bogaroveo* and Pandora *P. erythrinus* are less common and mainly found off Cornwall. The northern Ray's Bream, also known as Black Sea Bream *Brama raii,* belongs to a different family, Bramidae.

Breccia. A conglomerate rock formed by the compression during earth movements of many small fragments of rock; found on the coast, e.g. near Dawlish, south Devon.

Breck. A stretch of heathland in the region around Thetford in south-west Norfolk and Brandon in north-west Suffolk; the area has been known as Breckland since W. G. Clarke's book *In Breckland Wilds* (1925).

Breeze. A light wind, up to about 20 m.p.h.

Breeze Fly. A local name for the blood-sucking tabanid flies.

Brent. ⇨ Brent GOOSE.

Briar or Brier. Term applied to any prickly bush, but especially to roses. Sweet Briar *Rosa rubiginosa*, with sweet-smelling leaves, is a wild rose found mainly on calcareous soils in the south.

Brickearth. A fertile soil found especially in the lower Thames valley on top of the gravels of the second Interglacial period and probably deposited by winds.

Brier. ⇨ BRIAR.

Brill (Bothidae: *Scophthalmus rhombus*). A common left-sided flatfish, closely related to the turbot, but more oval in shape.

Brimstone Butterfly (Papilionina: *Gonepteryx rhamni*). The archetypal butterfly, the whole group being named after the brimstone – or butter – yellow wings of this common early spring species which is related to the Whites; widespread and common, adults spending the winter in hibernation, the only British butterfly to do so in the open; the larvae feed on buckthorn. The Brimstone Moth *Opisthograptis luteolata* is a common yellow geometrid.

Brisling. A fisherman's term, especially in Norway, for small sea fish, especially sprats, but including the young of other members of the Herring Family (Clupeidae).

Bristle Back or Hairy Back (Aschelminthes: Gastrotricha). A small class of tiny freshwater animals, closely allied to the rotifers, and named from the hairs or bristles on their backs.

Bristle Fern. ⇨ FILMY FERN.

Bristle-Tail (Arthropoda: Insecta). Two orders of tiny wingless insects, living among stones, logs, leaves, and all kinds of detritus. Two-pronged Bristle-tails (Diplura) are white and eyeless, with two long tail filaments; ten British species. The larger Three-pronged Bristle-tails (Thysanura) have eyes and three filaments; twelve British species, of which one, *Petrobius maritimus*, is common on the shore. ⇨ FIRE-BRAT, SILVER-FISH.

Bristle Worm (Annelida: Polychaeta). A class of marine annelid worms, with bristles borne in small sheaves on their segmented bodies. They are conveniently but unscientifically divided into the Errantia, which move or swim about freely, and the Sedentaria, which live in a tube of their own manufacture. The Errant Polychaetes include the catworm, ragworms (Nereidae) white ragworm, scale worms and sea-mouse; the Sedentary Polychaetes include the fan or peacock worms, lugworms, redthreads, ragworms (*Nerine*), the bioluminescent *Chaetopterus variopedatus*, and the rock-boring worms of the genus *Polydora:* some species make hard, pale calcareous tubes, but in others tubes are made only of mud or mucus.

Brit or Britt. Local name on the west coast for small fish such as terns feed to their young.

British Bryological Society. The national society for the study of mosses and liverworts, founded in 1896 as the Moss Exchange Club, changing in 1922; publishes *Transactions*.

British Deer Society. The national society concerned with the study and conservation of deer, founded in 1963; publishes *Deer*.

British Ecological Society. The national society for the study of ecology in the broadest sense, founded in 1904 as the British Vegetation Committee and assuming present name in 1913; publishes *Journal of Ecology*, *Journal of Animal Ecology* and *Journal of Applied Ecology*.

British Herpetological Society. The national society for the study and protection of reptiles and amphibians, founded in 1947; publishes *British Journal of Herpetology*.

British Lichen Society. The national society for the study and conservation of lichens, founded in 1958; publishes the *Lichenologist*.

British List. The list of birds proved to have occurred at least once in some part of the British Isles by natural means, and brought up to date from time to time by the British Ornithologists' Union, is widely known to ornithologists as 'the British List'.

British Mycological Society. The national society for the study of fungi, founded in 1896; publishes *Transactions.*

British Naturalists' Association. Founded in 1905, as the British Empire Naturalists' Association, to provide a national society for naturalists; has eighteen local branches, and publishes three times a year its journal *Country-Side.*

British Ornithologists' Club, founded in 1892 as a dining club for members of the British Ornithologists' Union; publishes a *Bulletin* largely concerned with taxonomy.

British Ornithologists' Union, founded in 1859, is the oldest national bird organization still in existence. It is concerned especially with museum and international aspects of ornithology, organizes expeditions, publishes periodical check lists of British birds, and *Ibis* quarterly.

British Phycological Society. The national society for the study of algae, founded in 1952; publishes the *British Phycological Bulletin.*

British Pteridological Society. The national society for the study of ferns and their allies, founded in 1891; publishes the *British Fern Gazette.*

British Trust for Ornithology. Founded in 1932 to promote the scientific study of birds in the field, especially among amateur bird watchers. Runs cooperative inquiries, notably the annual heron census, and the national bird ringing scheme, also coordinates the coastal bird observatories through the Bird Observatories Committee. Publishes *Bird Study.*

Brittle-Star (Echinodermata: Ophiuroidea). A class of starfish-like marine animals, but with a fairly clear division between the central disc or body and the, usually five, 'arms'. The common species on our shores include the Black Brittle-star *Ophiocomina nigra*, the Daisy Brittle-star *Ophiopholis aculeata* and the Sand Stars *Ophiura.*

Broad-leaved Tree. Trees other than conifers, not quite coterminous with deciduous trees because on the one hand the deciduous larch is a conifer and on the other a few broad-leaved trees such as holly are evergreen.

Broads. Shallow lakes and waterways extending out of the rivers of north-east Norfolk; they are now known to be flooded medieval peat workings.

Brome Grass (Gramineae: *Bromus*). A fairly large genus of long-awned grasses, mostly unimportant as fodder, the commonest being the ubiquitous Soft Brome *B. mollis* and Barren Brome *B. sterilis*, both of grassy and waste places. Two taller species are Upright Brome *B. erectus*, often dominant in chalk grassland, and Hairy Brome *B. ramosus*, frequent in hedge-banks and shady places. ⇨ FALSE BROME GRASS.

Brook. A small stream in England and Wales.

Brooklime (Scrophulariaceae: *Veronica beccabunga*). A blue-flowered speedwell, widespread and common in wet places.

Brookweed (Primulaceae: *Samolus valerandi*). A local white-flowered plant of damp places.

Broom (Papilionaceae: *Sarothamnus scoparius*). An unarmed shrub with trefoil leaves and brilliant golden yellow pea-flowers; widespread and frequent on acid soils.

Broomrape (Orobanchaceae: *Orobanche*). Curious leafless plants, yellow, reddish, or purple in colour, parasitic on the roots of other plants, many of our ten species being highly specific to one or two hosts. Common Broomrape *O. minor*, mainly on clovers and other small peaflowers, is the commonest and most widespread species, the only others which are not very scarce or local being the Knapweed Broomrape *O. elatior* and the the Ivy Broomrape *O. hederae*.

Brown Argus. ⇨ BLUE BUTTERFLIES.

Brown Butterflies (Papilionina: Satyridae). A family of eleven butterflies, all brown except the Marbled White, and including the abundant Meadow Brown *Maniola jurtina* of grassy places, the Hedge Brown, Small Meadow Brown or Gatekeeper *M. tithonus* of bushy places, the Speckled Wood, the Wall, the Ringlets, the Scotch Argus, and the Heaths.

Brown Earths. One of the three main climatic soil types in Britain (the others are Podsols and Blanket Bog Peats), typical of a mild climate and moderate rainfall as in the south and midlands, and normally supporting deciduous forest.

Brown Rot (Ascomycetes: *Sclerotinia*). Small fungi which cause ripe fruits to turn brown and become rotten, such as *S. fructigena*, which rots apples and plums, and *S. laxa*, which not only rots fruit but causes fruit blossom and young leaves to wilt and die.

Brown Seaweeds (Phaeophyta). A major division or phylum of marine algae, mostly brown or olive-green in colour, and the most conspicuous seaweeds around our shores. There are some very large species, including the kelps and thongweed, some medium-sized, including the wracks, and a number of smaller plants with no vernacular names, all included in a single class, the Phaeophyceae. ⇨⇨ PEACOCK'S TAIL, POD-WEED, SEA-LACE.

Browsing. Damage done by deer eating young shoots and buds off trees.

Bryology. The scientific study of mosses and liverworts.

Bryony. Our two bryonies are alike in being climbers with small greenish flowers and poisonous red berries, in being our sole native representatives of two great tropical plant families, and in being locally common in hedges and scrub, mainly in the south. Black Bryony *Tamus communis*, of the Yam Family (Dioscoreaceae), twines clockwise and has heart-shaped leaves. White Bryony *Bryonia dioica*, of the Cucumber Family (Cucurbitaceae), has ivy-shaped leaves and climbs anti-clockwise with tendrils.

Bryophyte (Bryophyta). A major grouping or phylum of green lower plants, comprising the mosses and liverworts.

Bryozoa. ⇨ SEA-MAT.

Buck. Male of fallow, roe and muntjac deer.

Buckbean. ⇨ BOGBEAN.

Buckie. ⇨ Common WHELK.

Buckler Fern (Aspidiaceae: *Dryopteris*). A group of larger ferns closely allied to Male Fern, of which the commonest are *D. dilatata*, widespread in woods, and *D. carthusiana*, more local in marshy woods and bogs. The Hay-scented Buckler *D. aemula*, found mainly in the north and west, smells like hay when dried.

Buckthorn (Rhamnaceae). Two species, both shrubs or small trees and found mainly in the south: Common or Purging Buckthorn *Rhamnus catharticus*, thorny, with black berries and preferring dry chalky soils, and Alder Buckthorn *Frangula alnus*, unarmed, with red berries and preferring wet acid soils; both, however, grow in the basic peaty fens of East Anglia. Sea-Buckthorn (Elaeagnaceae: *Hippophaë rhamnoides*) is a thorny, thicket-forming shrub of coastal districts, especially on sand dunes, with narrow silvery leaves and bright orange berries.

Buddleia or Butterfly Bush (Buddleiaceae: *Buddleia davidii*). Deciduous shrub, frequent in gardens, that has seeded itself widely on waste ground in recent years; highly attractive to butterflies in late summer.

Budgerigar (Psittacidae: *Melopsittacus undulatus*). ⇨ ACCLIMATIZATION.

Buff Cap (Basidiomycetes: *Hygrophorus pratensis*). A common edible toadstool growing in grassland in the latter half of the year.

Buff-Tip Moth (Notodontoidea: *Phalera bucephala*). A large Prominent, with buff tips to its wings, noted for its defoliating caterpillars, swarms of which may strip whole branches of trees of all their leaves; widespread and common.

Bug. A loose general term for wingless or apparently wingless insects, more properly applied to the Hemiptera, and especially to the Heteroptera.

Bugle (Labiatae: *Ajuga*). Three blue-flowered plants, two native and one introduced, much the commonest being the woodland *A. reptans*. ⇨⇨ GROUND-PINE.

Bugloss (Boraginaceae). Name of several blue-flowered plants of the Borage Family, especially the widespread and common Viper's Bugloss *Echium vulgare* and the arable weed Small Bugloss *Lycopsis arvensis*.

Bulb. A swollen underground stem, consisting of a series of fleshy scales or leaf-bases enclosing the bud that will produce the leaves and flowering stems the following year; especially in the Lily and Daffodil Families.

Bulb Fly (Cyclorrhapha: Syrphidae). Hover flies of the genera *Merodon* and *Eumerus*, whose larvae live in bulbs and can be serious pests to the bulb grower and gardener. The largest is Narcissus Fly *M. equestris*, which looks like a small bumblebee.

Bulbil. A small bulb or tuber at the base of a leaf-stalk, e.g. in Coralwort, or among the flowers, e.g. in some garlics.

Bulb Mite (Acari). The Large Bulb Mite *Rhizoglophus echinopus* (Astigmata) is found in the soil almost wherever narcissi are grown. The Bulb Scale Mite (Prostigmata: *Steneotarsonemus laticeps*) is a serious pest of bulbs, especially narcissus, living between their fleshy scales.

Bulin (Pulmonata: *Ena*). Two rather narrowly spiral land snails of drier places, *E. obscura* being the commoner and *E. montana* the larger.

Bullace. ⇨ PRUNUS.

Bullfinch (Passeriformes: *Pyrrhula pyrrhula*). Handsome finch, the cock with black cap and red breast, liable to do serious damage to buds of ornamental and fruit trees; widespread and locally common.

Bullhead (Scleroparei: *Cottus*). Four small spiny broad-headed fishes related to the gurnards, three marine and one freshwater species. The Miller's Thumb *C. gobio*, is frequent in freshwater streams and brooks. The two commoner marine ones, both frequent in rock pools, are the Father-lasher or Short-spined Sea-scorpion *C. scorpius*, which grunts when taken from the water, and the Long-spined Sea-scorpion *C. bubalis*. Armed Bullhead, ⇨ POGGE.

Bulrush (Cyperaceae: *Scirpus lacustris*). A tall aquatic rush-like sedge, often growing in midstream, with clusters of oval reddish-brown flower-spikes near the top of its leafless stem. The plant popularly known as bulrush, thanks to the Victorian painting of Moses in the bulrushes, is actually the great reed-mace *Typha*. There are several less common, or rare, allied bulrushes.

Bumblebee or Humblebee (Aculeata: *Bombus*). Social bees, usually coloured with some combination of black with yellow, red or white, named from their humming note and living in much smaller colonies, usually underground, than the honeybee, though with a similar social organization of queens (developed females), males (not called drones) and workers (undeveloped females); eighteen species the commonest including *B. terrestris*, *B. lucorum*, and *B. lapidarius*. ⇨⇨ CARDER BEE, CUCKOO BEE.

Bunt. A smut disease of wheat.

Bunting (Passeriformes: Fringillidae). Small seed-eating birds allied to the finches. Five British breeding species: Cirl Bunting *Emberiza cirlus*, rather local in south England; Corn Bunting *E. calandra*, our largest bunting, widespread but local; Reed Bunting *E. schoeniclus*, widespread and frequent in marshes and by fresh water, also known as pit sparrow and reed sparrow; Snow Bunting *Plectrophenax nivalis*, rare on Scottish mountain tops, locally frequent in winter on high moors and along the coast. Yellow Bunting ⇨ YELLOWHAMMER. Three regular passage migrants in varying numbers: Lapland Bunting *Calcarius lapponicus*, the Lapland Longspur of north America; Little Bunting *E. pusilla*, and Ortolan *E. hortulana*. Also six stragglers.

Bur. The dead fruiting heads of burdock and other plants, armed with hooked bristles that adhere to dogs and clothing,

Burbot or Eel Pout (Gadidae: *Lota lota*). A somewhat elongated fish, the only freshwater member of the Cod Family, confined to rivers flowing into the North Sea between Durham and Norfolk.

Burdock (Compositae: *Arctium*). Tall spreading, large-leaved plants with sprays of purple flowerheads surrounded by hooked bristles, which when dried form burs. Common or Lesser Burdock *A. minus* is widespread and common in woods and on waysides and waste ground. Greater Burdock *A. lappa* is largely confined to waysides and waste ground in England and Wales.

Bur Marigold (Compositae: *Bidens*). Two waterside plants, Trifid Bur Marigold *B. tripartita* and Nodding Bur Marigold *B. cernua* with normally unrayed button-like flowers, and flattened fruits armed with barbed bristles that adhere firmly to clothing; both are locally frequent in England and Wales, especially on the dried beds or shores of ponds and lakes.

Burn. A small stream or brook in Scotland.

Burnet (Rosaceae: *Poterium* and *Sanguisorba*). Plants with pinnate leaves and tight heads of small flowers: Salad Burnet *P. sanguisorba*, named from the former use of its cucumber-tasting leaves in salads, common in calcareous grassland; Fodder Burnet *P. polygamum*, an escape from cultivation; and Giant Burnet *S. officinalis*, frequent in damp meadows, with reddish-purple flowers.

Burnet Companion (Caradrinoidea: *Ectypa glyphica*). One of our few day-flying moths, named from being often seen in company with burnet moths.

Burnet Moth (Psychoidea: *Zygaena*). A group of seven small red-and-black day-flying moths, of which the commonest and most widespread is the Six-spot Burnet *Z. filipendulae*.

Burnet Rose. ⇨ ROSE, WILD.

Burnet Saxifrage. ⇨ SAXIFRAGE.

Burnt-Tip Orchid (Orchidaceae: *Orchis ustulata*). An attractive small orchid of calcareous turf in the south, resembling a miniature lady orchid.

Bur-Reed (Sparganiaceae: *Sparganium*). Aquatic perennial monocotyledons, with flowers in globular bur-like heads; four species, the commonest Branched Bur-reed *S. erectum*.

Burrel Fly. A local name for the blood-sucking tabanid flies.

Burrow. Tunnel excavated underground by rabbits and other animals.

Burrowing Shrimp. (Decapoda Reptantia: Callianassidae). Highly specialized crabs, which burrow in the muddy sand; the main species are *Callianassa subterranea* and *Upogebia stellata*.

Burrows. Extensive range of coastal sand dunes in Wales and W. England.

Burweed, Spiny. ⇨ Spiny COCKLEBUR.

Bury. An isolated rabbit burrow.

Burying Beetle (Staphylinoidea: *Necrophorus*). A genus of carrion beetles also known as sexton beetles from their singular habit of burying the carcases of small vertebrates such as mice and rolling their skinned flesh up into a ball to provide food for their larvae; *N. humator* is the commonest species.

Bush. A small shrub.

Bush Grass (Gramineae: *Calamagrostis epigejos*). A tall reed-like grass of woods and fens on heavy soils.

Bustard (Ralliformes: Otitidae). Large, fairly long-legged ground-living birds, related to the cranes rather than the waders. Three British species, all rare: the Great Bustard *Otis tarda*, used to breed but now a very rare straggler from Europe; the Little Bustard *O. tetrax*, rare visitor from Europe; Macqueen's or Houbara Bustard *Chlamydotis undulata*, very rare straggler.

Butcher Bird. ⇨ Red-backed SHRIKE.

Butcher's Broom (Liliaceae: *Ruscus aculeatus*). A singular plant, not known ever to have been used by butchers for brooms; the only shrubby monocotyledon native to Britain; locally frequent in woods and scrub, mainly in the south; flowering in early spring. Its tiny greenish-white flowers, which turn to red berries, are borne on the upper surface of what appear to be small oval spine-tipped leaves, but are actually flattened stems, known as cladodes.

Butterbur (Compositae: *Petasites*). Three plants with broad brush-like tufts of pink or white flowers appearing early in the year before the large broad leaves, much the commonest and the only native being Butterbur *P. hybridus* of roadsides and shady stream-sides, with leaves 1 – 3 ft broad. ⇨ WINTER HELIOTROPE.

Buttercup (Ranunculaceae: *Ranunculus*). Wild flowers with five shiny yellow petals and acrid poisonous juice; a dozen species, including the spearworts and wood goldilocks, the commonest being Meadow Buttercup *R. acris*, Creeping Buttercup *R. repens*, and Bulbous Buttercup *R. bulbosus*, all growing mainly in grassy places.

Butterfish. ⇨ GUNNEL.

Butterfly (Lepidoptera: Papilionina). The superfamily, formerly known as Rhopalocera, containing the butterflies, which are characterized by possessing club-tipped antennae and no mechanism for hooking the fore and hind wings together, as in moths. All butterflies but only a few moths are day-flying. There are sixty-eight British species, of which ten are only rare or occasional visitors, in eight families, of which the most important are Saturnidae, containing the 'browns'; Nymphalidae, including the fritillaries, the vanessids, the purple emperor and the white admiral; Lycaenidae, containing the blues, coppers and hairstreaks; Pieridae, including the whites, brimstone and clouded yellow; and Hesperiidae, the skippers.

Butterfly Bush. ⇨ BUDDLEIA.

Butterwort (Lentibulariaceae: *Pinguicula*). Insectivorous plants with purple two-lipped violet-like flowers, catching their insect prey with their sticky leaves which fold in over the victims; three species, the commonest being *P. vulgaris* of bogs and wet moors and heaths, mainly in the north and west.

Butt Rot. The decay of the lower part of a living tree.

Buzzard. (1) (Falconiformes: Falconidae). A large, broad-winged bird of prey. Two breeding species: The Common Buzzard *Buteo buteo*, not uncommon in parts of the north and west, especially Devon, and the Honey Buzzard *Pernis apivorus*, noted for feeding on the grubs of wild bees and wasps, which still nests irregularly, most often in the New Forest. The Rough-legged Buzzard *B. lagopus* is a scarce winter visitor from northern Europe. (2) A local name for noctuid moths.

By-the-wind-Sailor or Jack-Sail-by-the-Wind (Siphonophora: *Velella spirans*). An oceanic siphonophore, floating on the surface of the sea, like a blue jellyfish with a stiff 4-in. sail, and polyps hanging down. An erratic visitor from the open Atlantic, especially to our western coasts after prolonged south-westerly winds.

C

Cabbage (Cruciferae). Name given to several members of the Cabbage Family, including Wild Cabbage *Brassica oleracea*, the origin of the vegetable, a local sea-cliff plant, mainly in the south; Bargeman's Cabbage *B. campestris*, the wild form of the turnip, frequent on river banks; Dune or Isle of Man Cabbage *Rhynchosinapis monensis*, local on sandy shores in the north-west; and Warty Cabbage *Bunias orientalis*, a spreading alien with warty fruits.

Cabbage Moth (Caradrinoidea: *Mamestra brassicae*). A noctuid whose caterpillar is a pest of cabbages and other brassicas, though it will also feed on many other plants.

Cabbage Root Fly (Muscidae: *Erioischia brassicae*). A two-winged fly whose larva is an economic pest, damaging the roots of cabbages and allied vegetables.

Cabinet Beetle (Diversicornia: *Trogoderma versicolor*). An introduced pest of stored products.

Cacao Moth or Tobacco Moth (Pyraloidea: *Ephestia elutella*). A small pyralid moth, whose larvae are reckoned our most serious insect pest of stored cacao and tobacco, though they also eat other stored products.

Cachalot or Sperm Whale (Cetacea: *Physeter catodon*). An uncommon but regular visitor from the tropics to British seas, the largest of the toothed whales, much sought after for the spermaceti and ambergris which it produces.

Caddis-Fly or Sedge-Fly (Insecta: Trichoptera), Moth-like winged insects, several of which have fancy anglers' names, such as Halford's Welshman's Button; 188 species in ten families. Their aquatic larvae are noted for their habit of wrapping themselves in a protective silken tube, to which are attached fragments of stick, sand, stone and other detritus. They are valued by anglers because fish feed on both larvae and adults.

Caducous. Botanical adjective used of flowers, leaves or other parts of a plant which fall off early in their development.

Caespitose. Botanical adjective meaning tufted.

Cake Urchin. An order of sea urchins (Clypeasteroida).

Calamint (Labiatae: *Calamintha*). Three mint-scented aromatic plants with pale pinkish-lilac flowers, growing on banks on calcareous soils; *C. ascendens* is the most frequent.

Calcareous. Containing calcium in the form of chalk or lime.

Calcicole. A plant liking lime in the soil, appearing to require a high proportion of available calcium salts, e.g. the common rock-rose, salad burnet and traveller's joy.

Calcifuge. A plant disliking lime in the soil, calcium salts being apparently poisonous to it, e.g. most members of the Heath Family.

Calf. The young (first year) of red deer and Japanese sika deer.
Calf's Snout. ⇨ Lesser SNAPDRAGON.
Callunetum. A plant community dominated by ling heather *Calluna vulgaris*, usually a heath or moor. ⇨ –ETUM.
Calyx. The outer ring of components of a flower, often divided into separate sepals, usually green, but occasionally, as in the marsh marigold, brightly coloured and appearing like a corolla (petals). ⇨ PERIANTH.
Camberwell Beauty (Papilionina: *Nymphalis antiopa*) A large and handsome vanessid butterfly, with a white or pale yellow border, which occasionally visits us from Scandinavia but does not breed; it was first seen in the eighteenth century in Camberwell. It has been suspected that some may come as chrysalids by ship with loads of timber.
Cambrian Period. ⇨ GEOLOGICAL PERIODS.
Camm. A linear buttress of shingle on the inner side of the great shingle beach of Chesil Bank, Dorset, cut by a series of small gullies called 'cans'.
Camouflage. ⇨ CRYPTIC COLORATION.
Campanulaceae (Angiospermae: Dicotyledones). The Bellflower Family, characterized by its bell-shaped, usually blue, flowers, including harebell, the rampions and sheepsbit scabious, as well as the bellflowers.
Campanulate. Botanical adjective meaning bell-shaped.
Campion (Caryophyllaceae: *Silene*). A dozen native or established species, mostly with inflated calyxes and white or pink flowers; the commonest white-flowered are White Campion *S. alba*, arable and wayside weed; Bladder Campion *S. vulgaris*, in bare grassy places; and Sea Campion *S. maritima* on the shore; pink-flowered: Red Campion *S. dioica*, in woods and on hedge-banks and sea-cliffs. Moss Campion *S. acaulis*, also pink, grows on many mountain tops.
Can. A gully on Chesil Bank. ⇨ CAMM.
Canary. ⇨ SERIN.
Canary Grass (Gramineae: *Phalaris canariensis*). A constituent of bird-seed mixtures, now often found growing on rubbish tips and waste ground.
Candle-Snuff Fungus (Ascomycetes: *Xylaria hypoxylon*). A blackish club-shaped fungus, named from its appearance when covered with white spores, related to Devil's Fingers and growing on rotting tree stumps and logs.
Candytuft, Wild (Cruciferae: *Iberis amara*). White- or mauve-flowered annual of disturbed ground, largely confined to the Chilterns.
Cane. A country name for the female weasel, which is smaller than the male. ⇨ WHITRICK.
Canopy. The layer of vegetation in a woodland represented by the upper branches of the trees.
Capercaillie (Galliformes: *Tetrao urogallus*). A large woodland grouse, which became extinct in Scotland in the eighteenth century but was successfully reintroduced from Sweden about fifty years later and is now widespread in pinewoods, especially in the eastern Highlands.
Capillary. Botanical adjective meaning hair-like.

Capitate. Botanical adjective meaning head-like.

Capitulum. A compound flower-head, as in the Compositae (dandelions, daisies, etc.) usually surrounded by an involucre. The term also has various zoological meanings.

Caprimulgiformes. The order of birds including the nightjars (Caprimulgidae).

Capsid Bug (Heteroptera: Miridae). A large family of delicate-looking, somewhat elongated greenish or brownish plant bugs, including some predatory species. Some 200 British species, including the abundant Common Green Capsid *Lygocoris pabulinus*, sometimes a pest of fruit and vegetables; the notorious Apple Capsid *Plesiocoris rugicollis*, an orchard pest, now controlled by DDT; and the Black-kneed Capsid *Blepharidopterus angulatus*, an important predator of the red spider mite, whose destruction by DDT at the same time as the apple capsid has turned the mite, which is not affected by DDT, into a serious pest of apples. ⇨ BRACKEN BUG, FERN BUG, PLANT BUG.

Capsule. A dry fruit containing one or more seeds and splitting open in various ways, but not down one side (a follicle) or down two sides (a pod).

Carabid Beetle. ⇨ GROUND BEETLE.

Caradrinoidea (Lepidoptera: Heteroneura). The super-family of moths containing, the two families of tiger, ermine, and footman moths (Arctiidae, Hypsidae), the large noctuid family (Caradrinidae), the silver Y, snouts, red underwings and their allies (Plusiidae) and the tussock moths (Lymantriidae).

Carapace. The upper part of the shell of a tortoise or turtle, covering the animal's back; also used of the corresponding part of crabs and some other arthropods.

Caraway (Umbelliferae: *Carum carvi*). The white-flowered umbellifer whose aromatic seeds are used to flavour cakes, established in meadows in the north and more widespread as a casual. Whorled Caraway *C. verticillatum*, with thread-like leaf-segments in apparent whorls, is local in damp grassy places on acid soils in the west. Corn Caraway, ⇨ CORN PARSLEY.

Carboniferous Period. ⇨ GEOLOGICAL PERIODS.

Carder Bee (Aculeata: *Bombus*). Three of the smaller species of bumblebee, *B. muscorum*, *B. sylvarum* and *B. solstitialis*.

Cardinal Beetle (Heteromera: Pyrochroidae). Three species of beetle named from their bright red colouring, *Pyrochroa serraticornis* being perhaps the most frequent.

Cardinal Spider (Araneae: *Tegenaria parietina*). One of the larger of our three species of House Spider, with a body up to ¾-in. long, named from the legend that it once frightened Cardinal Wolsey at Hampton Court.

Carnation Grass (Cyperaceae: *Carex flacca* and *C. panicea*). Two common sedges whose greyish leaves recall those of carnations, *C. flacca* in grassland, especially calcareous, *C. panicea* in marshy spots.

Carnivora. An order of predatory, flesh-eating mammals, represented in Britain by eight natives: fox, badger, otter, wild cat, stoat, weasel, polecat and pine marten; and by two aliens: feral domestic cat and mink.

Carnivore. An animal or plant that feeds on flesh.

Carp (Pisces: Cyprinidae). The Carp *Cyprinus carpio* is a rather large, long-lived, freshwater fish, attaining weights up to 44 lb. in the wild in Britain; widespread in ponds as a result of medieval introductions. Golden Carp, King Carp and Mirror Carp are all varieties in captivity and often put into ponds; the Prussian Carp is another variety. The smaller Crucian Carp *Carassius carassius,* found in rivers and ponds in England, may also be introduced.

Carpal Joint. The wrist of a bird's wing, the joint which projects forward when the wing is at rest, also applicable to the corresponding limbs of mammals, reptiles and amphibians.

Carpel. One of the units making up the pistil, and consisting of an ovary, which is the future seed-box, usually surmounted by a stalk known as the style, at whose tip is the stigma, which receives the pollen to start the process of fertilizing the ovules inside the ovary and turning them into seeds.

Carpenter Bee (Aculeata: Apidae). A group of solitary bees which burrow into wood, especially rotting wood, to make their nests; *Ceratina cyanea* is our commonest species. ⇨ LEAF-CUTTER BEE, WOOL-CARDER BEE.

Carpet Fly. ⇨ WINDOW FLY.

Carpet Moth (Notodontoidea: Hydriomenidae). Name given to many geometrid moths, especially of this family, one of the best known being the Garden Carpet *Xanthorhoë fluctuata.*

Carpet Shell (Lamellibranchia: *Venerupis*). Four species, two widespread and two mainly southern, of marine bivalves, resembling smooth cockles, the commonest being the Pullet Carpet Shell *V. pullastra.*

Carr. A fen wood of which the typical tree is the alder. Common in East Anglia, especially the Broads, and found in many other areas.

Carragheen Moss or Irish Moss (Rhodophyta: *Chondrus crispus* and *Gigartina stellata*). A name given to two edible seaweeds on the south and west coasts of Ireland, where they are used in making soups, jellies and other gelatinous preparations, also in making agar. In the Hebrides they are used to thicken milk and make a custard-like dish. Carragheen is a place in Co. Waterford.

Carrion Beetle (Staphylinoidea: Silphidae). A family of carrion-eating beetles, with sixty species, the best known being the burying beetle.

Carrot, Wild (Umbelliferae: *Daucus carota*). A common white-flowered (but often with a single red or purple flower in the centre of the umbel) umbellifer of dry, especially calcareous grassland; the origin of the vegetable. Moon Carrot *Seseli libanotis* is an extreme rarity of chalk grassland in the south.

Carstone. A building stone quarried from the Lower Greensand in Bedfordshire and elsewhere.

Cartwheel Flower. ⇨ GIANT HOGWEED.

Caryophyllaceae (Angiospermae: Dicotyledones). The Pink Family, characterized by its jointed stems with undivided leaves, and including the campions, catchflies, chickweeds, stitchworts, pearlworts, sandworts and spurreys, as well as the pinks.

Casual. ⇨ ADVENTIVE.

Cat (Carnivora: *Felis*). The fierce Wild Cat *Felis sylvestris*, the only native member of the cat tribe, is confined to the Scottish Highlands. Elsewhere there are numerous feral Domestic Cats *F. catus*, and in Scotland the two are said to interbreed.

Catchfly (Caryophyllaceae: *Silene* and *Lychnis*). Name, derived from their sticky hairs, for some of the rarer campions, of which the most widespread is the Small-flowered Catchfly *S. gallica* of fields and waste places.

Catchment Area. The area within which all water drains into a single river system, separated from other catchment areas by a Watershed.

Caterpillar. The larva of a butterfly or moth.

Caterpillar Fungus. ⇨ VEGETABLE FLY.

Cat-Fish (Percomorphi: *Anarhichas*). Northern deep-water sea fishes, allied to and resembling the blennies and gobies; four British species, of which the only really frequent one is *A. lupus*, also known as Sea-cat or Wolf-fish. Not to be confused with the freshwater Catfishes *Silurus*, which are not native, though the Wels *S. glanis* is naturalized in a few waters.

Cat Ice. Thin ice forming over a pool or puddle, often separated from the water by an air space, and supposedly able to bear no greater weight than that of a cat.

Catkin. A long tassel-like spike of small flowers, especially in hazel, the oaks, the poplars and willows, and their allies.

Catmint (Labiatae: *Nepeta cataria*). White-flowered perennial with a pungent smell highly attractive to cats, like that of the allied purple-flowered plant of the same name commonly grown in gardens, *N. x faassenii*.

Catsear (Compositae: *Hypochaeris*). Three grassland plants with yellow dandelion-like flowers, much the commonest being Common Catsear *H. radicata*. ⇨⇨ MOUNTAIN EVERLASTING.

Catsfoot. ⇨ MOUNTAIN EVERLASTING.

Catstail. Alternative name for reed-mace, now chiefly used in North America. ⇨⇨ TIMOTHY.

Cattle, Wild White (Artiodactyla: *Bos taurus*). There are no genuinely wild cattle left in the British Isles; nor are the so-called 'wild white cattle' emparked at Chillingham, Northumberland, and in a few other parks and zoos the descendants of our former wild stock; the herd formerly at Chartley Park, Staffordshire, is now at Whipsnade Zoo.

Catworm (Polychaeta: *Nephthys hombergi*). A long (up to 8 ins.) pearly pink or white, active marine bristle worm, widely distributed and common on the shore and often used as bait: related to the white ragworm.

Caudata or Urodela (Chordata: Amphibia). The order of tailed amphibians containing the newts and the European but non-British salamanders (both Salamandridae).

Cauliflower Fungus (Ascomycetes: *Sparassis crispus*). An edible fungus superficially resembling a cauliflower, found at the base of pine stumps.

Cauline. Botanical adjective used of leaves growing on the stem of a plant, but not right at the base or immediately beneath a flower or inflorescence, the latter type of leaf being known as a bract.

Cave Research Group of Great Britain. The national body concerned with the scientific study of caves, including natural history and geology, founded in 1946; publishes *Transactions.*

Cavernicolous. Inhabiting caves.

Caving. ⇨ SPELAEOLOGY.

Celandine. Name given to two quite unrelated yellow-flowered plants because in ancient Greece both used to bloom at the time when the swallows (*chelidon* in Greek) arrived. Greater Celandine (Papaveraceae: *Chelidonium majus*) is a medicinal herb now widely established in hedge-banks in England and Wales. Lesser Celandine (Ranunculaceae: *Ranunculus ficaria*) is widespread and common in copses and hedge-banks in early spring.

Celery, Wild (Umbelliferae: *Apium graveolens*). A native umbellifer with greenish-white flowers, the origin of the garden vegetable, locally frequent in grassy places by the sea in the south, scarce inland.

Celery Fly (Cyclorrhapha: *Philophylla heraclei*). A gall fly well known as a pest of celery in gardens; its larvae are leaf-miners.

Cellar Beetle. ⇨ CHURCHYARD BEETLE.

Census. A numerical count of the population of a bird or other animal in any area. Censuses have been made of the populations of the heron, gannet, and great crested grebe among widespread and common British birds, and of the St Kilda wren and other rare birds.

Centaury (Gentianaceae: *Centaurium*). Low-growing but upright plants with pink flowers, closing except in full sunshine; six species, of which only Common Centaury *C. erythraea* and Slender Centaury *C. pulchellum* are at all widespread. There are also two yellow centauries, known as yellowworts.

Centipede (Arthropoda: Chilopoda). A class of myriapods, living under stones, logs, dead leaves, etc., mainly carnivorous. None of the forty or so British species have vernacular names: *Lithobius forficatus* is a common large reddish-brown species; *Scolioplanes maritimus* is common on the shore.

Cep. ⇨ BOLETUS.

Cephalochordata. A very primitive sub-phylum of the Chordata, containing the lancelet.

Cephalopod (Mollusca: Cephalopoda). A specialized and highly developed class of active, carnivorous marine molluscs, propelling themselves by a form of jet propulsion, and possessing eight or ten tentacles or arms, equipped with strong suckers. The squids and cuttlefishes belong to the sub-order Decapoda, and the octopuses to the sub-order Octopoda.

Cestoda. ⇨ TAPEWORM.

Cetacean. Member of the mammalian order Cetacea, the whales, porpoises and dolphins.

Chaetognatha. ⇨ ARROW WORM.

Chaetopoda. The name formerly given to the greater part of the Phylum Annelida (segmented worms) now split into the Polychaeta, the Oligochaeta, and the small class Myxostomaria containing parasitic marine worms.

Chafer (Lamellicornia: Melolonthinae). Vegetarian allies of the dung beetles, including the well known Cockchafers, the Summer Chafer *Amphimallon solstitialis* and the Garden Chafer or Bracken-clock *Phyllopertha horticola*. Rose Chafer, ⇨ ROSE BEETLE.

Chaffinch (Passeriformes: *Fringilla coelebs*). One of the two commonest British birds, a brightly coloured finch widespread in woods, hedgerows and gardens.

Chaffweed (Primulaceae: *Anagallis minima*). A tiny pink pimpernel, often under an inch high, growing in wet heathy places.

Chalcid Wasp (Parasitica: Chalcidoidea). A super-family of minute, mostly parasitic wasps, including the fairy-flies, parasitizing a wide variety of insects and their eggs, including some as small as aphids and scale-insects; some 1,564 British species. One of the best known species is *Pteromalus puparum*, a parasite of the cabbage white butterfly, laying its eggs in the newly formed pupae.

Chalk. A soft white very porous form of limestone, containing from 90 – 99 per cent of calcium carbonate. It contains the remains of small marine animals and shells and was laid down under the sea that covered the south-eastern half of England in the Cretaceous period, about 100 million years ago. The chalk dominates the south-eastern half of England, and gives the typical smooth rounded outline of the Downs, with many dry valleys. Beech is the characteristic tree, and the flora is rich in orchids and other colourful wild flowers.

Chalybeate Spring. A spring whose water contains a high proportion of iron.

Chamaephyte. A plant whose resting buds in winter are above the surface of the ground, but not more than a foot.

Chamomile (Compositae). Plants with daisy-like flowers: Wild Chamomile *Matricaria recutita*, a common aromatic weed of cornfields and waste ground; Corn Chamomile *Anthemis arvensis*, a less aromatic and less common weed; and Common Chamomile *Chamaemelum nobile*, a highly aromatic local plant of sandy grasslands, formerly much planted for lawns.

Chanterelle (Basidiomycetes: *Cantharellus cibarius*). A well-known yellow edible fungus, flat-topped but not of the true toadstool shape, locally frequent in broad-leaved woods, especially beech, and much esteemed by the comparatively few who eat edible fungi other than mushrooms. The False Chanterelle *Hygrophoropsis aurantiaca*, growing in coniferous woods and sometimes confused with it, is edible but not recommended.

Char (Salmonidae: *Salvelinus alpinus*). Non-migratory relatives of the trout, landlocked in cold and usually deep mountain lakes in North Wales, the Lake District, Scotland and Ireland. At least fifteen different races are recognized, including the torgoch and haddy.

Charadriiformes or Lari-Limicolae. A large order of mostly aquatic or marsh-loving birds, including the waders (Charadriidae, Scolopacidae), gulls and terns (Laridae), skuas (Stercorariidae), and auks (Alcidae).

Chardonneret Trap. A type of trap, operating on the deadfall principle, used by bird ringers to catch small passerines and named from the French word for goldfinch.

Charlock. ⇨ Wild MUSTARD.

Charm. A small flock or family party of goldfinches.

Charophyceae, Charophyta. ⇨ STONEWORT.

Chat. Name given in Britain to song birds of the genus *Saxicola* (stonechat, whinchat), and elsewhere to those of the genus *Oenanthe*, usually known in Britain as wheatears.

Cheesecakes. An old name for the common mallow.

Cheese Mite (Astigmata: *Tyrolichus casei*). A notorious pest of cheese, also infesting many other stored products. The allied *Carpoglyphus lactis* also occurs on cheese.

Cheese Skipper (Cyclorrhapha: *Piophila casei*). The larva of a small two-winged fly (Piophilidae), which is a common pest of stored products containing animal protein; the larva 'skips' by grasping its hind parts with its mouth and suddenly letting go.

Chelicerae. The biting jaws and poison fangs of spiders and other arachnids.

Chelonethi. ⇨ FALSE SCORPION.

Chelonia. Alternative name for the Testudines (tortoises etc.).

Chemotaxis. ⇨ TAXIS.

Chemotropism. ⇨ TROPISM.

Cherry (Rosaceae: *Prunus*). Three species grow wild: one tree, the widespread native Gean or Wild Cherry *P. avium*, ancestor of the garden sweet cherries, especially in beechwoods; and two shrubs, the mainly southerly Sour Cherry *P. cerasus*, probably originally bird-sown from sour and Morello cherries in gardens and orchards; and the mainly northerly Bird Cherry *P. padus*, with flowers in spikes like the garden Cherry-Laurel *P. laurocerasus*.

Chert. A grey, brownish or white flint-like stone found in chalk, limestone and Greensand districts, and composed of sand grains cemented together by silica.

Chervil (Umbelliferae). A name for several white-flowered umbellifers, of which much the commonest is Wild Chervil or Cow Parsley, which is followed in the hedge-banks by Rough Chervil *Chaerophyllum temulentum*, with purple-spotted stems. Bur Chervil *Anthriscus caucalis* is more local, mainly on sandy banks near the sea.

Chestnut, Horse (Hippocastanaceae: *Aesculus hippocastanum*). A tall deciduous tree from the Balkans, often planted as an ornamental tree for its handsome white flowers, and not infrequently naturalized. Its fruits, known as 'conkers', are superficially similar to edible chestnuts.

Chestnut, Sweet (Fagaceae: *Castanea sativa*). A tall deciduous catkin-bearing tree, native of the Mediterranean, often planted and readily regenerating itself on light sandy soils; believed by some to be native also in parts of England. Fruit is the edible chestnut.

Chestnut Worm or Purple Worm (Oligochaeta: *Lumbricus castaneus*). A fairly common and widely distributed earthworm of rich, moist soils, named from its chestnut or violet-brown colour. Also called marsh worm by anglers.

Chick. The small young of terrestrial nidifugous birds, especially of game birds and waders.

Chickweed (Caryophyllaceae: *Stellaria*). Four plants with broad leaves and deeply cleft white petals, especially Common Chickweed *S. media*, an abundant weed; also Water Chickweed *Myosoton aquaticum*, frequent by fresh water. Mouse-Ear Chickweed, ⇨ MOUSE-EAR.

Chicory. ⇨ SUCCORY.

Chiffchaff (Passeriformes: *Phylloscopus collybita*). Leaf warbler, noted for its monotonous song, *chiff-chaff*, which alone readily distinguishes it from the willow warbler in the field; summer visitor to woods and scrub north to the Forth, but mainly in the south.

Chilopoda. ⇨ CENTIPEDE.

Chimaera (Bradyodonti: Chimaeridae). Deep-water fishes, mainly of northern seas; three British species, the commonest being the rabbit-fish.

Chinaman's Hat (Prosobranchia: *Calyptraea chinensis*). A whitish limpet-like mesogastropod operculate mollusc, related to the bonnet and slipper limpets and local on the lower shore on our south-west coasts.

China Mark Moth (Lepidoptera: *Nymphula*). Three species of moths of the microlepidopterous family Pyralidae, named from the markings on their wings, and with aquatic larvae.

Chinchbug, European (Heteroptera: *Ischnodemus sabuleti*). An elongated blackish plant-bug, related to the serious cereal pest of North America; has spread widely over south-east England during the twentieth century, feeding on grasses.

Chine. A ravine on the coasts of Dorset, Hampshire and the Isle of Wight.

Chink Shell (Prosobranchia: *Lacuna*). Small intertidal operculate molluscs, closely related to and somewhat resembling the winkles, the commonest being the Banded Chink Shell *L. vincta*, yellow with red bands, widespread on seaweed.

Chironomid. ⇨ Non-biting MIDGE.

Chiroptera. ⇨ BAT.

Chiton or Coat-of-Mail Shell (Mollusca: Amphineura). Small primitive limpet-like molluscs, with eight-valved shells, able to curl up like pill-bug woodlice, and feeding on seaweeds. Representatives of seven genera occur in Britain, and some are common both between the tides e.g. *Lepidochitona cinereus*, and at greater depths, e.g. *Lepidopleurus asellus*.

Chives (Amaryllidaceae: *Allium schoenoprasum*). The well-known culinary herb, with tufts of greyish leaves and pink-purple flowers, which is a scarce native in Cornwall, South Wales and northern England, growing on cliffs and rock ledges.

Chlorophyceae, Chlorophyta. ⇨ GREEN ALGAE.

Chlorophyll. The green colouring matter of almost all plants, except fungi, and a few animals. ⇨ PHOTOSYNTHESIS.

Chlorosis (adj. Chlorotic). A condition of plants in which leaves, or parts of leaves, turn pale green or yellowish, due to a variety of causes, usually magnesium or other mineral deficiency, sometimes induced by an excess of calcium, when it is known as Lime Chlorosis.

Chondrophore. ⇨ SIPHONOPHORE.

Chordata. The most advanced phylum, or main grouping, of the Animal Kingdom, characterized by the possession, at least in the embryo or larval state, of a notochord, a skeletal rod lying lengthwise between the central nervous system and the gut. There are four subphyla, all of which are represented in the British fauna: Hemichordata (enteropneusts); Urochordata or Tunicata (sea-squirts); Cephalochordata (lancelets), and Vertebrata (mammals, etc.).

Chough (Passeriformes: *Pyrrhocorax pyrrhocorax*). An uncommon relative of the crows, with all-black plumage but red legs and bill, named from its call-note. Sometimes known as the Cornish chough, but now almost if not quite extinct in Cornwall, though still found in Ireland, Wales, the Isle of Man and the Inner Hebrides, mostly on sea cliffs. Also called the Red-billed Chough, to distinguish it from the yellow-billed Alpine Chough *P. graculus*, which has only once occurred in Britain.

Christmas Tree. ⇨ SPRUCE.

Chromosome. The protein threads on the nucleus of the cell on which the Genes are carried.

Chrysalis. Pupa, applied originally only to the brightly coloured pupae of certain butterflies.

Chrysalis Snail (Gastropoda: Pulmonata). A group of four small land snails, slightly resembling a moth chrysalis, related to the whorl snails, the commonest being *Lauria cylindrica* and the largest *Abida secale*.

Chrysophyta. A phylum or major division of small, mainly freshwater algae with three classes: the Xanthophyceae or Heterokontae, which somewhat resemble the green algae; the often golden-yellow Chrysophyceae, and the Bacillariophyceae or diatoms.

Chub (Cyprinidae: *Squalius cephalus*). A widespread and common surface-living river fish, closely related to the dace, preferring fast streams and favoured by coarse fishermen.

Churchyard Beetle or Cellar Beetle (Heteromera: *Blaps mucronata*). One of the nocturnal ground beetles, often found at night in cellars, stables and large kitchens.

Cicada (Homoptera: Cicadidae). A family of insects allied to the aphids and froghoppers, widespread in warmer climates but represented in Britain only by the rare *Cicadetta montana* in the New Forest.

Ciliate. (1) ⇨ INFUSORIAN. (2) Botanical adjective used of leaves, sepals or other organs with tiny hairs growing along their margins.

Cilium (pl. Cilia). A minute thread-like organ, moving in unison with other cilia to move the organism, e.g. a protozoan, or provide it with food.

Cinnabar Moth (Caradrinoidea: *Hipocrita jacobaeae*). One of our most striking moths, red and black, with equally striking warning coloration in its black-and-yellow banded caterpillars, which often completely defoliate ragwort and groundsel.

Cinquefoil (Rosaceae: *Potentilla*). Mostly low-growing, yellow-flowered plants with leaves palmately divided into leaflets, often five in number; a dozen species, including the common weed Creeping Cinquefoil *P reptans*, the purple-flowered Marsh Cinquefoil *P. palustris*, and the rare Shrubby Cinquefoil *P. fruticosa*. ⇨ SILVERWEED, BARREN STRAWBERRY, TORMENTIL. Least Cinquefoil *Sibbaldia procumbens* is a closely related mountain plant.

Circumscissile. Botanical adjective used of fruits which open to shed their seeds in such a way that the top comes off like a lid.

Cirripedia. ⇨ BARNACLE.

Cirrocumulus; Cirrostratus; Cirrus or Marestail Clouds. ⇨ CLOUD.

Cladoceran. ⇨ WATER FLEA.

Cladode. A green shoot that is flattened and leaf-like, as in butcher's broom.

Clam. ⇨ Great or Edible SCALLOP. Soft-shelled Clam, ⇨ GAPER.

Clap Net. A net pegged to the ground, which can be released by a spring to catch birds, formerly much used by bird-catchers.

Clary (Labiatae: *Salvia horminoides*). Purple-flowered relative of the scarce meadow sage, fairly frequent in grassy spots in the south.

Class. One of the groups in the Classification of living organisms, comprising one or more Orders. Classes are grouped into a Phylum (pl. phyla) in the Animal Kingdom or a Division in the Vegetable Kingdom.

Classification. All living things, plants, mammals, insects, etc., are classified in a hierarchy of similar groups. The lowest group is the Species; similar species are grouped into a Genus; similar genera into a Family; families into an Order; orders into a Class; classes into a Phylum (zoological) or Division (botanical). The system is not finalized and is constantly being adapted to fit new knowledge. Within the species there are Subspecies, Varieties and Forms. ⇨ TAXONOMY.

Clavate. Botanical adjective meaning club-shaped.

Clavicornia (Coleoptera: Polyphaga). A heterogeneous super-family of beetles, containing 470 species in twenty-two families, and including the ladybirds, various wood-boring beetles, the raspberry beetle, the grain beetles and the spider beetles.

Clay. A soil consisting of extremely fine particles, very retentive of moisture. It is sticky and swells when wet; hardens and shrinks on drying. Boulder Clay is a glacial deposit, also known as Till, often highly calcareous, and so called because it contains boulders carried considerable distances by the ice. Gault is a stiff rather calcareous clay, occurring between the Upper and Lower Greensands, sometimes directly underlying the chalk. London Clay underlies much of the London basin and is particularly thick and stiff. Oxford Clay, laid down in Jurassic times between 100 and 150 million years ago, stretches in a narrow band diagonally across the country from

Dorset to Yorks via Oxford. Weald Clay is a Cretaceous bed lying below the Lower Greensand.

Clay-with-Flints. A non-calcareous, sometimes acid deposit found on top of the chalk hills of the Chilterns, the North and South Downs and part of the Hampshire Downs. Oak, sometimes beech, is the typical tree of woods on the clay-with-flints. ⇨ LEACHING.

Clearwing (Tinaeoidea: Sesiidae). A small family of moths, with few or no scales on their wings, so that they look like bees or wasps, one being known as the hornet moth.

Cleavers. ⇨ GOOSEGRASS.

Cleaves. The two halves of a deer's hoof.

Cleg. The most commonly used vernacular name for the smaller blood-sucking tabanid flies (*Haematopota*).

Cleistogamous Flower. A flower that is self-pollinated and which therefore does not need to open to allow the pollen to be dispersed by insects or the wind.

Cleugh. Scots version of Clough.

Click Beetle or Skipjack (Diversicornia: Elateridae). A family of sixty-five beetles named for their power of leaping into the air while on their backs, making a clicking sound. The larvae of many click beetles, especially of the genera *Agriotes* and *Athous*, are the root-destroying pests of agriculture known as wireworms.

Clifden Nonpareil (Caradrinoidea: *Catocala fraxini*). A handsome large moth, resembling its relatives the red underwings but with blue-grey on the hindwings, formerly regarded as a rare visitor, but in recent years discovered to be breeding in Kent and Norfolk.

Climax. The final stage, or culmination, in a succession of natural plant communities in one area under a particular set of conditions, e.g. any bare piece of ground becomes colonized by small plants, which in turn are superseded by larger plants, which are followed perhaps by scrub and finally trees. Oakwood is the natural climax in much of Britain, but on chalk it is beechwood and on limestone ashwood. Climaxes can be held at any stage by biotic factors, such as grazing on chalk downland.

Climber. Plants that raise themselves off the ground by climbing up other plants, either by twining, by tendrils, by twisting leafstalks (traveller's joy) or by rootlets (ivy).

Cline. A situation in which certain characters in an animal or plant population tend to change gradually and continuously over a large area, as in Allen's, Bergmann's and Gloger's Rules, so that the populations at each end of the cline may be substantially different from each other.

Clint or Grike. A fissure in the surface of limestone rock. ⇨ LIMESTONE PAVEMENT.

Clitter. A jumbled collection of boulders, often very large, that have been eroded from the granite tors on south-western moorlands and fallen down the hillside.

Clone. A group of animals or plants of a precisely similar genetical make-up, because they have been produced either vegetatively or apomictically from a single plant, or asexually or parthenogenetically from a single animal.

Clothes Moth (Tinaeoidea: Tinaeidae). Four members of this large family of 'micros' are household pests, the most familiar being the Common Clothes Moth *Tineola bisselliella*, whose larvae attack woollen materials, furs and skins. ⇨ HOUSE MOTH, TAPESTRY MOTH.

Cloud. Water droplets condensed from the air and suspended at a high level; at ground level it is called Fog or Mist. Clouds are classified according to their shape and height. Cirrus clouds are of the wispy 'mares-tail' type (see below). Cumulus are the great woolly white clouds of a summer's day (see below). Stratus is the leaden sheet that spreads uniformly across the sky. 'Nimbus' signifies the presence of rain. High Clouds, at 25,000 to 30,000 ft are Cirrus, Cirrocumulus and Cirrostratus. Middle Clouds at 10,000 to 25,000 ft are Altocumulus and Altostratus. Low Clouds below 10,000 ft are Stratocumulus, Stratus, Nimbostratus and Cumulus. Cumulonimbus clouds occur up to 25,000 ft.

Cloud shapes

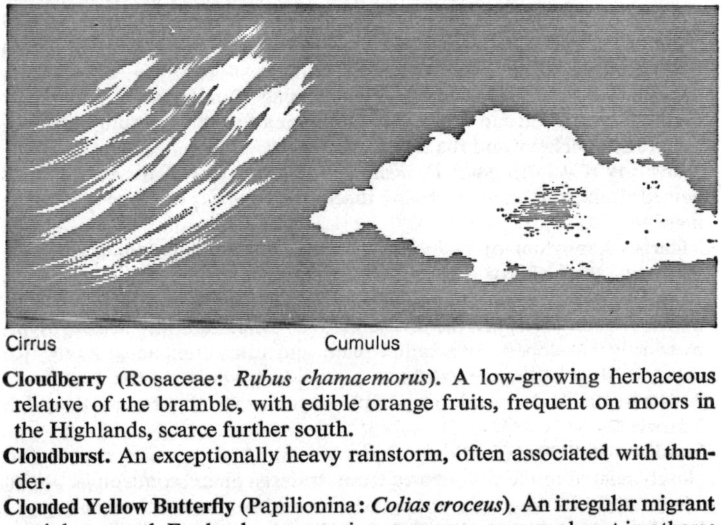

Cirrus Cumulus

Cloudberry (Rosaceae: *Rubus chamaemorus*). A low-growing herbaceous relative of the bramble, with edible orange fruits, frequent on moors in the Highlands, scarce further south.

Cloudburst. An exceptionally heavy rainstorm, often associated with thunder.

Clouded Yellow Butterfly (Papilionina: *Colias croceus*). An irregular migrant mainly to south England, common in some years, rare or absent in others; deep yellow in colour, related to the Whites. The Pale Clouded Yellow *C. hyale* is much less common, and Berger's Clouded Yellow *C. calida* has only quite recently been discovered to visit us at times and possibly even to over-winter as a hibernating larva in Kent.

Clough. A ravine or small, steep-sided valley in the north of England especially Lancashire and Cheshire; called Cleugh in Scotland. ⇨ DENE, GHYLL.

Clover (Leguminosae: *Trifolium*). A large genus of peaflowers with trefoil leaves, most of which have their small flowers in tight roundish heads and grow in grassy places. Much the commonest are White or Dutch Clover *T. repens*, grown by farmers as Wild White or Kentish Clover, and Red or Purple Clover *T. pratense*, also commonly cultivated. Alsike Clover *T. hybridum*, sometimes grown as a crop, is like a taller pink White Clover. Bokhara Clover is White Melilot grown as a crop; Calvary Clover is Black Medick. The small Haresfoot Clover *T. arvense* has elongated heads; Strawberry Clover *T. fragiferum* has flower-heads looking like tiny strawberries after they have flowered. ⇨ FENUGREEK, TREFOIL.

Club Fungus, Coral Fungus, or Fairy Club (Basidiomycetes: *Clavariaceae*). Small erect club-shaped or branched and coral-like fungi, growing on dead wood or the ground.

Clubmoss (Lycopsida: *Lycopodium* and *Selaginella*). An ancient group of plants superficially resembling mosses, but actually more closely allied to the ferns, having a vascular system. The most widespread of our six species are Fir Clubmoss *L. selago*, Stagshorn Clubmoss *L. clavatum*, and Lesser Clubmoss *S. selaginoides*.

Club-Rush (Cyperaceae). A name given to several members of the Sedge Family, the commonest being Bristle Club-rush *Scirpus setaceus*, a small plant of damp heathy places, and Sea Club-rush *S. maritimus*, frequent in brackish ditches.

Clupeidae (Pisces: Isospondyli). The Herring Family, an economically highly important group of sea fish, containing the herring, sprat, pilchard or sardine, anchovy and the two shads.

Cluster Fly (Cyclorrhapha: *Pollenia rudis*). A relative of the greenbottles named from its habit of clustering together in crevices or corners of buildings.

Cnidaria. A phylum of mainly marine aquatic animals, comprising the Hydrozoa (hydroids or sea-firs, siphonophores), Scyphozoa (jellyfish) and Anthozoa (sea-anemones and corals), formerly grouped with the Ctenophora (comb-jellies) as the Coelenterata. Cnidarians are usually symmetrically developed on a radial plan, and often look remarkably like flowers. Most of them go through both a free-swimming (medusa) stage and an attached (zooid or polyp) stage in the course of their life history.

Coal-Fish or Saithe (Gadidae: *Gadus virens*). A deep-water marine fish closely related to the cod, named from its dark colour, common in British seas, mainly in the north, young fish or cuddies coming close inshore.

Coal Measures. A deposit of Carboniferous times, especially found in Lancashire and Yorkshire, Nottinghamshire, Durham and Northumberland, forming soils with a poor and uninteresting flora. The coal was formed from forests which were swamped by sediment brought down by rivers.

Coat-of-Mail Shell. ⇨ CHITON.

Cob. (1) The male swan. (Female: pen.) (2) A cultivated hazel nut, especially Kentish cobs.

Cobweb. ⇨ SPIDER'S WEB.

Coccid. ⇨ SCALE INSECT.

Coccidiosis, A disease of mammals and birds caused by *Eimeria (Coccidia)*, a parasitic protozoon, in various internal organs, such as the liver. Often occurring in epidemic proportions in wild populations.

Cock. (1) Name for male bird, especially game birds and song birds. (Female: hen). (2) Sportsman's slang for woodcock.

Cockchafer or Maybug (Lamellicornia; *Melolontha*). Two large beetles, familiar for their bumbling flight at dusk in May and June; *M. melolontha* is the commoner. Their larvae can be destructive pests to farmers, who call them white grubs or rookworms.

Cockle (Lamellibranchia: *Cardium*). Marine bivalves with roundish triangular ribbed shells, the Common Cockle *C. edule* being often abundant on the shore and eaten as human food; other, mainly offshore, species are the Little Cockle *C. exiguum*, the Prickly Cockle *C. echinatum*, and the large Spiny Cockle *C. aculeatum*, known as Red Nose from the colour of its foot. The even more rounded Dog Cockle *Glycymeris glycymeris*, widespread and common on the shore, is related to the Noah's ark shell. ⇨ FRESHWATER COCKLE.

Cocklebur (Compositae: *Xanthium*). Green-flowered plants widespread as casuals, especially in areas where shoddy is used as manure. Spiny Cocklebur *X. spinosum* is more frequent than the Cocklebur *X. strumarium*, and has a variety of graphically descriptive names, such as spiny burweed, barbed-wire weed, and thistle weed.

Cockroach (Insecta: Dictyoptera). Flattened, superficially beetle-like insects, of which we have three native species (*Ectobius*) found in natural habitats in southern England and four or five aliens, inhabiting buildings; the Common Cockroach *Blatta orientalis* from tropical Asia, an abundant pest of old houses, popularly known as 'black beetle', despite its not very dark brown colour; the small German Cockroach *Blattella germanica*, known as 'the shiner', frequent in London and other cities; the large red-brown American Cockroach *Periplanata americana*, up to 1½ ins. long; its yellow banded Australian relative *P. australasiae*, and the South American *Pycnoscelus surinamensis*, all of which are liable to colonize zoos, hot-houses and other places where great heat indulges their tropical origins.

Cocksfoot Grass (Gramineae: *Dactylis glomerata*). An abundant fodder grass, with characteristic one-sided flower-heads of rounded or egg-shaped individual flower spikelets.

Cockspur or Gilt-tail (Oligochaeta: *Dendrobaena*). Two widely distributed and common earthworms of compost and manure heaps and similar places (*D. subrubicunda*) and under moss and stones (*D. rubida*).

Cock-Tail. ⇨ DEVIL'S COACH HORSE.

Cocoon. A protective case around the pupa, constructed by the caterpillars of moths and skipper butterflies before they pupate. Many cocoons, as of the non-British Silk Moth *Bombyx mori* and our own vapourer, are spun entirely of silk; in others, e.g. of the skippers, the silk is strengthened with leaves or fragments of food; yet others, made underground like those of the hawk-moths, are primarily constructed of a larval secretion and only lined with silk.

Cod (Gadidae: *Gadus callarias*). A large deep-water marine fish, abundant in northern British waters, especially the North Sea, and the second most valuable economic fish in the world. Young cod are known as codlings. The Poor Cod or Power *G. minutus*, also common in British seas, is the smallest of the cod's immediate relatives.

Codling. The young cod.

Codlin(g) Moth (Tortricoidea: *Ernarmonia pomonella*). A notorious pest of apples, named after the old-fashioned elongated apples called codlin(g)s.

Codlins and Cream. ⇨ Great WILLOW-HERB.

Cod-Worm (Nematoda: Anisakinae). A name given to several parasitic roundworms which infest cod, some of which are also found in grey seals; they are not harmful for human consumption if the fish is properly cooked.

Coelenterata. A name formerly given to the marine animals now classed in the phyla Cnidaria (jellyfish, sea-anemones, corals, etc.) and Ctenophora (comb-jellies). The known British species of these groups total about 265.

Coin Shell (Lamellibranchia: *Lepton squamosum*). A tiny marine bivalve, with flat translucent shells, which is uncommon around the burrows of the shrimp *Urogebia* in sand and shelly gravel on the south coast.

Cold-blooded. ⇨ POIKILOTHERMIC.

Cole, Coleseed. ⇨ RAPE.

Collembola. ⇨ SPRING-TAIL.

Colonist. ⇨ ADVENTIVE.

Colorado Beetle (Phytophaga: *Leptinotarsa decemlineata*). A North American leaf beetle, which has become a serious pest of potato crops in Europe. It invades Britain from time to time, and drastic official measures have to be taken to prevent it becoming established. The beetle is yellow with conspicuous black stripes, and its larva also orange-yellow with black markings; the presence of either should be promptly reported to the police.

Colour Phase. A regularly occurring variety involving a different colour to the animal's or plant's normal one, e.g. the melanic form of Montagu's harrier, the variety *valesina* of the silver-washed fritillary butterfly, and the white variety of the sweet violet.

Coltsfoot (Compositae: *Tussilago farfara*). A composite with rayed, yellow, superficially dandelion-like flowers appearing early in the year on leafless stems well before the horseshoe-shaped leaves which name the plant; formerly valued as a specific against coughs; widespread and often abundant on bare and waste ground, especially on clay.

Columbiformes. The order of birds containing the pigeons and doves (Columbidae) and the sand-grouse (Pteroclidae).

Columbine (Ranunculaceae: *Aquilegia vulgaris*). Handsome blue wild flower, closely resembling the garden plant, local in woods and scrub on limy soils in England and Wales.

Colymbiformes. Disused name for the order now known as Podicipitiformes (grebes).

Comb-Jelly (Ctenophora). A small phylum of pelagic marine jellyfish-like animals, formerly included with the true jellyfish and corals in the Coelenterata. They are small and rounded, swimming by means of eight rows of iridescent comb-like organs. Several species occur in offshore waters, or stranded in pools on the shore, but only two are common: the sea gooseberry and *Beroë cucumis*.

Comfrey (Boraginaceae: *Symphytum*). Rough-leaved plants with creamy-white, blue or purple flowers, the commonest being the white or purple Common Comfrey *S. officinale*, widespread by streams and in damp places, and its hybrid the blue or purple Russian Comfrey *S. x uplandicum*, an escaped fodder plant now widely established on roadsides.

Comma Butterfly (Papilionina: *Polygonia c–album*). A once rare and still rather local and southern vanessid butterfly, somewhat like a small tortoiseshell with raggedly edged wings and a white comma-shaped mark on the underwings, the c of the scientific name.

Commensal. Living together, used of members of different species, e.g. the house sparrow and man, inhabiting the same 'home', but not having much biological effect on each other, i.e. neither parasitic nor, in the strict sense, symbiotic.

Compass Plant. ⇨ Prickly LETTUCE.

Composite (Dicotyledones: Compositae). The largest family of flowering plants, including the daisies, dandelions, thistles and their allies, characterized by flowers which consist of a number, often large, of tiny flowers (florets) all packed tightly together in a button-like, brush-like or flattened head, or capitulum. These florets are of two kinds, tubular disk-florets and strap-shaped ray-florets, and composite flowers may consist of either or both kinds. Brush- or button-like flowers, such as thistles and groundsel, have tubular florets only, dandelion-like flowers have ray-florets only, and daisy-like flowers have both kinds. The fruits are often borne on a feathery appendage or pappus.

Conchological Society of Great Britain and Ireland. The larger and older of the two national societies for the study of molluscs, founded in 1874; publishes the *Journal of Conchology*.

Conchology. The study of molluscs and more especially of their shells.

Conditioning. In animal behaviour, the process of acquisition of the capacity to respond to a given stimulus with the reflex reaction proper to another stimulus when the two stimuli are applied concurrently for a number of times. (W. H. Thorpe).

Cone. The flower, and also usually the fruit of a coniferous tree, formed of overlapping spirally arranged scales. Though pine cones, for instance, are conical, many cones are not.

Conger or Conger Eel (Apodes: *Conger conger*). A large marine eel, with a similar early life history to the common eel, breeding in the deep waters of the Atlantic and passing through a leptocephalid stage. Males grow to 2 – 2½ ft. and females to 4 – 5 ft.

Conglomerate Rock. ⇨ PUDDING-STONE.

Conifer (Gymnospermae: Coniferae). Monoecious trees or shrubs, all the native species (Scots pine, yew, juniper) and naturalized British species being included in three families. Pinaceae, Taxaceae, and Cupressaceae. Though the name derives from the woody cone, the typical fruit of the Pine Family (Pinaceae), the fruits of both yew and juniper are in fact berry-like rather than cone-like; ⇨ ARIL. All the British species, however, have long and very narrow leaves, known as needles.

Conker. ⇨ Horse CHESTNUT.

Connate. Botanical adjective used of two leaves or other organs of a plant which have grown together and become joined, e.g. the leaves of yellow-wort.

Connivent. Botanical adjective used of two organs of a plant whose tips are closer together than their bases.

Conservation. The maintenance of the optimum flow of energy intake from the sun and outgoing through the growth and activity of animals and plants. Normally this is achieved by maintaining the natural succession of vegetation in any area, which by achieving the natural climax maximizes the flow of energy in the given climatic and physical conditions. The maintenance of the animal populations that have adapted themselves to the vegetation is also involved. Hence, if the natural animal and plant communities are more or less undisturbed, conservation means just preservation, though this does not preclude a limited amount of cropping of the annual increases. However, any cropping is bound in some degree to upset the natural balance, by depriving some predator or herbivore of its food supply. On the other hand, the great majority of habitats in the British Isles are either artificial or semi-natural, so that here conservation implies some degree of management, according to whether we are striving to restore, for instance, a mixed deciduous wood to high oak forest, or to maintain an artificial habitat such as an oak coppice woodland or an East Anglian mown fen. This management may involve the elimination or control of surplus unwanted animal or plant populations, e.g. hawthorn scrub on a chalk down, or rats on a marine island.

Conservation Corps. An organization of mainly young people giving time voluntarily to do work of clearing and maintenance on local and national nature reserves and sanctuaries. The national corps, founded in 1959, is run by the Council for Nature, and many naturalists' trusts run local ones.

Convolute. Botanical adjective meaning coiled or rolled up.

Coombe Rock. A deposit of chalky rubble in the valleys of the North and South Downs, caused by Solifluction in the Ice Age.

Coot (Ralliformes: *Fulica atra*). Widespread and common water bird, especially on larger ponds, lakes and reservoirs, also in estuaries in winter. Has white forehead, hence phrase 'as bald as a coot'.

Copepod (Crustacea: Copepoda). A large sub-class of small, and often tiny, crustaceans, so abundant in the sea that they are almost certainly the most numerous animals on earth; they form up to 90 per cent of the marine zooplankton. One of the best known marine copepods is *Calanus finmarchicus*, a favourite food of the herring. There are also some fresh-water copepods, including the well known *Cyclops*. Some copepods are parasitic.

Copper Butterflies (Papilionina: *Lycaena*). Two bright coppery-red butterflies closely related to the blues: the widespread and common Small Copper *L. phlaeas*, and the very rare Large Copper *L. dispar*, which became extinct in its Fenland habitat about 1848 but has been successfully reintroduced at Woodwalton Fen, Huntingdonshire.

Coppice with Standards. A method of exploiting oakwoods in which all the trees except a rather open network of oaks, the standards, are felled, leaving plenty of space for hazels and other underwood to grow and be cut or coppiced at intervals of ten to fifteen years, to provide hurdles, stakes, etc.

Coprolite. The fossilized dung of reptiles and fishes, occurring in certain geological formations, e.g. the junction of the Chalk Marl and the Gault in Hertfordshire and Bedfordshire, as hard nodules containing much phosphate of lime; in some districts these were formerly used as fertilizers.

Coprophagy. The eating of animal dung.

Coprophilous. Growing in or inhabiting animal dung.

Copse. A small wood.

Coraciiformes. A somewhat heterogeneous order of birds containing the kingfishers (Alcedinidae), bee-eaters (Meropidae), rollers (Coracidae) and hoopoes (Upupidae).

Coral (Cnidaria: Anthozoa). Several orders of this class of marine coelenterate animals, which also contains the sea-anemones, are called corals, but only a few species occur on our coasts. The reef-building corals are exclusively tropical, those most closely related to them being the true or hard corals, known as madrepores, and including the Devonshire cup coral and the scarlet and gold star coral. The Soft Corals (Sub-class Octocorallia) are soft and anemone-like, and sometimes referred to as sea-anemones. Our three principal species are dead men's fingers, the beautiful green *Corynactis viridis*, and the local *Cerianthus lloydii*, which has an elongated tube like some species of sea-anemone and belongs to a different sub-class (Ceriantipatharia).

Coral Fungus. ⇨ CLUB FUNGUS.

Coral Necklace (Caryophyllaceae: *Illecebrum verticillatum*). Small prostrate plant named from the appearance of the tiny white flowers along the stem; rare in damp places in the south.

Coralroot or Coralwort (Cruciferae: *Cardamine bulbifera*). A woodland plant somewhat resembling lady's smock, with dark purplish wart-like bulbils at the base of the leaves; local and mainly in the Chilterns and the Weald. Coralroot Orchid (Orchidaceae: *Corallorhiza trifida*) is a small leafless saprophytic orchid, with yellowish flowers, very local in moist spots in the north. Spurred Coralroot, ⇨ GHOST ORCHID.

Corbie. The Scottish name for the raven.

Cordate. ⇨ LEAF-SHAPES.

Cord Grass or Rice Grass (Gramineae: *Spartina x townsendii*). A locally abundant grass of mudflats, often planted in estuaries to fix soft mud; a natural hybrid of the uncommon native *S. maritima* and the North American *S. alterniflora*, which appeared in Southampton Water about 1870, and still grows there.

Coriaceous. Botanical adjective used of leaves and other organs with a rather tough, leathery texture.

Coriander (Umbelliferae: *Coriandrum sativum*). A white-flowered umbellifer, smelling nauseously of bed-bugs, casual on waste ground.

Corixid Bug (Heteroptera: Corixidae). Rather elongated water bugs; thirty-three species, one of the commonest being *Corixa punctata*, widespread in still or slow-flowing water. ⇨ WATER SINGER.

Corkwing. ⇨ Baillon's WRASSE.

Corm. A short swollen underground storage stem, arising on top of the previous year's growth. ⇨ TUBER.

Cormorant (Pelecaniformes: *Phalacrocorax carbo*). A large, mainly blackish sea-bird, nesting on cliffs around most of the British and Irish coasts; also frequenting fresh water in some districts, but not breeding there except in Ireland and on one loch in Galloway. The closely related shag is sometimes called the green cormorant.

Corncrake or Land Rail (Ralliformes: *Crex crex*). Like a small, slim gamebird, shy and retiring in long grass, whence it makes its rasping 'crake' call. Summer visitor, formerly widespread but now rather local in the north and west and in Ireland.

Cornel, Dwarf (Cornaceae: *Chamaepericlymenum suecicum*). A small moorland plant with conspicuous white petal-like bracts; locally frequent in the Highlands, rare elsewhere in the north.

Cornflower or Bluebottle (Compositae: *Centaurea cyanus*). A relative of the knapweeds, with beautiful bright blue flowers, much decreasing in cornfields and now perhaps commoner as a garden escape.

Corn Fly. Ribbon-footed Corn Fly is a name for gout fly.

Cornish Moneywort (Scrophulariaceae: *Sibthorpia europaea*). A small delicate creeping plant with tiny pink and yellow two-lipped flowers, named from its small rounded leaves; local in moist woods in the south, especially Cornwall.

Cornsalad or Lamb's Lettuce (Valerianaceae: *Valerianella*). Five species of former salad plants with flattish button-like heads of tiny pale lilac flowers, growing in bare and disturbed ground, especially cornfields, the two commonest being *V. locusta* and *V. dentata.*

Corolla. The most conspicuous component of most flowers, lying inside the sepals, and usually brightly coloured, but occasionally absent or modified to form nectaries; often divided into separate petals. ⇨ PERIANTH.

Corrie. A steep-sided circular hollow on a mountainside in Scotland, deriving from the Gaelic word for cauldron. Often the slopes have a good arctic-alpine flora.

Corvidae (Aves: Passeriformes). The song-bird family that includes the crows, including the raven, rook, chough, magpie and jay.

Corydalis. ⇨ FUMITORY.

Corymb. ⇨ INFLORESCENCE.

Cossoidea (Lepidoptera: Heteroneura). A small superfamily of fairly large moths with wood-boring caterpillars, including the goat and leopard moths (Cossidae).

Cotoneaster (Rosaceae: *Cotoneaster*). Low, mostly spreading shrubs with small oval or roundish leaves, pink or white flowers and red berries. One very rare native growing on limestone rocks near Llandudno; three frequently bird-sown garden escapes, also mainly on chalk and limestone:

Khasia Berry *C. simonsii*, Rockspray *C. microphyllus*, and Wallspray *C. horizontalis*.

Cotton-Grass or Bog Cotton (Cyperaceae: *Eriophorum*), Four sedges growing in wet boggy conditions and named from their silky white fruiting heads; *E. angustifolium* and harestail are the commonest.

Cotton Spinner. ⇨ NIGGER.

Cotton Weed (Compositae: *Otanthus maritimus*). A silvery downy yellow-flowered plant of the seashore, now confined to one beach in Co. Wexford.

Cotyledon. Seed-leaf, the first leaf of the seedling to appear, usually differing from subsequent ones. ⇨ DICOTYLEDON, MONOCOTYLEDON.

Couch Grass (Gramineae: *Agropyron*). A group of five grasses, somewhat resembling wheat, to which they are closely related. Common Couch Grass or Twitch *A. repens* is a pestilential weed, owing to its creeping rhizomes. Bearded Couch Grass *A. caninum* is frequent in damp shady places. Sea Couch Grass *A. pungens* and Sand Couch Grass *A. junceiforme* are maritime, the latter sometimes helping to form small dunes.

Coumarin. An aromatic substance found in certain grasses, e.g. sweet vernal grass, and some other plants, and which produces the characteristic scent of hay, especially when the plants are dried.

Council for Nature, founded in 1958, is the national body representing the voluntary element in the natural history movement, and has more than sixty-four national and 264 local natural history societies in its membership of 440. It runs the national Conservation Corps, and its news bulletin *Habitat* appears monthly.

County Naturalists' Trust. ⇨ NATURALISTS' TRUST.

Courser, Cream-coloured (Charadriiformes: *Cursorius cursor*). A tern-like or swallow-like wader, a straggler from the Mediterranean region.

Covert (often pronounced *cover*). A small wood left or planted as a harbourage for foxes or game.

Covey. A small flock or family party of partridges.

Cowbane (Umbelliferae: *Cicuta virosa*). A highly poisonous white-flowered umbellifer, very local by fresh water.

Cowberry (Ericaceae: *Vaccinium vitis-idaea*). Evergreen undershrub allied to the bilberry, with edible red fruits, growing on moors and in woods on acid soils in the north and west.

Cowrie (Prosobranchia: *Trivia*). Two small mesogastropod marine operculate molluscs are our only representatives of the great array of large and variegated tropical cowries, viz. the widespread and common European Cowrie *T. monacha* (*Cypraea europaea*), with three brownish-purple spots on the characteristically shaped oval shell, and the slightly smaller and less common unspotted *T. arctica*.

Cowslip (Primulaceae: *Primula veris*). A widespread and common fragrant yellow spring flower of grasslands, especially on limy soils, closely related to the primrose, with which it hybridizes to form the False Oxlip.

Coypu (Rodentia: *Myocastor coypus*). A large aquatic South American rodent, the nutria of commerce, which escaped from fur farms during the 1930's and established itself in various places, notably the Norfolk

Broads. During the late 1950's coypus began to spread widely over eastern England, damaging both crops and river banks, resulting in a successful official campaign both to reduce their numbers and restrict their spread.

Crab (Malacostraca: Decapoda Reptantia). Broad, clawed marine crustaceans, whose abdomen and tail are much reduced and tucked out of sight under the body. The commonest species is the typically dark green Shore Crab *Carcinus maenas*; others include the large Edible Crab *Cancer pagurus* of rocky shores; the Fiddler or Swimming Crabs *Portunus*, of which the best known is the red-eyed Velvet Crab *P. puber;* the small Hairy Crab *Pilumnus hirtellus;* the remarkable soft-bodied Hermit Crabs *Eupagurus*, of which the best known is the Common Hermit Crab or Soldier Crab *E. bernhardus* which inhabits whelk or (when young) winkle shells, only its head and claws emerging; the yellowish Masked Crab *Corystes cassivellaunus;* the tiny Pea Crab *Pinnotheres pisum;* the small, almost round-bodied Porcelain Crabs *Porcellana;* and the strikingly long-legged Spider Crabs (Maiidae), of which the largest and best known is the Spiny Spider Crab *Maia squinado* which often invades lobster pots in the south.

Crab Apple (Rosaceae: *Malus sylvestris*). The true wild crab, with its hairless leaves and very sharp-tasting fruits, is less common than wilding orchard apples.

Crab Spider (Araneae: Sparassidae and Thomisidae). Two families of spiders named from their curved legs, giving them a crab-like stance and sideways movements.

Crag. (1) A steep, irregularly shaped rock or outcrop of rock. (2) A series of shelly deposits in East Anglia, divided into Coralline, Red and Norwich Crags, and highly fossiliferous.

Crake (Ralliformes: *Porzana*). Small moorhen-like marsh birds; four British species, mostly rare: Spotted Crake *P. porzana*, scarce passage migrant, occasionally breeding; Baillon's Crake *P. pusilla*, rare visitor from Europe, has bred; Little Crake *P. parva*, rare visitor from Europe; and Carolina Crake or Sora Rail *P. carolina*, rare straggler from America. ⇨ CORNCRAKE.

Crambid. ⇨ GRASS MOTH.

Cramp Ball (Ascomycetes: *Daldinia concentrica*). A common, almost spherical red-brown fungus, later turning black, growing on dead wood from ash and other broad-leaved trees, and formerly regarded as a specific against cramp if carried in the pocket.

Cranberry (Ericaceae: *Vaccinium oxycoccos*). An undershrub with pink flowers and edible red berries used to make jam and tarts; increasingly local as bogs and wet heaths are drained, and now mainly in the north and west. Small Cranberry *V. microcarpum* is confined to the Highlands.

Crane (Ralliformes: *Megalornis grus*). A large, long-legged marsh bird; an occasional passage migrant *en route* to and from its Scandinavian breeding grounds and its southern winter quarters; formerly bred in the Fens.

Crane Fly or Daddy-long-legs (Nematocera: Tipulidae). A family of nearly 300 species of primitive, long-legged, long-winged, two-winged flies; the larvae of many species, abundant in grassland, are known as leather-jackets and can be a serious economic pest.

Cranesbill (Geraniaceae: *Geranium*). A genus of mainly pink-flowered plants of grassy places and hedge-banks, named from the long 'bill' of their fruits; some, such as as the blue Meadow Cranesbill *G. pratense*, the blue-purple Wood Cranesbill *G. sylvaticum* and the blood-purple Bloody Cranesbill *G. sanguineum* with large showy flowers. Dovesfoot Cranesbill *G. molle* and Cut-leaved Cranesbill *G. dissectum* are the two commonest small-flowered species. ⇨ HERB ROBERT, LITTLE ROBIN.

Crawfish (Decapoda Reptantia: *Palinurus vulgaris*). A large marine crustacean, also known as Rock or Spiny Lobster, but distinguished from the true and squat lobsters by its spiny carapace, its much smaller claws and its tail not being tucked underneath; frequent in the south and west.

Crayfish (Decapoda Reptantia: *Astacus pallipes*). Our largest freshwater crustacean, resembling a miniature lobster, to which it is closely related; widespread but decreasing in calcareous rivers and streams.

Creek. A narrow inlet of the sea.

Creeping Jenny (Primulaceae: *Lysimachia nummularia*). A prostrate yellow-flowered plant, frequent in woods and damp places, also grown in gardens; related to the loosestrifes and yellow pimpernel.

Crenate. ⇨ LEAF-SHAPES.

Cress. A name given to many white-flowered crucifers, e.g. Bittercress; Hoary Cress, ⇨ HOARY PEPPERWORT; Penny-cress; Rock-cress; Shepherd's Cress *Teesdalia nudicaulis*, a local plant of bare ground; Smith's Cress *Lepidium heterophyllum*, frequent in grassland in the west; Swine's Cress, ⇨ WART-CRESS; Thale Cress *Arabidopsis thaliana*, a common weed; Wart-cress; Watercress; and Winter-cress; also to the yellow-flowered Yellowcresses. The Garden Cress, often seen on waste ground, is *Lepidium sativum*.

Crested Dogstail (Gramineae: *Cynosurus cristatus*). A very common grass of meadows and pastures, named from the supposed resemblance of its one-sided flower-spike to a dog's tail.

Cretaceous Period. ⇨ GEOLOGICAL PERIODS.

Cribellum. An extra silk-spinning organ possessed by some spiders.

Cricket (Insecta: Orthoptera). Three families of grasshopper-like insects with long antennae. Of the true Crickets (Gryllidae) we have three species: the common alien House Cricket *Acheta domesticus*, whose incessant chirruping comes from bakehouses, rubbish dumps and other warm places, and two scarce natives, the Field Cricket *Gryllus campestris* and the Wood Cricket *Nemobius sylvestris*. There are ten native species of Bush Crickets (Tettigoniidae), of which the largest and best known is the Great Green Grasshopper *Tettigonia viridis*, mainly in the south; another green species *Meconema thalassina* sometimes comes to light in houses in the autumn. The nocturnal Mole-Cricket *Gryllotalpa gryllotalpa*, named from its burrowing habits, is our sole representative of the Gryllotalpidae. It is a remarkable looking insect, 2 ins. long, and makes a churring sound like a nightjar; local in the New Forest and elsewhere in the south.

Crinoidea. ⇨ FEATHER STAR.

Crocus (Iridaceae: *Crocus*). There are no native wild crocuses in Britain, but both Spring *C. purpureus* and Autumn Crocuses *C. nudiflorus* are naturalized in places, the latter often confused with meadow saffron.

Crossbill (Passeriformes: *Loxia*). Large finches named from the strikingly crossed mandibles of their beaks, adapted to feeding on the cones of pine, spruce and other coniferous trees. The Common Crossbill *L. curvirostra* is an irregular visitor in some numbers from the Continent, often staying to breed and in some districts, such as the Breckland of East Anglia, permanently established. The Scottish Crossbill *L. pityopsittacus* breeds locally in the Scottish Highlands. The Two-barred or White-winged Crossbill *L. leucoptera* is a rare straggler.

Crosswort (Rubiaceae: *Cruciata chersonensis*). A yellow-flowered relative of the bedstraws, widespread and common in grassy and bushy places, but local in Scotland and absent from Ireland.

Crotties. The heaps of deer droppings.

Crow (Passeriformes: Corvidae). Large song or perching birds, with songs of low quality; referring in the widest sense to all members of the Crow Family (Corvidae), but usually applied more narrowly to the genus *Corvus*, comprising in Britain the Raven, Rook, Jackdaw, Carrion Crow and Hooded Crow. The two latter are now regarded as geographical races of one species, *C. corone*. The carrion crow is all black, like the rook and raven, and is resident throughout Great Britain, except in north and west Scotland. The hooded crow, black with a grey body, is resident in north and west Scotland and Ireland and a decreasing winter visitor to the rest of Britain, especially on the east side.

Crowberry (Empetraceae: *Empetrum nigrum*). A heath-like undershrub with small pink flowers and black berries, common on moors in the north and west. *E. hermaphroditum* is similar but less common on higher ground.

Crowfoot. An old term for the buttercups, now most often used for the Water Buttercups or Water Crowfoots *Ranunculus aquatilis* agg., a group of white-flowered aquatic species of ponds and rivers, with finely dissected submerged leaves and/or lobed floating ones.

Crucifer (Dicotyledones: Cruciferae). A member of the large Cabbage Family, characterized by its four petals and sepals, both arranged crosswise, and including various cresses, mustards, rockets and scurvy-grasses.

Crustacean (Arthropoda: Crustacea). The most primitive class of arthropods, distinguished by their two pairs of antennae, mostly living in the sea, where they form a substantial part of the Plankton, but a few forms also in fresh water and on land. The principal sub-classes are the Branchiopoda (water fleas), Ostracoda (plankton), Copepoda (plankton), Branchiura (fish-lice), Cirripedia (barnacles), and Malacostraca (lobsters, crabs, shrimps, woodlice). The first stage of the more primitive crustaceans is known as a nauplius larva. Over 1,850 species of crustaceans are known in Britain.

Cryptic Coloration. Development by an animal of a colour pattern which conceals it against its background as, e.g., certain geometer caterpillars look like brown twigs, and many moths closely resemble the tree trunks or other surfaces on which they rest by day. ⇨ WARNING COLORATION.

Cryptogam. The old Linnean term for all non-flowering plants, as opposed to the Phanerogams or flowering plants, still preserved in the term Vascular Cryptogams used to cover the ferns, horsetails, club-mosses and their allies, commonly included in floras with the flowering plants because they share with them a water-conducting (vascular) system in their tissue.

Cryptostigmata (Arachnida: Acari). An order or sub-order of mites, living especially in the upper layers of woodland soils, and feeding on algae, fungi and decaying leaves and wood. ⇨ SOIL MITE.

Crystalwort (Hepaticae: *Riccia fluitans*). One of our two free-floating aquatic liverworts, frequent in still or slow-moving lowland eutrophic waters.

Ctenophora. ⇨ COMB -JELLY.

Cuckoo (Cuculiformes: *Cuculus canorus*). Widespread and common summer visitor to Britain, named from its persistent call, and noted for its singular habit of laying its eggs in other birds' nests and leaving its young to be reared by the fosters. The young cuckoo normally ejects its hosts' own eggs or young from the nest, thus securing for itself the fosters' undivided attentions and food supply. Three other species of cuckoo are rare stragglers.

Cuckoo Bee (Aculeata: *Psithyrus*). Six species of parasitic bees which resemble and are closely related to bumblebees, the queen cuckoo bee killing the queen bumblebee and laying her eggs in her adversary's nest, to be reared by the bumblebee workers.

Cuckoo Fish. ⇨ BOAR FISH.

Cuckoo Flower. Name given to several spring flowers that start to bloom about the time of the cuckoo's arrival, including lady's smock.

Cuckoo Pint. ⇨ LORDS AND LADIES.

Cuckoo-Spit. The spume or foam produced by the nymphs of frog-hoppers to protect them against desiccation while they are sucking plant juices; often very common on plants in summer. The froth is produced by discharging air through a valve in the abdomen to mix with liquid issuing from the anus.

Cuculiformes (Vertebrata: Aves). The order of birds including the cuckoos (Cuculidae).

Cuddy. The young coal-fish.

Cudweed (Compositae: *Filago* and *Gnaphalium*). A group of ten plants covered with silvery white down, with tiny brush-like composite flowers arranged either in tight roundish heads or elongated leafy spikes, mainly growing in rather bare places. The commonest species are Marsh or Wayside Cudweed *G. uliginosum* of damp places, Common Cudweed *F. germanica* and Small Cudweed *F. minima* of sandy fields, and Heath Cudweed *G. sylvaticum* of heaths and open woods. Dwarf Cudweed *G. supinum* is frequent on mountains in the Highlands.

Cudworm. Old country name for a pill woodlouse.

Culm. The flowering stem of a grass.

Culm Measures. Carboniferous rocks in Devonshire which contain thin bands of impure anthracite or 'culm'.

Cumacean (Crustacea: Cumacea). Minute bottom-living, somewhat scorpion-like marine crustaceans, which sometimes ascend into the Plankton: seventy-two British species.

Cumulonimbus; Cumulus ⇨ CLOUD.

Cuneate. ⇨ LEAF-SHAPES.

Cup Fungus (Ascomycetes: Discomycetes). A group including the morels and elf-cups.

Curlew (Charadriiformes: *Numenius arquata*). A large wader with a long curved bill, breeding commonly on moors and in marshy meadows, especially in the north and west, and abundant on the shore in winter. Named from its main call-note. The now almost extinct Eskimo Curlew *N. borealis* is a rare straggler from Arctic America, ⇨ STONE-CURLEW, WHIMBREL.

Currant (Grossulariaceae: *Ribes*). Small bushes similar to those cultivated as soft fruits. Three native species: Black Currant *R. nigrum* with acid black fruits and aromatic leaves; Red Currant *R. rubrum* with acid red fruits; these two grow widely in fens and damp woods; Mountain Currant *R. alpinum*, with tasteless red fruits and smaller leaves, local in limestone woods in the north.

Cushat or Cushie-Doo. Scots name for the woodpigeon.

Cushion Star (Asteroidea: Poraniidae). A starfish family.

Cuspidate. ⇨ LEAF-SHAPES.

Cuticle. The outermost layer of both higher plants and some invertebrate animals, secreted by the epidermis, which is the outermost cellular layer of the skin, but of very different structure in different organisms.

Cut Grass. ⇨ RICE GRASS.

Cutlass Fish. ⇨ HAIR-TAIL.

Cuttlefish. (Cephalopoda: Decapoda). Marine molluscs with four pairs of short and one of long tentacles; their flat, white, internal shell or cuttle-bone is often found washed up on the shore. The Common Cuttlefish *Sepia officinalis*, up to 1 ft. long, is widespread, but commonest on the south coast, frequently inshore in quiet bays or estuaries in summer. *S. elegans*, about 5 ins. long, is less common. The Little Cuttlefish *Sepiola atlantica*, our smallest and commonest cephalopod, is only 1–2 ins. long and frequent on sandy shores. Two species commoner in the north are *Sepietta oweniana* and *Rossia macrostoma*.

Cyanophyceae, Cyanophyta. ⇨ BLUE-GREEN ALGAE.

Cyclorrhapha (Insecta: Diptera). The largest of the four sub-orders of two-winged flies, including the bee-lice (Braulidae), bluebottles, flesh, warble and deer bot flies (Calliphoridae), carrot rust fly (Psilidae), cheese skippers (Piophilidae), dung flies (Cordiluridae), frit fly (Chloropidae), fruit flies (Trypetidae, Drosophilidae), gall flies (Trypetidae), house and other muscid flies, hover flies (Syrphidae), kelp flies (Coelopidae), leaf miners (Agromyzidae), marsh flies (Sciomyzidae), nest flies (Neottiophilidae), parasite flies and sheep bot fly (Larvaevoridae) and stilt-legged flies (Tylidae).

Cyclostome (Marsipobranchii: Cyclostomata). A sub-class of primitive fishes with cartilaginous skeletons, but no limbs, ribs, jaws or scales, comprising two orders: the lampreys (Hyperoartii: Petromyzontidae) and the hag-fishes (Hyperotreti: Myxinidae).

Cygnet. The young swan.

Cyme. An inflorescence in the form of a forking cluster. ⇨ DICHASIUM.

Cyphal (Caryophyllaceae: *Cherleria sedoides*). A cushion-forming mountain plant of the Highlands with small yellow-green flowers.

Cyperaceae (Angiospermae: Monocotyledones). The Sedge Family, including, besides the true sedges *Carex*, the cotton-grasses, bulrushes, spike-rushes and club-rushes.

Cyprinidae (Pisces: Ostariophysi). The most important family of freshwater fish, whose members have scaly bodies but naked heads; including the barbel, bleak, two breams, two carp, chub, dace, goldfish, gudgeon, minnow, roach, rudd and tench.

Cyst. A wall secreted around the resting stages in the development of many animals, including the membrane or bladder enclosing a parasitic larva, maintaining it in a temporarily dormant state.

Cytology. The scientific study of cells in animals and plants.

D

Dab (Heterosomata: Pleuronectidae). Name given to several small right-sided flatfish, especially to *Limanda limanda*, smallest of the pleuronectids, and abundant in sandy bays. The Long Rough Dab *Hippoglossoides platessoides* is like a small halibut, and also has a northern distribution. Lemon Dab and Smear Dab, ⇨ Lemon SOLE; Pole Dab, ⇨ WITCH.

Dabchick or Little Grebe (Podicipitiformes: *Podiceps ruficollis*). Our smallest grebe, resident and common on ponds and small lakes.

Dace (Cyprinidae: *Leuciscus leuciscus*). Resembles the chub in most respects, but is smaller; rare and introduced in Ireland.

Daddy-Long-Legs. ⇨ CRANE FLY.

Daddy-Long-Legs Spider (Araneae: *Pholcus phalangoides*). A long-legged spider inhabiting houses in the south of England, its web often suspended between walls and ceiling.

Daffodil or Lent Lily (Amaryllidaceae: *Narcissus*). Two native species: Wild Daffodil *N. pseudonarcissus*, locally frequent in damp meadows in England and Wales: and Tenby Daffodil *N. obvallaris*, confined to West Wales.

Daisy (Compositae: *Bellis perennis*). One of the most familiar wild flowers, widespread and often abundant in lawns and other short turf, and the type of the composites with both rays and disc florets. The much larger Ox-eye Daisy or Moon Daisy *Chrysanthemum leucanthemum* is also widespread and common in grassy places. Michaelmas Daisies (*Aster* spp.), with purple rays, are common garden escapes on waste and marshy ground.

Dame's Violet (Cruciferae: *Hesperis matronalis*). A familiar garden plant, with fragrant purple or white flowers, whose name means 'violet of Damascus'; often naturalized by streams and waysides.

Damsel Bug (Heteroptera: Nabidae). Somewhat inappropriately named predatory bugs; a dozen British species, the commonest being *Nabis rugosus*. The young larvae of the Ant Damsel Bug *Himacerus mirmicoides* are remarkably like small black ants.

Damsel Fly (Insecta: Odonata). Fifteen species of small dragonfly, notably the Common Blue Damsel Fly *Enallagma cyatherigum* and the Red-eyed Damsel Fly *Erythromma najas*.

Dandelion (Compositae: *Taraxacum officinale*). Familiar yellow-flowered weed, with flowerheads all consisting of ray-florets, and milky juice (latex) in stems and leaves; named from the jagged shape of its leaves, supposed to resemble lions' teeth (French: *dent de lion*). The fruits are attached to feathery parachutes, making the well known 'dandelion clocks', also found on hawkbits and goatsbeard.

Danewort or Dwarf Elder (Caprifoliaceae: *Sambucus ebulus*). A stout plant with large flat heads of small white flowers, resembling a herbaceous elder shrub (which is what it is) or an outsize umbellifer; growing in widely scattered colonies, mainly on roadsides, reputedly on the sites of former battles with the Danes.

Darkie Charlie (Pleurotremata: *Scymnorhinus licha*). A Mediterranean shark that occasionally strays to British waters.

Darnel (Gramineae: *Lolium temulentum*). A now rare weed, somewhat like the closely related rye-grass.

Darter. ⇨ DRAGONFLY.

Darwinism. The theory of evolution by Natural Selection first propounded by Charles Darwin (1809–82) in his *Origin of Species* (1859).

Dawn Chorus. The great volume of song with which song-birds herald the dawn, especially in May and early June. The amount of song at this time, especially in Britain from blackbirds, song thrushes and robins, far surpasses that at any other time of day. The chorus begins around 80–100 minutes before dawn, and rises to its fullest volume from about half an hour to an hour before the sun rises. There is a similar but less striking outburst of song at dusk.

Dead Men's Fingers (Anthozoa: *Alcyonium digitatum*). A colonial soft coral, widespread on our rocky coasts, and sometimes growing to 6–9 ins. high; named from its pink fleshy appearance when removed from the water, in which it may vary from dirty white through yellow and pink to orange.

Dead-Nettle (Labiatae: *Lamium*). Aromatic plants with harmless nettle-like leaves; six species, the best known being White Dead-nettle or White Archangel *L. album*, a common wayside plant; Red Dead-nettle *L. purpureum*, a very common weed of cultivation flowering all the year round; and Henbit. ⇨⇨ YELLOW ARCHANGEL.

Deal-Fish. ⇨ VAAGMAR.

Dean. ⇨ DENE.

Death Cap (Agaricaceae: *Amanita phalloides*). One of the most lethal of our poisonous toadstools, the cause of most deaths reported from fungus poisoning; it is common in broad-leaved woods, and has a pale yellowish- or olive-green cap, with a white 'skirt' around the stem immediately below it. False Death Cap *A. citrina*, easily confused with both death cap and destroying angel, is harmless but tastes unpleasant. Some of its close relatives are edible, such as the blusher and the two grisettes, but others, notably the destroying angel and fly agaric, are also highly poisonous.

Death's Head. ⇨ HAWK-MOTH.

Death Watch Beetle (Clavicornia: *Xestobium rufovillosum*). A furniture beetle, notorious for the knocking sound made by the pupae tapping their heads against the pupal chamber and naturally often heard in the quiet of the night by those watching at the bedsides of dying people in old wooden houses. The larvae need the presence of certain wood-destroying fungi, which are most often found in damp wood, and can do immense damage to timber construction.

Debris Bug (Heteroptera: *Lyctocoris campestris*). A predatory bug, allied to the bedbug, found in various natural and artificial accumulations of vegetable debris, and occasionally attacking man.

Decapoda. (1) (Crustacea: Malacostraca). An important order, characterized by five pairs of 'walking legs', which contains all the larger crustaceans: divided into the Natantia (shrimps and prawns) and the Reptantia (crabs, lobsters, crawfish, crayfish, burrowing shrimps). (2) (Mollusca). The sub-order of Cephalopoda containing the squids and cuttlefishes.

Deceiver (Basidiomycetes: *Laccaria*). Two woodland toadstools, the Deceiver *L. laccata* and the edible Amethyst Deceiver *L. amethystina*.

December Moth (Bombycoidea: *Poecilocampa populi*). A relative of the eggars, named from its being on the wing at the end of the year.

Deciduous Plant or Tree. One which sheds its leaves annually, usually in the autumn, as a protection against winter conditions.

Decumbent. Botanical adjective used of prostrate stems which turn upwards towards the tip.

Decurrent. Botanical adjective used especially of leaves prolonged down on to the stem in the form of a 'wing'.

Decussate. Botanical adjective used when opposite pairs of leaves are arranged alternately down the stem at right angles to each other.

Deer (Artiodactyla: Cervidae). Ungulates bearing antlers as distinct from horns; in the British species only the males have them. Only two species are native: the Red Deer *Cervus elaphus* and the Roe. The four well established alien species are the Fallow Deer *Dama dama*, the Sika, the Muntjac, and the Chinese Water Deer *Hydropotes inermis*, which is no bigger than a large dog. The red deer (male, stag; female, hind; young, calf) occurs as a native in the Scottish Highlands, the Lake District and Exmoor, and as an introduced or escaped species elsewhere. The fallow deer (male, buck; female, doe; young, fawn) is found in woodland throughout Great Britain; normally spotted in summer only, but the menil variety is spotted throughout the year. The Chinese water deer (male, buck; female, doe; young, kid) is the least common of our introduced deer and occurs mainly in the counties around Bedfordshire.

Deer Fly. ⇨ Deer KED.

Deer Grass (Cyperaceae: *Scirpus caespitosus*). An abundant leafless sedge of moors and bogs, mainly in the north and west, somewhat resembling the spike-rushes and unrelated to the grasses.

Deflexed. Botanical adjective used of leaves, sepals or other organs which are markedly bent downwards.

Dehiscent. Botanical adjective used of fruits which split open to release the seeds.

Deltoid. Botanical adjective meaning shaped like the Greek letter Δ.

Demersal. Living on the bottom (of the sea), as opposed to pelagic (living in the upper layers of the sea).

Dene or Dean. A small usually steep-sided valley in the north of England, especially Yorkshire, also in south-east England. ⇨⇨ CLOUGH, GHYLL.

Dene Hole. A deep hole in chalky districts, with steep or almost vertical sides, of disputed origin, but probably a primitive chalk mine, similar to the Neolithic flint mines at Grimes Graves in the Norfolk Breckland.

Denizen. ⇨ ADVENTIVE.

Density-dependent Factors. In the study of animal populations, the factors which may vary according to the density of population; e.g., access to food supply, disease, reproductive rate, mortality, but not climatic or other external factors which are unaffected by the growth or decline of the population itself.

Dentate, Denticulate. Botanical adjectives meaning toothed and minutely toothed respectively, especially of leaves.

Depression. A low-pressure weather system, in which the air flows in an anti-clockwise direction, producing cloudy conditions with rain or mist.

Dermaptera (Arthropoda: Insecta). ⇨ EARWIG.

Desmid (Chlorophyceae: Conjugales). A large group of freshwater green algae, in some respects like diatoms, common on the surface of ponds and pools, often forming a green film, and on mud and aquatic plants.

Destroying Angel (Agaricaceae: *Amanita virosa*). A highly poisonous and lethal white toadstool, closely related to the death cap, growing, fortunately uncommonly, in broad-leaved woods.

Devil-Fish or Manta Ray (Pleurotremata: *Mobula giorna*). A giant tropical and Mediterranean ray, sometimes attaining 30 ft across, which has very rarely occurred in British seas; named from its 'horns' or head lobes, it is actually quite harmless.

Devil's Coach Horse or Cock-tail (Staphylinoidea: *Staphylinus olens*). A beetle, common in gardens and elsewhere, well known for its threat display of erecting the hinder part of its body and opening wide its mandibles.

Devil's Fingers (Ascomycetes: *Xylaria polymorpha*). A club-shaped blackish fungus, related to candle-snuff fungus, growing in tufts on the stumps of broad-leaved trees, especially beech.

Devonian Period. ⇨ GEOLOGICAL PERIODS.

Dew. Small drops of water condensed from the water vapour in the atmosphere on to vegetation and other low-lying objects, when radiation of heat from the earth's surface at night has lowered the temperature of the air near the ground to the point, known as the dew point, at which it is saturated with water vapour.

Dewberry (Rosaceae: *Rubus caesius*). A smaller and usually prostrate relative of the bramble, with similar edible purple-black fruits; widespread and frequent.

Dewpond. A shallow pond in a relatively waterless district, such as the chalk downs, which has a specially prepared bottom to catch and retain as much as possible of the rainfall; not, despite its name, substantially replenished by dew.

Dew Worm. An angler's name for the common earthworm *Lumbricus terrestris*.

Dextral. Of snails coiling right-handed.

Diapause. A state of dormancy or quiescence during the development of insects and other invertebrates.

Diatom (Chrysophyta: Bacillariophyceae). A class of microscopic, one-celled, solitary or colonial freshwater or marine algae (also found in damp soil), with hard, siliceous cell-walls, which occur in many, often beautiful forms, and are a favourite subject for microscopists. Some species form a brown scum on stones and other algae. Their skeletons may form deposits of the valuable mineral diatomite at the bottom of lakes.

Dichasium. A cyme with the branches approximately opposite and equal.

Dicotyledon (Angiospermae: Dicotyledones). The division of the flowering plants whose seedlings have two seed-leaves (cotyledons), whose leaves are usually broad, often stalked and nearly always net-veined, and whose flower parts are usually in multiples of four or five. The principal dicotyledonous families are the Ranunculaceae (buttercups), Cruciferae (cabbage, etc.), Caryophyllaceae (pinks and campions), Papilionaceae (pea-flowers), Rosaceae (roses, etc.), Umbelliferae (carrot, etc.), Scrophulariaceae (figworts etc.) Labiatae (two-lipped flowers), and Compositae (daisy, dandelion etc.).

Dictyoptera (Arthropoda: Insecta). The order containing the cockroaches; allied to the Orthoptera.

Digger Wasp. ⇨ SOLITARY WASP.

Digitate. Botanical adjective meaning arranged like the fingers of a hand.

Dike. ⇨ DYKE.

Dimorphic. Having two or more sharply contrasted forms. ⇨ POLYMORPHISM.

Dingle. A small wooded steep-sided valley, or a large wooded ravine, mainly in the west.

Dinoflagellate. ⇨ FLAGELLATE.

Dioecious. Adjective used of an animal or plant with male and female reproductive organs in different individuals i.e. unisexual.

Diploid. An organism with the chromosomes paired in the cell nucleus. ⇨ POLYPLOIDY.

Diplopoda (Arthropoda). ⇨ MILLIPEDE.

Diplura (Arthropoda: Insecta). ⇨ Two-pronged BRISTLE-TAIL.

Dipper (Passeriformes: *Cinclus cinclus*). Aquatic song bird with a conspicuous white breast, named from its habit of bobbing or dipping while standing on a rock in a stream. The only British bird which actually walks under water on the stream bed; frequent on fast-flowing streams in the north and west. Also known as water ouzel.

Dip-Slope. The long gentle slope of a line of hills where the underlying rock strata are not horizontal but tipped at an angle. The short steep slope on the other side is the Scarp.

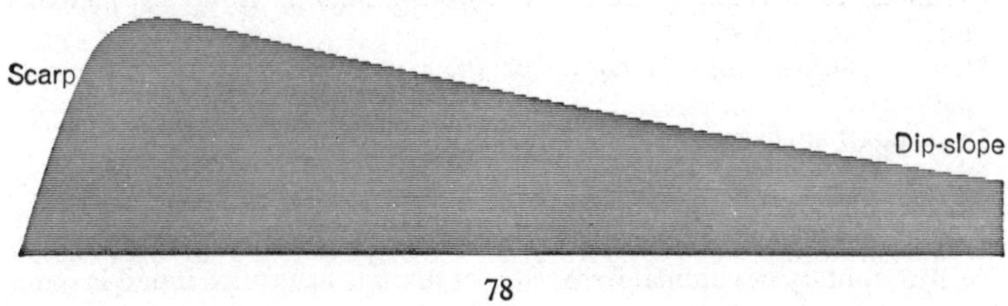

Diptera (Arthropoda: Insecta). The true or two-winged flies; about 5,200 species in 1,132 genera and three sub-orders; Nematocera, including the mosquitoes, midges and crane-flies; Brachycera, including the clegs; and Cyclorrhapha, including the house flies and bluebottles. Their two wings are the main distinguishing feature of the Diptera, their hind-wings being reduced to two small organs called halteres. Most Diptera have maggot larvae.

Dishwasher. ⇨ Pied WAGTAIL.

Displacement Activity. In animal behaviour, the performance of a behaviour pattern out of the particular functional context of behaviour to which it is normally related; e.g. 'false-preening' in birds. A displacement activity may result from the activation by one or more Drives of behaviour appropriate to another drive (W. H. Thorpe).

Display. A method of communication in animals, especially birds, which involves either the showing off of certain conspicuous physical features, e.g. the tail of a peacock or the red breast of a robin, and/or the ritual performance of certain actions, such as bowing or turning the head, and/or the uttering of certain sounds, in particular bird song. In birds display may include threat display, to warn off rivals, territorial display, to maintain the ownership of a territory, and courtship display, to facilitate pairing and mating.

Distichous. Botanical adjective used of leaves, flowers and other organs arranged in two exactly opposite rows.

Ditrysia. Alternative name for the Heteroneura.

Dittander (Cruciferae: *Lepidium latifolium*). Tall, white-flowered, broad-leaved saltmarsh plant, mainly in the south-east.

Divaricate. Botanical adjective, used of stalks and other organs which diverge at a wide angle.

Diver (Gaviiformes: *Gavia*). Elongated diving birds, ill-adapted to land to which they only come to breed; marine in winter, freshwater in summer. Three regular species: two breeding in Scotland, the Red-throated Diver *G. stellatus* and the Black-throated Diver *G. arcticus;* and one winter visitor, the Great Northern Diver *G. immer;* also one rare visitor, the White-billed Diver *G. adamsii.*

Diversicornia (Coleoptera: Polyphaga). A rather heterogeneous superfamily of beetles, with 248 species in nineteen families including the glow-worms, soldier and sailor beetles, skipjacks or click beetles, and the various pests of stored products belonging to the Dermestidae.

Diving Birds. The principal groups of diving birds in the British avifauna are the divers, grebes, auks, cormorants and some ducks. These dive from the surface of the water. Gannets and terns dive from the air.

Division. A major grouping in botanical classification, equivalent to a Phylum in zoology.

Dock (Polygonaceae: *Rumex*). A large group of mostly weedy plants, closely related to the sorrels, with spikes of small greenish flowers and undivided leaves; known as dockens in the north. The commonest species are Curled Dock *R. crispus* and Broad-leaved Dock *R.*

obtusifolius, both official noxious weeds; Clustered Dock *R. conglomeratus* and Wood Dock *R. sanguineus*, both in woods and damp grassy places; and the 6-ft Great Water Dock *R. hydrolapathum*, frequent by fresh water. Fiddle Dock *R. pulcher* is named from the shape of its leaves.

Docken. A northern name for docks.

Dodder (Convolvulaceae: *Cuscuta*). Parasitic plants with reddish leafless stems twining anti-clockwise and heads of small fragrant pinkish flowers: Common Dodder *C. epithymum* mainly on gorse and heather, the local Greater Dodder *C. europaea* mainly on nettles.

Doe. The female in fallow, roe and muntjac deer.

Dogfish or Hound (Selachii: Pleurotremata). Half a dozen small sharks that occur in British waters are so named, the commonest being the Lesser Spotted Dogfish or Rough Hound *Scyliorhinus caniculus.* Also common or frequent are the Greater Spotted Dogfish or Nurse Hound *S. stellaris,* the Smooth Hound *Mustelus mustelus* and the Piked Dogfish or Spur Dog *Acanthias vulgaris,* the two latter both viviparous.

Dog's Mercury (Euphorbiaceae: *Mercurialis perennis*). A common carpeter of woods in early spring, with catkins of tiny greenish flowers blooming in February and March. Annual Mercury *M. annua* is a locally frequent weed of cultivation.

Dogwood (Cornaceae: *Thelycrania sanguinea*). Deciduous shrub noted for its red twigs and common in England and Wales, especially on chalk and limestone; flowers white, berries black.

Dolerite. A volcanic rock, composed of augite and felspar, usually occurring in dykes or sills; the crystals are smaller than in gabbro because they cooled more quickly. ⇨ WHINSTONE.

Dolphin (Cetacea: Delphinidae). A family of small toothed whales with a conspicuous beak-like snout. A dozen species are recorded from British waters, four of them rare. Of those popularly named dolphin, the commonest off our southern coasts are the small Common Dolphin *Delphinus delphis*, which often follows ships, and the larger Bottle-nosed Dolphin *Tursiops truncatus;* and off our northern coasts the White-beaked Dolphin *Lagenorhynchus albirostris* and the White-sided Dolphin *L. acutus.* ⇨⇨ BLACKFISH, KILLER WHALE, NARWHAL and PORPOISE. Risso's Dolphin ⇨ GRAMPUS.

Dominant. Adjective used of the most numerous and usually the tallest plant in a community or vegetational complex, e.g. beech trees in a beechwood, heather on a heather moor or reeds in a reed-swamp.

Door Snail (Pulmonata: *Clausilia* and *Marpessa*). Four species of land snail, with elongated shells, the commonest being *C. rugosa;* unlike other land snails they are sinistral (coiling left-handedly).

Dor Beetle (Lamellicornia: *Geotrupes stercorarius*). A black dung beetle, often known as the Lousy Watchman, from its being infested with the mite *Gamasus coleopterum.*

Dormouse (Rodentia: Muscardinidae). The Dormouse *Muscardinus avellanarius,* like a large mouse with a squirrel's tail, now somewhat local and uncommon, is the only native British mammal that truly hibernates. It feeds mainly on wild fruits and is quite harmless to the farmer and forester. In the Chilterns there is also a small but well established population of the Continental Fat or Edible Dormouse *Glis glis*, first introduced in 1902. They sometimes make a nuisance of themselves by entering houses to hibernate and feed on stored fruit.

Dotterel (Charadriiformes: *Charadrius morinellus*). A small plover, breeding rather locally on high mountain tops in Scotland and northern England, and once not infrequent at regular halting places on migration further south. Ring dotterel is an old name for the ringed plover.

Dove (Columbiformes: Columbidae). Short-billed, seed-eating land birds, with songs composed mainly of 'coos' and 'turs'. Four British breeding species: Collared Dove *Streptopelia decaocto*, rapidly colonizing the British Isles after a headlong advance over Europe from Balkans in past thirty years; first bred in Norfolk 1955; resident and often found near chicken runs; Ring Dove, an old name for the woodpigeon; Rock Dove *Columba livia*, the origin of both domestic and feral pigeons, now confined to cliffs in far north and west of British Isles; Stock Dove *C. oenas*, widespread and frequent; Turtle Dove *S. turtur*, summer visitor, mainly to the south. The dove most often kept by bird-fanciers is the Barbary Dove *S. risoria*, which sometimes escapes.

Dowitcher or Red-breasted Snipe (Charadriiformes: *Limnodromus*). North American wader, two forms of which are occasional stragglers to Britain on migration *L. scolopaceus* being the more frequent.

Down. The soft natal plumage of newly born birds, also present under the feathers of ducks, geese and certain other birds, when it is sometimes used as a nest lining, notably by the eider duck.

Downlooker Fly or Downhill Fly (Brachycera: *Rhagio scolopacea*). One of the snipe flies, so named by fly-fishers, who use it for trout, because of its habit of resting head downwards on a tree trunk.

Downs. Open, largely treeless chalk hills in southern England – Kent, Surrey, Sussex, Hampshire, Wiltshire, Dorset and Berkshire. The characteristic short turf is not natural, but the result of sheep-grazing, and later when that became uneconomic rabbit-grazing. Since myxomatosis drastically reduced the rabbits many areas have become covered with scrub. Because of the grazing the typical chalk flora of the downs is low-growing.

Dragonet (Percomorphi: *Callionymus*). Two smallish, 10 – 12 in. shallow-water marine fish, of striking appearance with fantastically shaped back fins and handsomely variegated colouring. The Common Dragonet *C. lyra* is much commoner than the Spotted Dragonet *C. maculatus*.

Dragonfly (Insecta: Odonata). A fairly primitive order of carnivorous winged insects, passing through an aquatic nymph stage, the adults also usually frequenting the neighbourhood of fresh water; forty-three British species. Roughly speaking the larger dragonflies are known as Hawkers, from their habit of hawking around the edges of ponds, the medium-sized ones are Darters, and the small ones Damsel Flies. Few have genuine English names, except for the Demoiselle *Agrion virgo* and the Emperor

Anax imperator, one of our largest species. Dragonfly nymphs are also called naiads.

Drake. The male duck.

Drawmoss. Scottish name for cotton-grass.

Drepanoidea (Lepidoptera: Heteroneura). The superfamily of moths containing the hook-tips.

Drey. A squirrel's tree nest.

Drift. ⇨ GLACIAL ACTION.

Drill (Prosobranchia: Muricidae). Marine operculate molluscs allied to the whelks, but with more pointed and sharply sculptured shells, named from their habit of drilling through their victims' shells. The Common Drill or Sting Winkle *Ocenebra erinacea* feeds on bivalves and barnacles, and used to be a serious pest of east-coast oyster-beds till it was exterminated by the hard winters in the 1940s. The American Oyster Drill *Urosalpinx cinerea* is a hardier and so more serious alien pest of oyster-beds in Essex and Kent.

Drinker Moth (Bombycoidea: *Philudoria potatoria*). A common large reddish-brown eggar moth, named from the dew-drinking habit of its large brown furry caterpillars, often seen on grass stems in spring.

Drive. In animal behaviour, the complex of internal and external states of behaviour and stimuli leading to a given behaviour. (W. H. Thorpe).

Drone. The males of social bees and wasps.

Drone Fly (Cyclorrhapha: *Eristalis tenax*). A large hover-fly looking remarkably like a drone honeybee; its larva is of the 'rat-tailed maggot' type.

Dropwort (Rosaceae: *Filipendula vulgaris*). The downland counterpar of meadowsweet, confined to calcareous turf, mainly in England.

Drosophila. ⇨ FRUIT FLY.

Drumlin. An oval-shaped mound of boulder clay left by an ice-sheet and elongated in the direction of the flow of the ice. One explanation is that the ice was slowed up by the quantity of debris and the main flow went between these mounds. They are found particularly in northern England, southern Scotland and northern Ireland.

Drupe. Fruit with the seeds enclosed within a stone or pip, the whole with a fleshy surround e.g. plum, blackberry. ⇨ BERRY.

Dryad's Saddle (Basidiomycetes: *Polyporus squamosus*). One of the larger bracket fungi, up to a foot across.

Dry Rot (Basidiomycetes: *Serpula* [*Merulius*] *lacrymans*). A fungus with a honeycomb-like fruit-body anything up to 3 ft across, which destroys wood in damp situations, where there is little movement of air, and can cause immense damage to old buildings. Two other fungi which cause dry rot are the Wet-rot Fungus *Coniophora cerebella* and *Poria vaillantii*.

Dubhlochan. ⇨ LOCHAN.

Duck. The smaller waterfowl of the family Anatidae; males are drakes, females ducks. The British breeding species are divisible into seven species of dabbling or surface-feeding ducks; four species of diving ducks, including the Tufted Duck *Aythya fuligula*, whose drakes are conspicuously black and white; two species of saw-bill, and one of shelduck. Four more species of diving duck, including the sea-going Long-tailed Duck *Clangula hyemalis*, and one saw-bill, the Smew, are regular winter visitors. A dozen more species, including the strikingly patterned Harlequin Duck *Histrionicus histrionicus* from Iceland, are irregular or casual visitors, among them seven stragglers from across the Atlantic. Call Duck, a miniature breed of Mallard, sometimes escaping. Cuddy Duck or St Cuthbert's Duck, local names of the eider in Northumberland.

Duck Decoy. A specially designed trapping mechanism for ducks, with a star-shaped pool of water, the points of the star, known as 'leads' or 'pipes', curving away from the centre and ending in a trap. The ducks are led towards the trap by the intermittent sight of a dog moving behind staggered screens, which they follow in curiosity. Duck decoys are little used in Britain nowadays for their original purpose of catching ducks for the table; one famous decoy, Borough Fen in north Northamptonshire, is exclusively used to trap ducks for ringing.

Duckweed (Lemnaceae: *Lemna*). Very small, free-floating aquatic monocots, rarely flowering in Britain, whose leaf-like fronds have rootlets attached; often carpeting ponds, much relished by ducks and also eaten by herbivorous fish. The Common Duckweed *L. minor* is the commonest of our four species. Least Duckweed *Wolffia arrhiza*, uncommon in the south, with egg-shaped fronds 1mm. across, is our smallest flowering plant, though it has never actually flowered in Britian.

Duke of Argyll's Tea-Plant (Solanaceae: *Lycium halimifolium* and *L. chinense*). Low shrubs, with purple-and-yellow nightshade-like flowers, often planted and sometimes escaping into hedges, etc. Named from a supposed nurseryman's mistake when a former Duke of Argyll ordered tea-plants.

Dulse (Rhodophyta: *Dilsea carnosa*). A common medium-sized red seaweed, said to have been eaten by Cornish fishermen. The true Edible Dulse, however, is the allied *Rhodymenia palmata*, at one time a staple diet on the west coasts of Scotland and Ireland, and still sometimes chewed by fishermen.

Dumble. A small gorge made by a stream in the Keuper marls of Nottinghamshire.

Dun. The dull-coloured winged penultimate (sub-imaginal) stage of mayflies.

Dun-Bar (Caradrinoidea: *Cosmia trapezina*). A noctuid moth noted for the cannibalistic habits of its black-dotted green larvae.

Dune. An accumulation of sand blown by the wind, usually on the coast, occasionally inland, e.g. in the Breckland. Sand blown from shoals offshore that are exposed by the tide or from sloping beaches builds up into a hummock at any point where it meets an obstacle. Plants help to build up the dunes by continuing to push up shoots through the new sand as it arrives, sending out new roots and so maintaining the dunes by holding the sand surface. Marram grass (*Ammophila*), which spreads by underground runners, is the most efficient and common of these binding plants. If their growth is disturbed the wind may get in and a 'blow out' results, shifting large volumes of sand. The grass tufts also provide shelter for other plants to grow between them. A fixed or grey dune is one with a fairly complete covering of vegetation, as opposed to the bare mobile white or yellow dunes. A foredune is a small dune in front of the main range, on the shore often dominated by the Sand Couch Grass *Agropyron junceiforme.*

Dung Beetle (Polyphaga: Lamellicornia). Beetles, allied to the chafers, with the unlovely habit of living on animal dung; one of the best known is the dor beetle. The Horned Dung Beetle *Typhaeus typhaeus* is noted for its long horns.

Dung Fly (Cyclorrhapha: *Scopeuma*). Predatory two-winged flies (Cordiluridae), whose larvae feed in animal dung. The yellow dung flies which assemble in such large numbers on cow-pats are the males of the Common Yellow Dung Fly *S. stercorarium* awaiting the arrival of females seeking to lay their eggs in the dung.

Dunlin (Charadriiformes: *Calidris alpina*). The commonest small wader on the shore in winter, also breeding on moors in the north and west.

Dunnock. ⇨ HEDGESPARROW.

Dun Sentinel (Prosobranchia: *Assiminea grayana*). A tiny operculate snail of brackish waters on the east coast, especially the Thames estuary.

Dust Devil. ⇨ WHIRLWIND.

Dutch Rush (Equisetaceae: *Equisetum hyemale*). An uncommon horsetail with unbranched overwintering stems.

Dyke. (1) In Scotland, a dry stone wall used as a field boundary. (2) In England, a ditch or small watercourse, or alternatively a grassy bank or wall built as a flood defence on the coast or in the Fens, the last often spelt 'dike'. (3) A vertical or steeply inclined sheet of igneous rock formed during the Alpine earth movements, when molten lava forced its way up through cracks in the older rock, and then cooled and solidified. Where lava forced its way horizontally, the result is a Sill. Dykes may be of any thickness.

E

Eagle (Falconiformes: Falconidae). A very large, broad-winged bird of prey. The only breeding species is the Golden Eagle *Aquila chrysaëtos*, almost confined to the Highlands of Scotland, where about 150 pairs breed. Until about 1908 the White-tailed or Sea Eagle *Haliaëtus albicilla* also bred in Scotland, but it is now only a rather irregular winter visitor and passage migrant. The Spotted Eagle *A. clanga* is a rare straggler from eastern Europe.

Eaglet. The young eagle.

Eagre. The bore on the River Trent.

Early Brown (Plecoptera: *Protonemoura meyeri*). A common stone-fly, emerging early in the year. Several other small brown stone-flies are also called 'browns'.

Early Purple Orchid (Orchidaceae: *Orchis mascula*). The earliest orchid to flower, and one of the commonest and most widespread in woods and grassland; leaves spotted and flowers purple, whence formerly called, e.g. in Shakespeare, long purples.

Earth. A fox's burrow.

Earth-Ball (Basidiomycetes: *Scleroderma*). Brown woodland fungi, resembling unopened puff-balls or earth-stars; *S. aurantium* is the commonest species, especially around birches.

Earth Fan (Basidiomycetes: *Thelephora terrestris*). A brown fan-shaped fungus of pinewoods and heaths.

Earth-Nut. ⇨ PIG-NUT.

Earthquake. Movements or tremors of the earth's crust are infrequent in Britain and rarely severe enough to damage buildings. In the Middle Ages several recorded earthquakes did fairly severe damage, e.g. to Lincoln Cathedral in 1185 and Wells Cathedral in 1248, but in modern times there has only been one earth movement severe enough to damage sound buildings. The Great Essex Earthquake of 22 April, 1884, felt over a wide area, from Somerset to the Belgian coast, and from the Isle of Wight north to Lincolnshire, damaged thirty-one churches and chapels, and 1,213 other buildings in Essex. The area of main damage was from Wivenhoe to Peldon, and around Abberton, where hardly a building is said to have been unscathed; Langenhoe Church was wrecked.

Earth Sciences. The study of the physical characteristics of the Earth, especially geography, geology, geomorphology and geophysics.

Earth-Star (Basidiomycetes: *Geastrum*). Brown woodland fungi, at first resembling puff-balls, which then split open to resemble a star or flower; *G. triplex*, which favours beech woods, is one of the larger and commoner species.

Earth-Tongue (Ascomycetes: *Geoglossum*). A group of tongue-shaped fungi, usually black or olive-green, growing directly from the soil in woods or grassland.

Earthworm (Oligochaeta: Lumbricidae). A family of twenty-five terrestrial oligochaete worms, with bristles borne singly; often abundant in the soil, where they may number millions per acre. They perform a most important function in aerating the soil, and moving nutrient salts upwards to the surface. Attempts by gardeners to destroy them in order to have a lawn free from Worm Casts are therefore highly misguided. The commonest makers of casts on lawns are *Allolobophora longa* and *A. nocturna.* Our largest and one of our commonest species is *Lumbricus terrestris*, which can be a foot long, and is known to anglers under various names, such as dew worm, flat-tail, squirrel-tail and twachel. It has a U-shaped burrow, from which it half-emerges at night while feeding on leaves and other decaying vegetable matter, plugging the entrance by day with leaves or twigs. ⇨ BRANDLING, CHESTNUT WORM, COCKSPUR, MARSH WORM.

Earwig (Insecta: Dermaptera). Small brown skulking insects, with rarely seen but well developed wings, and a pair of forceps at the hinder end of the body. The young pass through a nymphal stage. There are five native and two or three introduced species, the only widespread and common one being the only too well known Common Earwig *Forficula auricularia.* Earwigs are named from a supposed propensity to enter the human ear, but there is no evidence that this occurs as anything but the rarest chance.

Easter-Ledge. A local name for bistort in the Lake District, where its young leaves are eaten as Easter-ledge pudding.

Ebracteate. Botanical adjective meaning without bracts.

Ecad. A form modified by its habitat.

Ecdysis. The periodic shedding or sloughing of the old skin, made necessary by a hard exoskeleton which will not expand, revealing a new soft one underneath, which rapidly hardens, especially in reptiles, amphibians and insects (usually in immature stages) and other arthropods.

Echinoderm. The Echinodermata are a phylum of marine animals with a radially symmetrical ground-plan, moving with the aid of suckers called 'tube-feet' and operated hydrostatically by the animals' water vascular system. They include the Classes Crinoidea (feather-stars), Holothuroidea (sea cucumbers), Echinoidea (sea urchins), Asteroidea (starfishes) and Ophiuroidea (brittle stars). Some 170 British species are known.

Echinoidea. The sea urchins.

Echiuroidea. A small phylum of sausage-shaped, worm-like, marine animals, formerly included in the abandoned phylum Gephyrea. *Thalassema neptunei* is uncommon on our south and west coasts.

Eclipse. The moult plumage of ducks, which become flightless for a time owing to the loss of their wing quills.

Ecological Niche. The place in the biological community occupied by one or more organisms; or a distinctive habitat into which a pre-adaptively specialized animal or plant can fit exactly; e.g., the introduced little owl found a vacant ecological niche in England for a small insectivorous diurnal bird of prey, and the introduced ivy-leaved toadflax a niche on old walls.

Ecology. The scientific study of the relationships between living animals and plants and their environment, animate and inanimate.

Ecosystem. A unit of vegetation together with all the animals associated with it, and all the physical and chemical components which make it a recognizably self-contained habitat. Three main groups of factors affect an ecosystem: the effects of living organisms are the *biotic* factors, e.g. grazing animals on vegetation; the effects of the soil and physical environment are the *edaphic* factors; and the weather provides the *climatic* factor. The anthropogenic factor is the human aspect of the biotic factor, e.g. felling of trees.

Ecotone. A border zone between two habitats, such as the edge of a wood, where woodland blends into grassland or scrub.

Ecotype. A group of plants of the same species adapted to a particular habitat and capable of interbreeding with other ecotypes of the same species where their habitats overlap.

Ectoparasite or Exoparasite. A parasite living outside its host, e.g. a louse on a human head.

Ectoprocta. ⇨ SEA-MAT.

Edaphic Factor. ⇨ ECOSYSTEM.

Edible Fungi. Although numerous British fungi are edible, and many of them excellent eating, only a few of the more distinctive species are at all commonly eaten in Britain, owing to fear of confusion with poisonous fungi. The main species currently eaten are the field and horse mushrooms, blewits, chanterelles, and, especially in the north of England, morels. During the last century English truffles were also eaten, but the art of discovering these is now almost lost. Many quite erroneous ideas exist as to tests of the safety of eating fungi, such as that they are safe if they peel and have a ring round the stem; in fact the death cap responds to both these tests. It is therefore essential, before gathering fungi to eat in the fields or woods, to be familiar with the poisonous species likely to be confused with those you find. ⇨ POISONOUS FUNGI.

Edward Grey Institute for Field Ornithology. Founded in 1933 by the British Trust for Ornithology to administer the fund raised in memory of Lord Grey of Fallodon, the Institute is now part of the Department of Zoological Field Studies at Oxford University. It specializes in bird population and migration problems.

Eel (Apodes: *Anguilla anguilla*). A snake-like fish with a most remarkable life history. All British eels are born in the Sargasso Sea, whence the larvae, formerly known as Leptocephalids, make their way across the Atlantic. When three years old, they change to eel-shape and ascend the rivers as Elvers, to spend five or six years there as Yellow Eels, or in the case of well developed specimens with large mouths, Frog-mouthed Eels; females may attain a length of 3–4 ft. Finally, they assume a silvery breeding dress and as Silver Eels return to the sea to die. There is currently a controversy as to whether any European eels actually return to the Sargasso Sea to breed, or whether they die *en route*, the fresh supply of eels for Europe being maintained by the surplus stock of the American eels, which also breed in the

Sargasso Sea. Conger Eel, ⇨ CONGER. Horn-eel, ⇨ PIPE-FISH. Sand eels are unrelated.

Eel-Grass or Grass-wrack (Zosteraceae: *Zostera*). Three species of grass-like marine monocotyledons, the only flowering plants which actually grow in the sea, and a favourite food of brent geese and wigeon.

Eel-Pout or Viviparous Blenny (Percomorphi: *Zoarces viviparus*). One of our few viviparous marine fishes, usually 5–8 ins. long, allied to the blennies, common on our northern and eastern shores. ⇨ BURBOT.

Eelworm (Aschelminthes: Nematoda). Tiny free-living and parasitic round-worms, some species free-living in the soil or fresh water, others endoparasitic on plants or insects. These latter include the causes of many well-known plant diseases, such as the Potato-root Eelworm *Heterodera rostockiensis*, the Sugar-beet Eelworm *H. schachtii*, the Root-knot Eelworm *Meloidogyne* and the stem, bulb, and leaf eelworms of the genus *Ditylenchus*; some of these are aliens imported along with the introduced crops.

Eft. Old country name for newt.

Egg. The earliest stage of animal development, hatching internally in, e.g., mammals and some reptiles, which are born alive, and externally in birds and most other animals. Some animals, including birds, lay their eggs in specially constructed nests. An invertebrate's egg is usually referred to as an ovum (pl. ova). The Egg-tooth is the horny knob at the tip of the bill of a new-born young bird, which it discards shortly after it has used it to crack the eggshell on hatching. Eggshells are mainly composed of calcium carbonate, and vary in thickness from $\frac{1}{20}$ mm. in a goldcrest to $\frac{3}{4}$ mm. in a mute swan.

Eggar Moth (Bombycoidea: Lasiocampidae). A group of large, mostly brown moths, with furry larvae, closely related to the lackeys, the best known being the Oak Eggar *Lasiocampa quercus*, the Fox Moth, the Drinker, the Lappet and the December Moth.

Eggs and Bacon. A vernacular name for birdsfoot trefoil.

Eglandular. Botanical adjective meaning devoid of glands.

Egret (Ardeiformes: Ardeidae). Species of heron, usually rather small; three are irregular visitors or stragglers to Britain, the south European Little Egret *Egretta garzetta* being the most often seen.

Eider (Anatidae: *Somateria mollissima*). Large diving sea duck, lining its nest with very soft down from its breast which is taken for eiderdowns; known as cuddy duck or St Cuthbert's duck in Northumberland. Common breeder on the coast of Scotland, more local in northern England and northern Ireland; increasing in winter in south England and Wales. The King Eider *S. spectabilis* is a rare straggler from the Arctic.

Elder (Caprifoliaceae: *Sambucus nigra*). A widespread and common shrub with large flat heads of small white flowers, turning to purple-black berries from which elderberry wine is made; leaves pinnate, with a somewhat unpleasant smell. Red-berried Elder *S. racemosa* is well established in parts of south and east Scotland. Dwarf Elder, ⇨ DANEWORT. Ground-Elder is unrelated.

Elecampane (Compositae: *Inula helenium*). A tall composite with yellow daisy-like flowers, formerly valued medicinally and still sometimes growing on roadsides and near houses.

Elephant-Tooth Shell. ⇨ TUSK SHELL.

Elf-Cup or Cup Fungus (Ascomycetes: Pezizales). A group of small cup-shaped fungi, of which the best known species are the Orange Elf-cup or Orange-peel Elf-cup *Peziza aurantia*, frequent on lawns, paths and bare ground, and the Scarlet Elf-cup *P. coccinea*, locally common on dead sticks and branches in winter and early spring.

Elleck. ⇨ Red GURNARD.

Elm (Ulmaceae: *Ulmus*). Tall deciduous hedgerow and woodland tree. Three species in Britain: Common Elm *U. procera*, Smooth-leaved Elm *U. carpinifolia* and Wych Elm *U. glabra*. Dutch and Huntingdon Elms are commonly planted hybrids, and Cornish Elm is a form of the smooth-leaved with branches sweeping upwards. Dutch Elm Disease, produced by the fungus *Ceratostomella ulmi* and carried by the elm bark beetle *Scolytus scolytus*, causes extensive die-back but does not usually kill the elm; known in Britain since at least 1927.

Elm Gall Bug (Heteroptera: *Anthocoris gallarum-ulmi*). A flower bug which lives in the curled leaf-galls on elms produced by the aphid *Eriosoma ulmi*, on which it feeds.

Elver. The young common eel at three years old, when it first arrives in our rivers from the sea.

Elytron (Pl. Elytra). The forewing of a beetle, modified into a hard cover for the functional hindwing.

Emarginate. (1) Botanical adjective used of a leaf or other organ that is slightly notched at the tip. (2) Also used in ornithology to denote a feather which has one web reduced so as to produce a tapering effect.

Emerald Moth (Notodontoidea: Geometridae). Nine handsome pale green geometrid moths; the Large Emerald *Geometra papilionaria* is one of the commonest and most widespread. The Light Emerald *Campaea marginata* belongs to another geometrid family, Selidosemidae.

Emperor Moth. (Notodontoidea: *Saturnia pavonia*). A large moth with a conspicuous eye on each wing, the males flying commonly in day-time in spring in moorland districts, where the handsome large bright green caterpillars may be found on heather and other plants in the summer.

Empid. A family of robber flies.

Enchanter's Nightshade (Onagraceae: *Circaea*). White-flowered woodland plants, unrelated to the nightshades and obscurely named after the enchantress Circe: *C. lutetiana* common in woods and gardens, and *C. alpina* rare in the west. The hybrid between them is commoner than *alpina*.

Enchytraeid Worm. ⇨ WHITE WORM.

Endemic Species. A species of animal or plant confined to a particular country, region or island, and having, so far as is known, originated there. The British Isles have few endemic species because they have only comparatively recently been separated from the Continent. The Scots primrose is one.

Endogenous. Growing inside an organism.

Endoparasite. A parasite living inside its host, usually in the gut, e.g. a tapeworm.

Endoprocta. ⇨ SEA-MAT.

Engraver Beetle. ⇨ BARK BEETLE.

Ensiform. Botanical adjective meaning sword-shaped.

Enteropneust. ⇨ ACORN WORM.

Entire. Botanical adjective used of leaves or other organs whose edge is not toothed or cut.

Entomogenous. Growing on insects.

Entomology. The scientific study of insects.

Entomophily. The pollination of plants by insects.

Entoprocta. ⇨ SEA-MAT.

Eocene Period. ⇨ GEOLOGICAL PERIODS.

Ephemeral. A plant that completes its life cycle more than once in one year, producing more than one crop, e.g. groundsel.

Ephemeroptera. ⇨ MAYFLY.

Epicalyx. A calyx-like arrangement of bracts or stipules, lying just outside or beneath the true calyx.

Epichile. Front part of the lip or labellum of certain orchid flowers.

Epigeal. Botanical adjective meaning above the ground.

Epilimnion. The warm upper layer of water of a lake. ⇨ SEICHE.

Epiphyte. A plant growing on another plant, but not parasitic upon it, e.g. mosses, lichens and polypody fern on trees.

Epizoic. Living on or attached to the body of another animal, e.g. a barnacle or a limpet shell.

Equinox. The time of year at which the periods of daylight and darkness are both approximately twelve hours, and when the sun rises exactly in the east and sets exactly in the west. The spring or vernal equinox falls on or about 21 March, and the autumnal equinox on or about 22 September.

Equisetaceae (Pteridophyta: Sphenopsida). The family of vascular cryptogams containing the horsetails.

Ergot (Ascomycetes: *Claviceps*). A small black fungus parasitic on the seed-heads of grasses, including the cultivated cereals, especially rye. It is highly poisonous, producing constriction of the blood vessels at the extremities, so that hands, feet and limbs become gangrenous and drop off. The disease was known in the Middle Ages as Holy Fire and was prevalent in wet seasons, when the fungus was abundant and was ground into the rye flour. Another form of the disease produced convulsions and death. *C. purpurea* is the ergot of cultivated cereals and many wild grasses; a smaller species *C. microcephala* is common on purple moor-grass, reeds and mat-grass.

Ericaceae (Angiospermae: Dicotyledones). The family of flowering plants containing the heaths, heathers, bilberry and related under-shrubs.

Ermine. A stoat in its white winter pelage.

Ermine Moth (Caradrinoidea: Arctiidae). Three relatives of the tiger moths, with 'woolly bear' caterpillars, named from having black spots on a white or buff ground: White Ermine *Spilosoma lubricipeda* and Buff Ermine *S. lutea* are common; Water Ermine *S. urticae* is local in marshes and fens.

The Small Ermines (Tinaeoidea: *Yponomeuta*) are 'micros' with similar colouring and small web-forming colonial caterpillars.

Erose. Botanical adjective used of leaves or other organs which look as though they have been slightly gnawed.

Erratic. An isolated rock brought by glaciers and ice-sheets, often for a considerable distance from its place of origin, and left when the ice melted; e.g. blocks of granite from Shap Fell are found in many parts of the northern Pennines.

Eryngo, Field (Umbelliferae: *Eryngium campestre*). A now scarce white-flowered relative of the sea holly, also named Watling Street Thistle from once having grown there.

Erythrism (Adj. Erythristic). A colour variation of animals, in which black or brown is replaced by bright chestnut-red, especially in certain birds and their eggs.

Escape. An alien animal or plant that has escaped from captivity or cultivation into the wild and established itself; often wrongly written 'escapee'.

Escarpment. Scarp, but usually used of the more elongated scarps, such as the western edges of the Chilterns and Cotswolds.

Esker. A long, narrow, grass-covered ridge of sand and gravel deposited by a stream running under a glacier and left behind when the ice melted. Eskers are numerous and easily seen in central Ireland.

Ethology. The scientific study of the behaviour of animals in their natural environment, a comparatively modern development in the study of animal behaviour, initiated especially by Konrad Lorenz and carried on most notably by Niko Tinbergen.

-etum. A suffix used in plant ecology, attached to the Latin generic name of a plant to denote a community dominated by that plant, e.g. a heather *Calluna* moor is a Callunetum. Different types of community dominated by the same plant or tree may be distinguished by the addition of a Latin adjective. Thus a beechwood *Fagus* on chalk or clay-with-flints is a Fagetum rubosum, from its field layer being dominated by brambles *Rubus*, while a beechwood on sand is Fagetum ericetosum from the heathy nature of the ground vegetation. The species dominating the community may also be designated by the use of the genitive of the specific name, e.g. a wood of the common oak *Quercus robur* is Quercetum roboris.

Euglenophyta. A phylum or major botanical division into which botanists put certain flagellates which they regard as algae but zoologists class as protozoans of the sub-class Phytomastigina.

Euphorbiaceae (Angiospermae: Dicotyledones). The family of flowering plants containing the spurges and mercuries.

Euryhaline. Tolerating a wide range of saline conditions, as opposed to Stenohaline.

Eurytopic. Of an organism with a wide range of distribution, as opposed to Stenotopic.

Eutrophic. Rich in the basic salts that are plant nutrients; used of lakes, peat etc.; the opposite of Oligotrophic.

Evening Primrose (Onagraceae: *Oenothera*). Tall garden plants with rather pale yellow flowers, fragrant at night, several species of which have escaped from gardens and established themselves on waste ground and sand dunes; primrose-like only in their flower colour. The two most frequent species are *Oe. biennis* and *Oe. erythrosepala.*

Evergreen. A plant that keeps its leaves throughout the year, but usually for a given limited period, e.g. Scots pine three years. Our principal native evergreen trees and shrubs are Scots pine, yew, juniper, holly and ivy.

Evolution. The development of living organisms by a process of cumulative changes in successive generations, variations produced by mutation and other genetic mechanisms combining with selective adaptation to changing environments to produce different species and sub-species from a common original stock. ⇨ NATURAL SELECTION.

Exogenous. Growing inside an organism.

Exoparasite. ⇨ ECTOPARASITE.

Exoskeleton. A hard external structure supporting an animal, as in many insects and molluscs, and the bony plates of turtles and tortoises.

Exotic Animals and Plants. ⇨ ALIEN.

Exserted. Botanical adjective used mainly of stamens that protrude from the mouth of a two-lipped or tubular flower, especially in the mints.

Exstipulate. Botanical adjective meaning devoid of stipules.

Extrorse. Botanical adjective used of anthers which open towards the outside of a flower.

Eyas. The young falcon.

Eyebright (Scrophulariaceae: *Euphrasia officinalis*). Common hemiparasitic plant with white or purple flowers, formerly used medicinally as an eye lotion; very variable and divided into two dozen microspecies.

Eyrie. The nest of an eagle or a peregrine, usually placed on a more or less inaccessible rock ledge.

F

F_1 and F_2 Generations. In genetics the offspring produced by the parental generation (P_1) of animals and plants: F_1, first filial generation; F_2, the offspring of F_1, the second filial generation.

Fairy Club. ⇨ CLUB FUNGUS.

Fairy-Fly (Chalcidoidea: Mymaridae). A family of minute chalcid-type wasps with very delicate wings, laying their eggs in the eggs of other insects.

Fairy Ring. Ring or arc of darker coloured grass in meadows and pastures, caused by certain fungi, notably the Fairy Ring Champignon or Scotch Bonnets *Marasmius oreades*, a close relative of the horsehair fungus, and the parasol and St George's mushrooms. These rings may be a good many yards across and hundreds of years old; they are produced by a perennial mycelium continually growing outwards, each year's growth dying off. The darker grass is a richer growth fertilized by the nitrates produced by chemical changes which the mycelium stimulates in the soil. Many other fungi, such as common and horse mushrooms, grow in outward circles without producing visible rings on the turf.

Fairy Shrimp (Branchiopoda: *Chirocephalus diaphanus*). An inch-long transparent fresh water crustacean of the order Anostraca, living in temporary pools which dry up in summer; swims on its back; not closely related to the true shrimps.

Falcate. Botanical adjective meaning sickle-shaped.

Falcon (Falconiformes: *Falco*). Long-winged birds of prey, mostly feeding on birds and much favoured in falconry. Male, tiercel; female, falcon; young, eyas. There are four British breeding species, the Peregrine, Kestrel, Hobby, and Merlin; and three more or less rare stragglers, the northern Gyr Falcon *F. rusticolus* (including the Iceland and Greenland Falcons), the southern Red-footed Falcon *F. vespertinus*, and the Lesser Kestrel.

Falconiformes. The diurnal birds of prey, characterized by the hooked beaks and strong powers of flight that enables them to secure their carnivorous diet. The British breeding species (all Falconidae) include one eagle, two buzzards, the sparrowhawk, one kite, three harriers, the osprey and four falcons. Two vultures (Aegypiidae), two eagles, one buzzard, the goshawk, one kite, one harrier and three falcons are visitors, mostly very rare.

Falconry or Hawking. The use of trained hawks or falcons in pursuit of game or other birds or mammals. Falconry is still actively pursued in Britain today, and there is a Falconers Club of Great Britain. The most popular hawks are the peregrine, gyr falcon and goshawk.

False Brome Grass (Gramineae: *Brachypodium*). Two common grasses, long-awned like the true bromes: Slender False Brome *B. sylvaticum*, common in woods and hedge-banks, and Heath False Brome or Tor Grass.

False-Scorpion (Arachnida: Pseudoscorpiones). An order of small arachnids, 1 – 5 mm. long, superficially resembling minute scorpions or lobsters on account of their relatively large claw-like palps, which contain poison glands; two dozen British species, mainly found under stones, bark or other places where vegetable debris accumulates; *Dactylochelifer latreilli* and *Chthonius tetrachelatus* live on the shore. *Chernes nodosus* often makes use of Phoresy.

Family. A group in the classification of living organisms, consisting of related Genera. Related families are grouped into an Order. In botany the family name usually ends in *-aceae;* in zoology in *-idae.*

Fan Worm (Polychaeta: Sabellidae). Marine bristle worms living in tubes of mud projecting from the surface; the fans of feathery tentacles protrude from the tube only under water, the worm withdrawing well inside its tube at low tide. ⇨ PEACOCK WORM.

Fasciation. The fusion together of stems or branches in a plant to make an abnormally thick growth. The cause is obscure.

Fastigiate. Of trees or shrubs with a narrowly conical outline, e.g. as in the Lombardy poplar, due to all the branches growing straight upwards.

Fat-Basket. A receptacle made of two small pieces of wire-netting clasped together in which fat and other food for birds, especially tits, is hung up.

Fat Hen (Chenopodiaceae: *Chenopodium album*). The commonest of our native goosefoots, a weed of cultivation.

Father-Lasher. A marine species of bullhead.

Fault. A crack in the earth's crust, where the rocks have split and one side has fallen below the other so that the strata are discontinuous. There are several notable faults in the Scottish Highlands including the Great Glen with its chain of lochs including Loch Ness. A Rift Valley, such as the Central Lowlands of Scotland, has a fault on each side.

Fauna. The animal population of a place or period.

Fauna Preservation Society. Founded in 1903 as the Society for the Preservation of the Fauna of the Empire, the Society is concerned with the conservation of wildlife throughout the world; in the British Isles it is specially concerned with mammals, reptiles and amphibians. Its journal *Oryx* is published three times a year.

Fawn. The young (first year) of fallow and muntjac deer.

Feather. The characteristic component of the plumage or body covering of birds; a growth from the skin which is ultimately derived from the reptilian scale. For nomenclature of feathers, ⇨ p. 31.

Feather Mite (Astigmata: Analgesidae). A family of mites living among the feathers of birds, and living, like the fur mites of mammals, on the waste products of the skin.

Feather Star (Echinodermata: Crinoidea). A primitive group of marine animals, with ten feathery 'arms' attached to a small central body. The Rosy Feather Star *Antedon bifida*, which ranges from rosy red to deep purple, yellow and orange, occurs off our south-west coasts.

February Red (Plecoptera: *Taeniopteryx nebulosa*). A rather local stone-fly emerging between January and April.

Felt. Country name for the Fieldfare.

Felwort. ⇨ Autumn GENTIAN.

Femur (pl. Femora). The third joint in the legs of insects and spiders.

Fen. Waterlogged ground with a peat soil that may be alkaline, neutral or slightly acid, in contrast to Bog in which the soil is very acid and the vegetation consequently quite different. The Fens of eastern England have mostly been drained.

Fennel (Umbelliferae: *Foeniculum vulgare*). A fairly common large yellow-flowered umbellifer, with almost hair-like leaf-segments, used as a herb, and growing mainly in grassy places near the sea. Hog's Fennel (*Peucedanum officinale*) is very rare in Kent and Essex.

Fen Orchid (Orchidaceae: *Liparis loeselii*). A small orchid with yellow-green flowers. Very local in fens in East Anglia and dune slacks in South Wales.

Fenugreek (Leguminosae: *Trifolium ornithopodioides*). A small, inconspicuous and rather local clover with few flowers in the head.

Feral. Living wild. Several former domestic animals are now feral in Britain, among them the goat, pigeon and mute swan, and in some districts the cat.

Fermentation. The anaerobic decomposition of organic matter, especially carbohydrates (but not proteins, whose decomposition is called Putrefaction), by minute living organisms such as yeasts and bacteria to form simple organic chemicals such as alcohols, other organic chemicals, and carbon dioxide with the consequent liberation of energy.

Fern (Pteridophyta: Pteropsida). A plant that reproduces by means of spores, growing mainly in damp places and on rocks or walls. The great majority of the forty-seven British species fall in the Order Filicales of the sub-class Leptosporangiatae, and bear their spores on the back of the leaves (fronds), but a few bear them on special fertile stems that resemble the flowering stems of the seed-plants, viz. Royal Fern (Osmundaceae) and Adderstongues and Moonwort (Eusporangiatae: Ophioglossales).

Fernbug (Heteroptera: *Bryocoris pteridis*). A widespread and common capsid bug feeding on ferns, especially male and lady ferns.

Ferret. An albino domesticated form of the polecat, used for hunting rabbits in their burrows. Polecat-ferrets are dark hybrids between the ferret and the polecat, and are feral in several places, including the Isle of Mull.

Fescue (Gramineae: *Festuca*). Several of our commonest and most valuable forage grasses belong to this genus, notably Sheep's Fescue *F. ovina* and Red Fescue *F. rubra; F. pratensis* and *F. arundinacea*, two taller species, are also common in grassland. *F. vivipara* of moorlands in the north has green shoots instead of flowers. Grasses of several allied genera, such as *Vulpia* and *Catapodium*, are also given book names including fescue.

Feverfew (Compositae: *Chrysanthemum parthenium*). A medicinal herb with yellowish foliage and daisy-like flowers, widely established on walls and waste ground.

Fever Fly (Nematocera: *Dilophis febrilis*). A misnamed small black hairy two-winged fly, related to the gnats; other members of the genus are also loosely so called.

Fewmets. The individual droppings of deer.

Field Centre. A building, often with residential facilities, equipped with laboratories and other facilities for field studies, and staffed by teachers and demonstrators. ⇨ FIELD STUDIES COUNCIL.

Field Club. ⇨ NATURAL HISTORY SOCIETY.

Fieldfare (Passeriformes: *Turdus pilaris*). A large thrush with a grey-blue back often associating with redwings; a widespread and frequent visitor from northern Europe. Still known as Felt in some country districts.

Field Layer. One of the four levels of vegetation recognized by plant ecologists: the tree layer, the shrub layer, the field or herb layer and the moss or ground layer.

Field Studies. Scientific studies based on observation and/or experiments in the field, as distinct from the museum or laboratory.

Field Studies Council. The national body concerned with the organization of field courses in biological, geographical and related subjects in residential field centres in England and Wales, founded in 1943; publishes *Field Studies*. ⇨ SCOTTISH FIELD STUDIES ASSOCIATION.

Fierasfer or Pearl-fish (Percomorphi: *Fierasfer*). Two species of Mediterranean sea fish, with the singular habit of sheltering inside sea-cucumbers and feeding on the plankton brought in by their hosts' breathing; adults are rare in British waters, but young apparently less so.

Figwort (Scrophulariaceae: *Scrophularia*). Tall plants with inconspicuous two-lipped red-brown or yellow flowers; five species, the two commonest being Common Figwort *S. nodosa*, of woods and shady banks, and Water Figwort or Water Betony *S. aquatica*, of stream- and pond-sides. They are the only members of their family with the square stems otherwise associated with the labiates.

Filament. (1) The stalk of a flower's stamen, bearing an anther at its tip. (2) The string of cells of filamentous algae such as *Spirogyra*.

File Shell, or Gaping File Shell (Lamellibranchia: *Lima hians*). A swimming deep-water bivalve, whose oval white shell is covered with rough excrescences, and with long orange tentacles; occasional near low-water mark.

Filicales (Pteridophyta: Pteropsida). The order containing the typical ferns, in the families Hymenophyllaceae (filmy ferns), Dennstaedtiaceae (bracken), Adiantaceae (maidenhair fern and allies), Blechnaceae (hard fern), Aspleniaceae (spleenworts and hartstongue), Athyriaceae (lady fern and allies), Aspidiaceae (male, buckler and shield ferns), Thelypteridaceae (marsh fern and allies) and Polypodiaceae (polypodies); all bear their spores on the back of the fronds, which are usually two- or three-pinnate, only the hartstongue being undivided.

Filiform. Botanical adjective meaning thread-like.

Filmy Fern (Filicales: *Hymenophyllum*). Two delicate, moss-like small ferns, growing on damp rocks in the north and west: *H. tunbrigense* and *H. wilsonii*. The allied Killarney or Bristle Fern *Trichomanes speciosum* is now very rare in west Ireland.

Fimbriate. Botanical adjective used of leaves or other organs with fringed margins.

Fin. A group of spines linked by a membrane. For nomenclature of a fish's fins, ⇨ below.

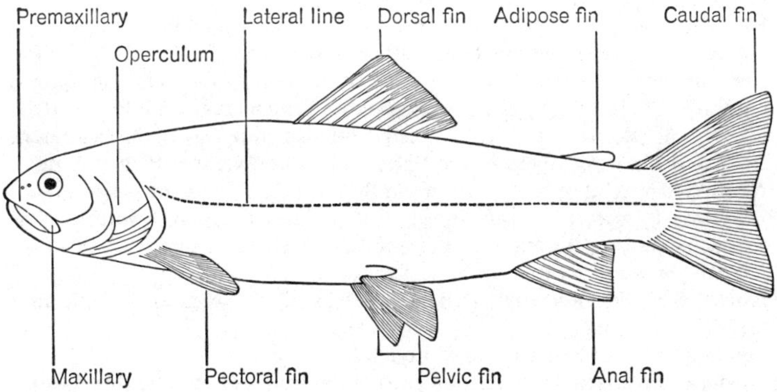

Finch (Passeriformes: Fringillidae). Small thick-billed seed-eating song birds, often flocking together in winter. Bramble Finch ⇨ BRAMBLING; Bullfinch; cardueline finch, a member of the genus *Carduelis*, viz. goldfinch, linnet, redpoll, siskin, twite; Chaffinch; Citril Finch *C. citrinella*, a very rare straggler; Goldfinch, Greenfinch, Hawfinch.

Finner. Seaman's name for the rorquals.

Finnock or Whitling. Scottish name for a sea trout on its first return from the sea.

Fir (Coniferae: Pinaceae). The true firs *Abies* are frequently planted, but nowhere naturalized, as the even more frequently planted Douglas Fir *Pseudotsuga menziesii* often is. Scotch Fir is an old-fashioned name for the Scots pine.

Fire, Effect of. In dry areas where fires occur more or less frequently, such as the heathy areas in the south and south-east, the plants are mainly limited to those that can survive the fires and those growing nearby that can quickly recolonize the area. Fires destroy tree seedlings and thus prevent the succession from heath to woodland. On grouse moors the heather is deliberately fired at regular intervals to produce new growth on which the grouse feed.

Fire-Brat or Baker's Brat (Thysanura: *Thermobius furnorum*). A bristle-tail found in bakehouses and similar warm places.

Firebug (Heteroptera: *Pyrrhocoris apterus*). A conspicuously coloured plant bug widespread in Europe but in Britain confined to one small island off the Devon coast; related to the cotton-stainers, pests of cotton fields.

Firecrest (Passeriformes: *Regulus ignicapillus*). A very small insectivorous song bird, resembling the goldcrest, but with white eyestripes, formerly miscalled fire-crested wren. Scarce winter visitor, mainly in the south; has recently begun to breed in Hampshire.

Fireweed. ⇨ ROSEBAY.

Firth. An inlet of the sea in Scotland, usually with low shores; often applied to the seaward end of an estuary, as with the Firths of Forth and Clyde on either side of central Scotland.

Fish (Chordata: Vertebrata). A group of 'cold-blooded', gill-breathing aquatic vertebrates, whose temperature varies according to that of their surroundings. They comprise four classes: Marsipobranchii (lampreys, hagfishes), Selachii (sharks), Bradyodonti (chimaeras) and Pisces (true or bony fishes). Anglers divide freshwater fish into Game Fish (salmon, trout and other members of the Salmonidae) and Coarse Fish (the rest). About 370 species of fish occur in the British Isles and their seas. ⇨ FIN.

Fish-Louse (Crustacea: Branchiura). A sub-class of tiny freshwater crustaceans, somewhat resembling copepods, which are parasitic on fish. *Argulus foliaceus* is our commonest species.

Fistular. Botanical adjective used of stems or stalks which are tubular and hollow.

Flag. ⇨ SWEET FLAG, YELLOW FLAG.

Flagellate (Protozoa or Flagellata). A large class of minute freshwater, marine and soil animals, moving by means of a flagellum (two in the Order Dinoflagellata) and so primitive that there is even some doubt whether they should be classed as animals or plants; ⇨ ALGAE. The most familiar of the freshwater ones are the spherical colonial *Volvox* and *Euglena*, which often forms a green scum on stagnant ponds and is one of those which many botanists classify as an alga. Some dinoflagellates occur on the shore, making green or brown patches on the sand (*Gymnodinium*), while others are found in the plankton, notably the phosphorescent *Noctiluca scintillans*. The most important parasitic flagellates are the Trypanosomes, the cause of sleeping sickness in tropical Africa.

Flagellum (Pl. Flagella). A minute thread-like organ, single or paired, performing similar functions to cilia, but not usually larger and not moving in unison.

Flamingo (Phoenicopteriformes: *Phoenicopterus ruber*). A large, very long-legged pink and white marsh bird, with a curiously bent 'Roman-nose' bill; rare as a genuinely wild visitor to Britain, and perhaps more often seen as a zoo escape.

Flash. A stretch of fresh water, usually produced by mining subsidences, especially in the salt districts of Cheshire and the Lancashire coalfield.

Flask Fungus (Ascomycetes: Pyrenomycetes). A group including the caterpillar fungi, candle snuff and dead men's fingers.

Flask Shell (Lamellibranchia: *Gastrochaena dubia*). A small piddock-like marine bivalve, which bores in rocks on our south and south-west coasts.

Flatbug (Heteroptera: Aradidae). A small flat plant-bug, living under bark and feeding on the mycelium of fungi; five British species, the most widespread being *Aradus depressus.*

Flatfish (Pisces: Heterosomata). Sea fish, highly adapted to bottom-living by being flattened laterally, with both eyes on one side, the lower side white, the upper coloured to harmonize with the bottom. Two main groups, the Bothidae with both eyes on the left, and the Pleuronectidae and Soleidae with both eyes on the right side of the body.

Flat-Tail. The common earthworm *Lumbricus terrestris.*

Flatworm (Platyhelminthes). The most primitive phylum of worms, with flattened, segmented bodies, the organs embedded in tissue instead of being contained in a body cavity. The principal classes are the free-living Turbellaria, which include the triclads and planarians, and the parasitic Cestoda (tapeworms) and Trematoda (flukes).

Flax (Linaceae: *Linum*). Plants with narrow leaves in pairs up the stem; three native species, the commonest being the white-flowered Fairy or Purging Flax *L. catharticum,* the two others blue-flowered and local. Cultivated Flax *L. usitatissimum* is a frequent casual or escape from cultivation.

Flax-Seed. ⇨ ALL-SEED.

Flea (Insecta: Siphonaptera, also known as Aphaniptera). A highly specialized order of wingless insects, whose adults are parasitic on the bodies of mammals and birds, whose blood they suck; most are specific to one or a few hosts. Of the forty-seven British species, fourteen live on birds, the largest being the Rock-dove Flea *Ceratophyllus columbae* and the smallest the House-sparrow Flea *C. fringillae.* The largest British flea is the Mole Flea *Hystricopsylla talpae,* and well known mammal fleas are the Human Flea *Pulex irritans,* the Dog Flea *Ctenocephalides canis* and the Rabbit Flea *Spilopsyllus cuniculi,* the carrier of myxomatosis.

Fleabane. The name of several yellow-flowered, daisy-like composites the commonest and most widespread being Common Fleabane *Pulicaria dysenterica* of damp grassy places, Blue Fleabane *Eri*[illegible] dry grassy places, and Canadian Fleabane *Conyza canadensis,* an introduced plant now abundant in waste places.

Flea Beetle (Phytophaga: Halticinae). A sub-family of leaf beetles, many of which can jump and are pests of farm crops, especially brassicas, e.g. the Turnip Flea Beetles *Phyllotreta* spp., also known as Turnip Fly, and *Chaetocnema concinna,* also known as Mangold Flea Beetle or Hop-flea; and the Potato Flea Beetle *Psylliodea affinis.*

Fleawort, Field (Compositae: *Senecio integrifolius*). A yellow-flowered ragwort, local in calcareous turf, mainly in southern England.

Fleet. An estuarine creek, especially, in the Thames estuary, one that has been cut off from the sea by land reclamation and become a landlocked freshwater channel.

Flesh Fly (Cyclorrhapha: *Sarcophaga*). Grey two-winged flies related to the bluebottles, whose larvae feed on carrion and other decaying animal matter.

Floccose. Botanical adjective meaning covered with soft white hairs, used especially of hawkweeds.

Flint. A hard siliceous deposit found as nodules in the Chalk.

Flixweed (Cruciferae: *Descurainia sophia*). A yellow-flowered weed of light soils, now mainly in East Anglia.

Flora. The plants of a region. Also the list of the plants of a particular area arranged in families and genera, with descriptions and sometimes also identification keys.

Floret. A small individual flower in a compound head, as in the composites, where the florets may be of two kinds, ray florets (strap-shaped) and tubular florets.

Florula. The flora of a small area, such as a single crag or pond.

Flote Grass, or Sweet Grass (Gramineae: *Glyceria*). Three grasses of freshwater margins and marshes, often growing right in shallow pools, with their leaves floating on the surface; *G. fluitans* is the commonest. Their sweet-tasting leaves are much relished by cattle.

Flounder or Fluke (Pleuronectidae: *Platichthys flesus*). A plaice-like, right-sided flatfish, characterized by its very white under surface and orange spots above that soon fade. Normally a fish of shallow seas, it also frequents estuaries and is the only flatfish to ascend rivers into fresh water. Pole Flounder, ⇨ WITCH.

Flour Mite (Astigmata: *Acarus siro*). The most important mite pest of flour and grain, also infesting many other stored products. The allied *Glycyphagus destructor* and *Goheria fusca* also infest grain and flour respectively. *Acarus* and *Glycyphagus* are both preyed on by the protostigmatid mite *Cheyletus eruditus*.

Flour Moth or Mill Moth (Pyraloidea: *Ephestia kühniella*). A small pyralid moth, whose larvae eat flour and other stored vegetable products and are reckoned our most serious insect pest of flour mills.

Flower. ⇨ FLOWERING PLANT.

Flower Bee. ⇨ MINING BEE.

Flower Bug (Heteroptera: Anthocorinae). Somewhat inappropriately named predatory bugs; a score of British species, the commonest being *Anthocoris nemorum*. ⇨ ELM GALL BUG.

The Parts of a Flower

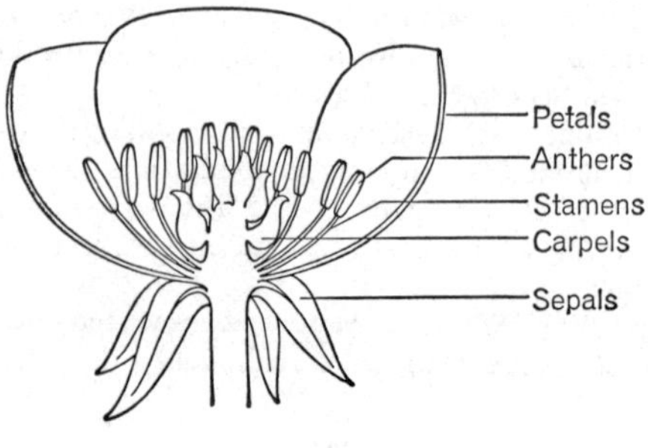

Flowering Plant (Spermatophyta: Angiospermae). The larger sub-phylum of the seed-plants, containing mostly herbaceous but also many woody plants, and divided into two large classes the Monocotyledons and the Dicotyledons; characterized by bearing their reproductive parts in flowers, which often have conspicuous petals or sepals surmounting the actual reproductive organs. ⇨ PISTIL.

Flower Fly. ⇨ HOVER-FLY.

Flowers of Tan (Myxomycetes: *Fuligoseptica*). A bright yellow slime fungus, named from being often found in tanyards, but also common in woodland.

Fluellen (Scrophulariaceae: *Kickxia*). Two toadflax-like weeds of calcareous soils: Sharp-leaved Fluellen *K. elatine* and Male or Round-leaved Fluellen *K. spuria.*

Fluke. (1) ⇨ FLOUNDER. (2) (Platyhelminthes: Trematoda). An order of small parasitic flatworms, including both endo- and exoparasitic species, many of which pass through more than one host on their way to their final destination. Much the best known is the notorious Liver Fluke *Fasciola hepatica,* whose larvae use water-snails, usually *Limnaea truncu-lata,* as their intermediate host, before emerging and encysting on blades of grass; these are eaten by sheep, in which the adults produce the disease called Liver-rot. Many trematodes live a largely freshwater existence, and may be found in fresh or brackish water while awaiting their next host.

Flush. A patch of wet ground, usually on a hillside, where the water flows diffusely and not in a fixed channel.

Fly. A general term for winged insects, but in common parlance often restricted to the two-winged flies or Diptera, and more especially to the house fly.

Flybug (Heteroptera: *Reduvius personatus*). A large, somewhat elongated assassin bug, noted for flying into lighted rooms on summer evenings; also known as Masked Bug from the fact that its larvae cover themselves with dust after each moult.

Flycatcher (Passeriformes: *Muscicapa*). Small song birds that feed by darting out to catch flying insects and returning to their perch. Five British species: Pied Flycatcher *M. hypoleuca,* with black and white plumage, summer visitor to woods in the west; Red-breasted Flycatcher *M. parva,* scarce passage migrant. Spotted Flycatcher *M. striata,* named from spotted breast of young birds, widespread and common summer visitor; also two rare stragglers.

Flying Bent. ⇨ Purple MOOR GRASS.

Flying-Fish (Pisces: *Exocoetus volitans*). A tropical fish able to glide in the air with the aid of elongated fins; very rare in British seas.

Fly Orchid (Orchidaceae: *Ophrys insectifera*). A local orchid of calcareous woods and scrub, related to the bee orchid, with flowers sufficiently like a certain small fly-like burrowing wasp to act as a lure to its males.

Fog. Cloud at ground level, consisting of water droplets condensed from the air and present in such numbers as to reduce visibility, and often mixed with impurities derived from atmospheric pollution. If visibility is less than one kilometre the low cloud is called fog, otherwise it is mist.

Fog and mist are of three main types: due to the contact of cold air with much warmer water; the contact of warm air with a cold surface; and the radiation of heat from the earth's surface into the atmosphere. The last is known as radiation fog.

Fog-Bow. A specialized form of Rainbow, which is white, not coloured, owing to the small size of the water droplets in fog.

Folkestone Ladies. Large billowy cumulus clouds, supposed to resemble the hair-do's of nineteenth century ladies at Folkestone, Kent.

Follicle. A dry fruit containing one or more seeds and splitting down one side only. ⇨ CAPSULE, POD.

Food Chain. A term devised by ecologists to show how all organisms are dependent on others in a natural community; e.g. an earthworm may feed on fallen leaves, a blackbird on the earthworm, and a sparrowhawk on the blackbird; or a minnow may eat a water-flea, a pike the minnow, a heron the pike, and a fox the heron. Each member of the hypothetical chain feeds on the one below and may be eaten by the one above; there are not often more than six links in such a chain, but there are often many side-links.

Footman Moth (Caradrinoidea: Lithosiinae). A sub-family related to the Tiger Moths, containing sixteen smallish species, the commonest being the Common Footman *Lithosia lurideola;* their caterpillars are very hairy. Two species of *Coscinia* in the sub-family Arctiinae are also so called.

Foraminifera (Protozoa: Rhizopoda). Minute shelled marine animals, remains of many of which occur on our beaches, while *Globigerina* contributes substantially to the muddy ooze on deep sea bottoms. The Foraminifera were immensely important in the building up of our limestone rocks.

Forb. A pasture herb.

Force. A Norse term still used for waterfall in the north of England, e.g. High Force in Teesdale.

Foreshore. The zone of the shore scarcely reached even by the highest normal spring tides, but subject to occasional disturbances by exceptional springs and having a sparse but specialized flora of such plants as sea rocket and saltwort.

Forest. An extensive area of land covered with trees, larger than a wood; originally an unenclosed open or woodland district preserved for hunting and usually owned by the Crown, e.g. the New Forest. The Forestry Commission calls blocks of woodland of 1,000 acres or more forests. High forest is the term used for mature woodlands when the tops of the trees close overhead to form a close canopy, and all trees are allowed to grow to their full height, as opposed to Coppice. A Deer Forest in the Highlands preserves the original meaning of the word, for it is largely treeless.

Forest Bug (Heteroptera: *Pentatoma rufipes*). A slightly greenish shieldbug widespread and common in woods and orchards.

Forester Moth (Psychoidea: *Procris*). Three small green day-flying moths, allied to the burnets; *P. statices* is the most widespread and frequent.

Forest Fly (Diptera: *Hippobosca equina*). A common parasitic fly of cattle and horses, known under this name in the New Forest in Hampshire.

Forget-Me-Not (Boraginaceae: *Myosotis*). Low-growing, blue-flowered plants; nine species, of which five grow in dry places, *M. arvensis* being the commonest, and four Water Forget-me-nots grow in wet places. *M. scorpioides* being the commonest. Early Forget-me-not *M. ramosissima*, the first to flower, is often only 1 – 2 ins. tall; the flowers of Changing Forget-me-not *M. discolor* open creamy yellow, change to blue-grey; Wood Forget-me-not *M. sylvatica* is also often grown in gardens.

Fork-Beard (Pisces: Gadidae). Two marine fishes of the Cod Family, named from the long forked filaments of one of their lower fins. The Greater Fork-beard or Forked Hake *Urophycis blennoides* is a deep-water fish appearing in British seas in winter. The Lesser Forkbeard, Tadpole-fish or Trifurcated Hake *Raniceps raninus* is a shallow-water fish of the east and south coasts.

Form. A hare's resting place in the grass, where it crouches in the day-time.

Form or Forma. In taxonomy, a taxon differing from other taxa by a single character.

Fossil. Animal or plant remains, or their outlines, preserved in the rocks, e.g. Ammonites.

Foumart. Old name for the polecat.

Fowl Mite. The commonest mite parasitic on birds, including domestic fowls, is the Red Mite (Mesostigmata: *Dermanyssus gallinae*) which sucks their blood at night, and will even bite man. The European or Northern Fowl Mite (M: *Ornithonyssus sylviarum*) sucks their blood day and night. The Fowl Cyst Mite (Astigmata: *Laminosioptes cysticola*), on the other hand, lives in the subcutaneous tissue, and produces small hard cysts.

Fox (Carnivora: *Vulpes vulpes*). The best known native mammalian predator, common and widespread throughout the British Isles.

Fox and Cubs or Grim the Collier (Compositae: *Hieracium brunneocroceum*). An orange-flowered hawkweed which has escaped from gardens and established itself, especially on railway banks in the north.

Foxglove (Scrophulariaceae: *Digitalis purpurea*). Common wild biennial flower with tall spires of purple-pink flowers, each shaped like the finger of a glove; avoiding calcareous soils.

Fox Moth (Bombycoidea: *Macrothylacia rubi*). A common large reddish-brown day-flying eggar moth, whose black furry caterpillar is strikingly marked with yellow rings.

Foxtail Grass (Gramineae: *Alopecurus*). Six grasses with sausage-shaped flower-spikes, similar to timothy, the commonest being the early-flowering Meadow Foxtail *A. pratensis*, Marsh Foxtail *A. geniculatus* of wet grassy places, and Slender Foxtail *A. myosuroides*, an arable weed.

Fragrant Orchid (Orchidaceae: *Gymnadenia conopsea*). A widespread and locally frequent pinkish-purple orchid of grassland turf and fens, noted for its sweet scent.

Frass. The droppings of larvae, especially of the caterpillars of butterflies and moths.

Fraying. Damage done to a tree by deer cleaning velvet off their antlers. Fraying Stock: the tree or bush used by deer for cleaning off velvet.

Freshwater Biological Association. The national body concerned with research into the biology of fresh and brackish water, founded in 1929; maintains laboratories at the Ferry House, Far Sawrey, Ambleside, Westmorland, and at East Stoke, Dorset; publishes an annual report.

Freshwater Cockle or Pea Mussel (Lamellibranchia: Sphaeriidae). A family of small freshwater bivalves, comprising four Orb-shell Cockles *Sphaerium*, of which *S. corneum* is the commonest and *S. rivicola* (up to 1 in.) the largest; and fifteen Pea-shell Cockles *Pisidium*, which are often found adhering to the feet of water-birds, *P. amnicum* (½ in.) being the largest and *P. cinereum* and *P. milium* the commonest and most widespread.

Freshwater Mussel (Lamellibranchia: Unionidae). A family of large freshwater bivalves with more or less oval shells, including the very similar Swan Mussel *Anodonta cygnea* and Duck Mussel *A. anatina*, which grow to 5 – 6 ins. long and the Swan Mussel occasionally to 9 ins.; the Pearl Mussel *Margaritifer margaritifer*, the most important pearl-producing freshwater mussel; and the Painter's Mussel *Unio pictorum*, so named because Dutch painters used to use the shells to hold their colours. Their larvae are called glochidia. The unrelated striped Zebra Mussel *Dreissena polymorpha* is an alien, which has become a pest owing to its habit of congregating in water mains and blocking them. Pea Mussel, ⇨ FRESHWATER COCKLE.

Freshwater Shrimp (Amphipoda: *Gammarus*). Small, slightly shrimp-like crustaceans of rivers and streams; much more closely related to the sand shrimps than to the true shrimps; much the commonest of our three species in Great Britain is *G. pulex*.

Frit Fly or Oat Fly (Cyclorrhapha: *Oscinella frit*). One of the best known economic insects, a tiny black two-winged fly (Chloropidae), which is a serious pest of cereals, especially oats. ⇨ GOUT FLY.

Fritillary or Snakeshead (Liliaceae: *Fritillaria meleagris*). A bulbous plant with handsome solitary variegated flowers, now very local in damp meadows in the south, especially in the Thames valley.

Fritillary Butterfly (Papilionina: Nymphalidae). A group of nine tawny brown butterflies, with black markings that give a chequer-board effect above (whence the name from Latin *fritillus*, a chess-board), and silver spots on the underwings; the commoner British species, many of whose larvae feed on violets, include the Dark Green Fritillary *Argynnis aglaia*, the Pearl-bordered Fritillary *A. euphrosyne* and the Silver-washed Fritillary *A. paphia*. The small and rather local Duke of Burgundy Fritillary *Hamearis lucina* is our sole representative of the family Riodinidae.

Frog (Salientia: *Rana*). Tailless amphibians, one native and two introduced species. The Common Frog *R. temporaria*, whose egg-masses (frogspawn) and tadpoles are decreasingly familiar in ponds in early spring, is widespread in Great Britain, but local in Ireland. The Edible Frog *R. esculenta*, from Europe, has established colonies which have lasted in parts of south and east England for up to 100 years, but have mostly died out in the end. The Marsh Frog *R. ridibunda*, recently introduced from Europe, appears well established in parts of Kent and Sussex. All frogs croak, but the voices of the marsh and edible frogs are louder and so more liable to provoke complaints.

Frog-Bit (Hydrocharitaceae: *Hydrocharis morsus-ranae*). Free-floating aquatic monocotyledon with white flowers, originally so named because supposed to be bitten by frogs.

Frog-Fish or Fishing Frog. ⇨ ANGLER.

Frog-Hopper (Homoptera: Ceropidae). A family of small jumping bugs whose nymphs produce Cuckoo-spit; named from their broad-headed, slightly frog-like appearance and their jumping proclivities. *Philaenus spumarius* is the commonest species.

Frog Orchid (Orchidaceae: *Coeloglossum viride*). A widespread but local small grassland orchid, whose green or reddish-green flowers appear remarkably like a small jumping frog.

Frond. The leaf or leaf-like structure of a lower plant, carrying spores, especially of a fern, a seaweed or a lichen.

Frost. A condition in which the air temperature is at or below the freezing point of water (32°F. or 0°C.), so that ice forms on puddles or pools of fresh water, and any moisture deposited on vegetation or other surfaces shows up as white rime or hoar frost, when the frost is known as a White Frost. Rime is deposited when very cold water-laden air deposits its surplus water vapour on terrestrial surfaces, Hoar Frost when very cold air comes into contact with dew already deposited. A Black Frost occurs when it freezes without such deposits.

Fruit. The ripened ovary of a seed-plant, containing the seeds. Dry fruits are called Achene, Capsule, Follicle or Nut. Fleshy fruits are Aril, Berry or Drupe.

Fruit Fly (Diptera: Cyclorrhapha). Two quite distinct families of two-winged flies are so called. The Drosophilidae are small yellowish or brownish flies, often frequent where fruit is being stored or vinegar or other fermented products being made, whence they are called Vinegar Flies; thirty-one species. Those of the genus *Drosophila* are now among the most famous insects in the world, being much used in genetical research on account of the large chromosomes in the salivary glands of their larvae and their short life-cycle. The Trypetidae include besides celery fly and gall flies, a group called Large Fruit Flies, whose larvae feed in fruits.

Fry. A collective term for young fish.

Fuchsia (Onagraceae: *Fuchsia magellanica*). A half-hardy deciduous shrub with distinctively shaped red and purple flowers, abundantly planted and naturalized on the west coast of Ireland, less common in eastern Ireland and on the west side of Britain.

Fugacious. Botanical adjective used of petals, leaves or other organs which fall quite soon after they are fully developed.

Fuller's Earth. An uncommon mineral deposit, hydrous silicate of aluminium, used in oil refining, the treatment of woollen cloth, and for toilet purposes.

Fulmar or Fulmar Petrel (Procellariiformes: *Fulmarus glacialis*). A superficially gull-like pelagic sea-bird that has spectacularly increased in the past century. Once a rare species of two northern islands, St Kilda and Foula, it is now an abundant breeder on our northern and western coasts, nesting also on many suitable cliffs on the south and east coasts.

Fumitory (Fumariaceae: *Fumaria*). Low-growing or scrambling plants with mostly pink tubular two-lipped flowers, named from their bluish foliage which gives them a smoky appearance (French, *fume-terre*) when the plants are growing *en masse*; ten species, mostly weeds of cultivation, but a few scramble over hedges; *F. officinalis* is much the commonest and most widespread. White Climbing Fumitory *Corydalis claviculata* is closely related but has tendrils. Yellow Fumitory *C. lutea* is a frequent garden escape on old walls.

Fungi. A very large phylum or major grouping of flowerless plants, varying greatly in size and shape from the one-celled yeasts through a wide range of mildews, moulds, smuts and rusts to the larger toadstools, including such monsters as the giant puff-ball and the larger bracket fungi. Their common characteristic is lack of chlorophyll—any green fungi are coloured by other substances – and they are all either Parasites or Saprophytes. Fungi consist of a mycelium which is a web of minute tubular threads known as hyphae and gives rise to much larger fruiting bodies, which may take the familiar toadstool shape, and produce spores. There are three classes of fungi, the microscopic Phycomycetes or Algal Fungi, the Ascomycetes or Sac Fungi, and the Basidiomycetes, or Club Fungi, together with the unclassifiable Imperfect Fungi. The common mushroom is by no means the only edible member of the larger fungi, though some species are also highly poisonous; ⇨ EDIBLE FUNGI, POISONOUS FUNGI. Many of the smaller fungi are of great economic importance, as the cause of plant diseases, such as potato blight *Phythium*, while others, such as *Penicillium*, are the source of highly valuable drugs.

Fungivorous. Feeding on fungi.

Funnel Cap (Basidiomycetes: *Clitocybe*). Toadstools with a somewhat funnel-shaped cap, especially the Common Funnel Cap *C. infundibuliformis* of woods and heaths and the edible Tawny Funnel Cap *C. flaccida*.

Fungus. An individual member of the Fungi.

Fur Mite (Astigmata: Listrophoridae). A family of mites living among the hair and fur of mammals, especially rodents, feeding on waste matter on the skin.

Furniture Beetle (Clavicornia: Anobiidae). A family of twenty-eight small wood-boring beetles, notorious for their destructiveness to wooden beams and furniture, and including the celebrated death watch beetle and the common household *Anobium punctatum* which accounts for four-fifths of all insect damage to timber in buildings.

Furunculosis. A serious bacterial disease of salmon, named from its characteristic 'furuncles' or boils, and often confused with much commoner salmon disease caused by the fungus *Saprolegnia ferox* forming white patches on sites of abrasions colonized by *Bacillus salmonis pestis*.

Furze. ⇨ GORSE; Needle Furze, ⇨ Petty WHIN.

Fusiform. Botanical adjective meaning spindle-shaped, cylindrical but tapering at both ends.

G

Gabbro. A volcanic rock consisting mainly of felspar and augite, the crystals being larger than in dolerite because it originally cooled more slowly.

Gadidae (Pisces: Anacanthini). The Cod Family; the largest and economically most important family of marine fish in British seas, whose members are commonly known as white fish, including the cod, haddock and whiting.

Gad Fly. A general name for the blood-sucking tabanid flies, apparently a misnomer, because the flies which actually cause cattle to gad about madly with tails erect are now believed to be the warble flies.

Gadwall (Anatidae: *Anas strepera*). A surface-feeding duck, resembling the female mallard; mainly a winter visitor, but breeding locally in Scotland and as an introduced bird in East Anglia and other parts of southern England.

Gale. A strong wind, over about 40 m.p.h. ⇨ BEAUFORT SCALE.

Galingale (Cyperaceae: *Cyperus longus*). A tall, handsome sedge, a very local native by fresh water in the south, but often planted by ornamental waters.

Gall. An abnormal growth of plant tissue, mostly caused by the activities of insects or mites, whose young find food and shelter inside it; eelworms and fungi can also produce galls. Plant galls are of two main types: simple galls, which involve only one plant organ, and compound ones, which involve two or more organs and are almost confined to buds. Simple galls are of two kinds: Felt Galls, which are excessively hairy patches of leaf surface and were formerly believed to be fungi, and Mantle Galls, which are cavities within excessive growth around the area punctured by the gall-maker. The mantle galls made by gall mites are further divided into Leaf-roll Galls, in which the edge of the leaf curls over, and Pocket Galls, often known as Nail Galls, in which the surface of the leaf makes a fold. Blister Galls are somewhat similar. Compound galls are subdivided into Bud Galls and Cluster Galls, or Witches' Brooms. The principal gall-making insects are Hymenoptera (gall wasps and some saw-flies), Diptera (gall midges and some other two-winged flies), and Hemiptera (aphids), together with the larvae of some moths and beetles. Many galls are made in oaks, including the Oak-apple, Spangle Galls, Currant Galls made by the gall wasp *Neuroterus lenticularis*, and Marble Galls made by another gall wasp *Adleria* (*Cynips*) *kollari*. Roses have Pea Galls made by *Rhodites* gall wasps, as well as the Robin's Pincushion or Moss Gall. The tubular nail galls on the leaves of lime trees and big-bud of black currants are both produced by mites, and the red Bean Galls commonly seen on willow leaves are due to the Bean-gall Saw-fly *Pontania proxima*.

Gallant Soldier (Compositae: *Galinsoga*). An infelicitous name invented for two white-flowered South American plants, *G. parviflora* and *G. ciliata*, which have been common weeds around London and elsewhere during the past fifty years, having probably escaped from Kew Gardens.

Gall Fly (Cyclorrhapha: Trypetidae). One group of this family of over seventy species of two-winged flies have larvae which live in plants and often produce Galls. ⇨ CELERY FLY, FRUIT FLY.

Gall Mite (Prostigmata: Eriophyina). Those eriophyid mites which produce Galls in their food plants, including the Big Bud Mite, Pear Leaf Blister Mite *Eriophyes pyri* and Blackberry Mite *Aceria essigi,* which also infests loganberries.

Gall Wasp (Parasitica: Cynipidae). A family of some 228 small hymenoptera, many of which produce Galls on plants in the course of their life history, while others are more normally parasitic like ichneumons; some even make use of other species' galls, either as true parasites or as inquilines. The best known galls produced by gall wasps are those on oaks and roses. On oak there are the Oak-apple and the marble, bullet or Devonshire galls made by *Adleria* (*Cynips*) *kollari*, which was accidentally introduced into Devon about 1830 and has since spread all over Britain. Many of the oak galls, however, are produced by alternate generations, e.g. currant and spangle galls by *Neuroterus quercus-baccarum*; silk-button and blister galls by *N. numismalis*, cherry and violet egg galls by *Dryophanta scutellaris* and artichoke or hop galls and hairy catkin galls by *Andricus fecundatrix.* On roses the best known hymenopterous galls are the robin's pincushion and the rose and spiked pea galls, produced by *Rhodites eglanteriae* and *R. nervosus* respectively.

Game. Hares, pheasants, partridges, grouse, blackgame and bustards are all game according to the Game Act of 1831. A licence is required to shoot woodcock, snipe, quail, landrail, conies (rabbits) and deer, but these are not legally game. For purposes of compensation for damage, deer are included as game in the Agricultural Holdings Act, and there are some similar anomalies. The term is widely used in popular speech to cover any animal or bird pursued for sport.

Game Bird (Galliformes). The order of birds including the grouse, ptarmigan and capercaillie (Tetraonidae), partridges, pheasants and quail (Phasianidae), much sought after by sportsmen. They are mostly ground-living and all are ground-nesting birds, but some may roost or feed in trees. Male is called cock; female, hen.

Gamopetalous. Botanical adjective, used of petals which are joined into a corolla tube, at least at the base.

Gander. The male goose.

Gannet (Pelecaniformes: *Sula bassana*). A large white sea-bird with black wing-tips, nesting in often substantial colonies (gannetries) on ten or twelve islands around the coasts of Scotland, Wales and Ireland, with a single very small one on the mainland cliffs at Bempton, East Yorkshire. The best known gannetries are those on the Bass Rock in the Firth of Forth and Ailsa Craig in the Firth of Clyde.

Gaper (Lamellibranchia: *Lutraria* and *Mya*). Four species of fairly large marine bivalve, with more or less oval shells gaping at both ends, found on sandy and muddy shores; two species of *Lutraria*, also called Otter Shells, *L. lutraria* being the larger (up to 6 ins.) and commoner; and two of *Mya*, the edible Sand Gaper *M. arenaria*, known in America as the Soft-shelled Clam, being the larger (up to 5 ins.) and commoner.

Gapes. ⇨ GAPEWORM.

Gapeworm (Nematoda: *Syngamus trachea*). A bright red roundworm causing gapes disease in poultry and many wild birds, especially rooks and starlings, by lodging in the windpipe (trachea).

Garden Snail (Gastropoda: Pulmonata). The name given to two widespread and common land snails, both common in gardens and much eaten by song thrushes at their 'anvils': *Helix aspersa* (⇨ SNAIL) and the highly variable *Cepaea nemoralis*, also found in hedgerows and woods, especially among nettles.

Garden Spider (Araneae: *Araneus diadematus*). One of the commonest of the orb-web builders, it is by no means confined to gardens.

Garfish (Pisces: *Belone belone*). Eel-like fish, with jaws elongated into a remarkable beak, much longer than the pike's, but not related to either eel or pike; arrives off our western coasts in shoals at the same time as the mackerel. Also known as Garpike, Sea Pike and, from the colour of its bones, as Green-bone.

Garganey or Garganey Teal (Anatidae: *Anas querquedula*). A small surface-feeding duck, a local summer visitor to southern England.

Garlic or Leek (Amaryllidacae: *Allium*). A genus of bulbous plants characterized by their distinctive garlicky smell, round heads of pink, purple or white flowers, and usually narrow, often cylindrical and hollow leaves. The commonest and most widespread of our eight native species is Ramsons, followed by the pestilential weed of arable land, Crow Garlic *A. vineale*, the rest being local, scarce or rare, among them Chives and the Wild Leek *A. ampeloprasum*, rare outside Cornwall, which may grow 6–7 ft tall. Hedge Garlic (Cruciferae: *Alliaria petiolata*) is a common white-flowered garlic-scented spring-flowering plant of hedge-banks, also known as Garlic Mustard and Jack by the Hedge.

Garpike. ⇨ GARFISH.

Gastropod or Univalve (Mollusca: Gastropoda). A large and varied class of asymmetrical molluscs, mainly vegetarian, marine and equipped with a single shell, usually spiralling to the right. Three sub-classes: the mainly marine operculates or Prosobranchia (limpets, winkles, whelks, etc.), the marine Opisthobranchia, mainly consisting of the shell-less sea-slugs, and the freshwater and terrestrial Pulmonata, including freshwater and land snails and land slugs.

Gastrotricha. ⇨ BRISTLE BACK.

Gatekeeper. ⇨ Small Meadow BROWN BUTTERFLY.

Gattorugine. ⇨ Tompot BLENNY.

Gault. ⇨ CLAY.

Gause's Principle or Law. The now well established fact that two species cannot exist together in the same habitat if they have identical ecological requirements, unless there is a superabundance of food and other resources of the environment. This was restated by the Russian ecologist Gause in 1934, but because he was by no means its first discoverer, it has been suggested that it should be renamed the competitive exclusion principle.

Gaviiformes. ⇨ DIVER.

Gean. ⇨ CHERRY.

Gene. The basic physical unit of inheritance of animals and plants; genes are arranged in pairs of chromosomes, one in each pair deriving from each parent. Each gene controls one or more physical characters. ⇨ ALLELE.

Genera. Plural of genus.

Genetics. The scientific study of heredity and variation, i.e. of the principal mechanism of natural selection and evolution.

Geniculate. Botanical adjective meaning bent like a knee.

Genotype. The basic genetic construction of an individual organism.

Gentian (Dicotyledones: Gentianaceae). Low-growing plants with handsome flowers, purple in the five *Gentianella*s, of which the commonest is Autumn Gentian or Felwort *G. amarella*, of calcareous turf and dunes; and brilliant blue in the three *Gentiana*s, of which the two most gorgeous are the scarce Marsh Gentian *G. pneumonanthe* and the rare Spring Gentian *G. verna*, confined to upper Teesdale and western Ireland.

Gentle. Anglers' term for maggot used as bait, also applied to white worms.

Genus (Pl. Genera). The next to the smallest division in the biological Classification of living things, comprising a group of similar Species. Closely related genera are grouped into a Family. ⇨⇨ NOMENCLATURE. This is an artificial grouping, the species being the only natural one.

Geo. A small narrow inlet of the sea on a cliffed coast in the far north of Scotland, especially the Shetlands.

Geography. The study of the surface of the earth, and the relationship of its physical features and climate to the distribution of the flora, fauna and human population.

Geological Periods (See table overleaf).

Geology. The scientific study of the rocks of the earth's crust.

Geometer or Geometrid (Heteroneura: Notodontoidea). A group of moths with 'looper' caterpillars, formerly all included in the Geometridae but now split into six families, including the Sterrhidae (waves), the Geometridae proper (emeralds), Hydriomenidae (carpet and pugs), and Selidosemidae, which includes the peppered moth, the magpie moth and the various thorn moths.

Geomorphology. The study of the evolution of land forms, or the arrangement and form of the earth's crust, e.g. how valleys are carved out of hills and mountains by the slow action of weather and water.

Geophysics. The science of the physical aspects of the earth's structure and atmosphere.

Geophyte. A perennial plant whose buds overwinter beneath the surface of the soil.

Geotaxis. ⇨ TAXIS.

Geotropism. ⇨ TROPISM.

Gephyrea. A miscellaneous grouping of marine worm-like animals, formerly

Geological Periods

Era	*Period or System*	*Approximate Duration in Years*	*Cycles of Earth Movement*
Quaternary	Recent (Holocene) Pleistocene		
Tertiary or Cenozoic ***Often written Cainozoic or Kainozoic***	Pliocene	65,000,000 (Pliocene to Eocene)	
	Miocene		
	Oligocene		Alpine
	Eocene		
Secondary or Mesozoic	Cretaceous	55,000,000	
	Jurassic	30,000,000	
	Rhaetic	10,000,000	
	Triassic	30,000,000	
Primary or Palæozoic	Permian	30,000,000	
	Carboniferous	60,000,000	Armorican
	Devonian	40,000,000	
	Silurian	30,000,000	Caledonian
	Ordovician	50,000,000	
	Cambrian	100,000,000	
Pre-Cambrian or Eozoic	Pre-Cambrian		Charnian

regarded as a phylum, but now commonly split by zoologists into the Phyla Sipunculoidea and Echiuroidea and the Class Priapulida of the Phylum Aschelminthes.

Geraniaceae (Angiospermae: Dicotyledones). The family containing the Cranesbills *Geranium* and Storkbills *Erodium*, named from the long pointed 'beak' of their fruits.

Germander (Labiatae: *Teucrium*). Labiates with only a lower lip; four species, three rare with pinkish-purple flowers and the common Wood-sage.

Ghost Larva or Glass Larva. The transparent aquatic larva of the phantom midge.

Ghost Moth (Homoneura: *Hepialus humuli*). One of the Swift Moths, whose white males fly around together in the dusk each appearing to be dangling from a string like a ghostly marionette. The larvae feed on roots.

Ghost Orchid or Spurred Coralroot (Orchidaceae: *Epipogium aphyllum*). A remarkable leafless saprophytic orchid, which flowers irregularly, usually after wet springs, in a few shady beech-woods in the Chilterns, and perhaps still also in a few oak-woods in the Welsh Border.

Ghyll. A small, steep-sided valley in the north of England, especially in Cumberland and the Lake District, also in Yorkshire, where it is often spelt Gill.

Gill. ⇨ GHYLL.

Gillaroo. A red-spotted Irish lake trout.

Gill Fungus. ⇨ TOADSTOOL.

Gilt Head. (1) ⇨ Baillon's WRASSE. (2) A name for the earthworms otherwise known as Cockspur and Brandling.

Gipsy Moth (Caradrinoidea: *Lymantria dispar*). A tussock moth which became extinct in Britain about 1850, but has developed into a serious pest in North America as a result of accidental introduction.

Gipsywort (Labiatae: *Lycopus europaeus*). Common white-flowered plant of freshwater margins, non-aromatic but closely related to the mints.

Glabrous. Botanical adjective meaning devoid of hairs or down. Glabrescent means becoming glabrous in the normal course of development.

Glacial Action. Glaciation during the Ice Age, which ended in Britain about 10,000 years ago, has produced two main effects in the landscape. On the higher ground much rock was swept bare of soil by ice-sheets and glaciers, leaving characteristically rounded contours; sometimes scratches can be seen where sharp stones or boulders have been ground across the rock surface by the moving ice. On the lower ground the soil so removed has been laid down in a thick deposit of glacial drift, usually of boulder clay, sand or gravel. ⇨⇨ DRUMLIN, ESKER.

Gladden. Norfolk name for reed-mace.

Gladdon, Stinking Iris or Roast-beef Plant (Iridaceae: *Iris foetidissima*). A pale grey-blue and yellow wild iris, named from the sickly sweet smell of its leaves when crushed, locally common in scrub, especially near the sea in the south.

Gladiolus, Wild (Iridaceae: *Gladiolus illyricus*). Handsome reddish-purple wild flower resembling garden gladioli, now confined to the New Forest.

Gland (adj. Glandular). A small sac secreting liquid, often aromatic, sunk in or protruding (often on a stalk) from the surface of a leaf, sepal or other part of a plant.

Glass Larva. ⇨ GHOST LARVA.

Glasswort or Marsh Samphire (Chenopodiaceae: *Salicornia*). A complex group of plants, with no leaves, rudimentary flowers and fleshy cylindrical stems, eaten as Samphire; widespread and common on saltmarshes and on coastal mud.

Glaucous. Dull greyish green or blue.

Glen. A valley in the Highlands of Scotland, usually less broad than a Strath.

Gley Horizon or Gley Soil, also spelt Glei. A soil horizon coming immediately above the subsoil or parent rock in brown earths; it is grey-green in colour, often mottled with reddish-brown, produced by the reduction and oxidation of iron salts due to the alternate downward percolation of rain water in wet weather and upward movement of moisture stored in the clay subsoil during dry weather; the mottling develops when the water-level falls and air gets in.

Globe Flower (Ranunculaceae: *Trollius europaeus*). A handsome globe-shaped yellow wild flower, somewhat buttercup-like, locally frequent in the north and west.

Gloger's Rule. The tendency for pigmentation in warm-blooded species to become paler as mean temperature decreases, i.e. going northward in the northern hemisphere.

Glochidium. The larva of a freshwater mussel, parasitic on fish in its early stages.

Glow-Worm (Diversicornia: Lampyridae). Two beetles, one *Lampyris noctiluca* frequent, the other *Phosphaenus hemipterus* rare, noted for the light-producing organs with which the larva-like females attract the male at night; these organs are luminous, producing light from the oxidation of a compound called luciferin and reflecting it by means of minute urate crystals, not phosphorescent. ⇨ BIOLUMINESCENCE, PHOSPHORESCENCE. *L. noctiluca* is widespread but local in grassy and heathy places.

Glume. A small bract at the base of a grass flower; ⇨GRASS. Glumaceous means glume-like.

Gnat (Nematocera: Culicidae). A family of delicate two-winged flies, the larger of which (Culicinae) are now usually called mosquitoes, although our commonest mosquito *Culex pipiens* is known as the Common Gnat. The larva of the Phantom Midge *Chaoborus crystallinus*, which is actually a gnat, is transparent and known as the Phantom, Ghost or Glass Larva. Winter Gnats (N.: Trichoderidae) are like small crane flies, do not bite humans, and are chiefly noted for their dancing swarms to be seen in winter, as well as at other times of year; ten British species. Fungus Gnats (N.: Mycetophilidae) are a large group (470 species) of small gnatlike two-winged flies, the males often dancing in swarms, and the larvae feeding on fungi and various kinds of decaying matter.

Gneiss (Pronounced 'nice'). A form of granite or similar rock in which the constituent mineral crystals of quartz, felspar and mica are arranged in bands instead of being mixed indiscriminately.

Goat, Wild (Artiodactyla: *Capra hircus*). Herds descended from the many goats kept in Britain during earlier times are found wild in many hill districts in Wales, northern England and Scotland.

Goat Moth (Cossoidea: *Cossus cossus*). A widespread large grey-brown moth, named from the smell of its wood-boring larva, which may take three to four years to complete its growth.

Goatsbeard (Compositae: *Tragopogon pratensis*). A common grassland plant with narrow grass-like leaves and yellow dandelion-type flowers, called Jack-go-to-bed-at-noon, from opening only in the morning.

Goat's Rue (Leguminosae: *Galega officinalis*). A garden escape with pinkish lilac or white flowers, often growing on waste ground

Goatsucker. ⇨ NIGHTJAR.

Goblet, The (Basidiomycetes: *Clitocybe cyathiformis*). An edible woodland fungus with a cup-shaped cap.

Goby (Percomorphi: Gobiidae). Small fishes of rocky and sandy shores, two lower fins united to form a sucker, mostly less than 5 ins. long. Thirteen British species, the commonest being the 2–3-in. Common Goby *Gobius minutus*, and the 2-in. Painted Goby *G. pictus*. The Black *G. niger* and common gobies live also in brackish estuarine water. The largest species is the Giant Goby *G. capito* of the Cornish coast, attaining 9 ins. The uncommon 1-in. Diminutive Goby *G. scorpioides*, the smallest British fish, is also one of the smallest fish in the world. Forster's Goby *G. forsteri* was recorded off Plymouth as new to science in 1958. The Transparent Goby *Aphia minuta*, common on our western shores, is both tiny and transparent.

Godwit (Charadriiformes: *Limosa*). Large waders, like smaller curlews but straight-billed, with handsome reddish breeding plumage. The Black-tailed Godwit *L. limosa* has one regular British breeding locality and also nests occasionally in the north of Scotland; it is also a frequent passage migrant and an increasing winter visitor. The Bar-tailed Godwit *L. lapponica* is frequent on the shore in winter.

Goldcrest (Passeriformes: *Regulus regulus*). The smallest British bird, a 3½ in. insectivorous song bird, somewhat resembling a tiny leaf warbler, but with an orange or yellow crest. Widespread and frequent in woods, especially associated with conifers. Formerly miscalled golden-crested wren.

Goldeneye (Anatidae: *Bucephala clangula*). A diving duck, widespread and frequent in winter, but not breeding. The tufted duck is sometimes also called goldeneye, because of its yellow eyes.

Golden Plusia (Caradrinoidea: *Polychrisia moneta*). A noctuid moth related to the Silver Y, which has spread all over Great Britain since it was first noted in Hampshire in 1890.

Golden-Rod (Compositae: *Solidago virgaurea*). A common plant of heathy places with a spike of yellow flowers. Two garden species, *S. canadensis* and *S. gigantea* are frequent escapes on to waste ground.

Goldfinch (Passeriformes: *Carduelis carduelis*). A small brightly coloured finch, whose recovery from scarcity to comparative frequency in many areas especially in the south is one of the great successes of the bird protection movement in Britain. Often feeds on thistles. Collective noun: a charm.

Goldfish (Cyprinidae: *Carassius auratus*). A popular ornamental fish, often liberated into ponds, when it is liable to revert to its natural greenish-brown, the brilliant gold and silver forms being entirely domesticated. Feral, naturally coloured goldfish are not uncommon in ponds round London and elsewhere.

Goldilocks (Compositae: *Crinitaria linosyris*). A rare yellow-flowered relative of the sea aster, confined to limestone cliffs in the west. Wood Goldilocks (Ranunculaceae: *Ranunculus auricomus*) is a widespread and frequent woodland buttercup, often with either ragged or no petals.

Gold-Sinny (Labridae: *Ctenolabrus rupestris*). One of the less common wrasses.

Good Friday Grass. A vernacular name for field woodrush, from its starting to flower about Easter.

Good King Henry (Chenopodiaceae: *Chenopodium bonus-henricus*). Our only perennial goosefoot, formerly eaten as a spinach-like green vegetable, and still surviving on many roadsides and field borders.

Goosander (Anatidae: *Mergus merganser*). A large hole-nesting diving duck, the largest native saw-bill. Breeds by rivers in Scotland and extreme northern England; widespread also in the south in winter.

Goose (Anatidae). Large semi-aquatic birds, mainly winter visitors to British Isles: male, gander; female, goose; young, gosling. Five species of grey geese *Anser:* Greylag Goose *A. anser*, ancestor of domestic geese and the only native breeding goose, nesting very locally in north Scotland, more numerous as winter visitor from Iceland; White-fronted Goose *A. albifrons*, named from its white forehead, mainly on the west coast, including the neighbourhood of the Wildfowl Trust's collection on the Severn estuary; Pink-footed Goose *A. arvensis brachyrhynchus*, the commonest wild goose generally, the large form known as the Bean Goose *A. a. arvensis* now being very local; also two rare visitors, the Lesser White-fronted Goose *A. erythropus*, sometimes with flocks of white-fronts, and the Snow Goose *A. caerulescens*, of which the Blue Goose is a colour phase. Four species of black goose *Branta:* Brent Goose *B. bernicla*, winter visitor, mainly on east coast; Barnacle Goose *B. leucopsis*, winter visitor, mainly in north-west; the introduced Canada Goose *B. canadensis*, naturalized and breeding on lakes and ponds in various parts of England; and the very rare Red-breasted Goose *B. ruficollis*.

Gooseberry (Grossulariaceae: *Ribes uva-crispa*). Low spiny bush with globular green fruits, widespread and frequent in woods and hedgerows; native but also bird-sown from gardens.

Goosefoot (Chenopodiaceae: *Chenopodium*). A genus of weeds with tiny green flowers in leafy or leafless spikes, the commonest being Fat Hen, Red Goosefoot *C. rubrum*, which often has reddish leaves, Many-seeded

Goosefoot *C. polyspermum*, and Good King Henry. The scarce and decreasing Stinking Goosefoot *C. vulvaria* smells repulsively of stale fish. ⇨ SOWBANE.

Goosegrass or Cleavers (Rubiaceae: *Galium aparine*). A close relative of the bedstraws, which straggles over vegetation, clinging to it and to clothing with the tiny prickles on its stems, leaves and fruits; widespread and common.

Gorge. A narrow ravine or small valley flanked by high rocky cliffs.

Gorse (Papilionaceae: *Ulex*). Spiny evergreen shrubs with fragrant yellow flowers, growing on heaths and downs. Three species: the common and widespread Gorse, Furze or Whin *U. europaeus*, flowering all the year but mainly in spring; and two smaller species, both confined to acid soils and flowering in late summer. Western Gorse *U. gallii*, mainly in the west, and the still smaller and often almost prostrate Lesser Gorse *U. minor*, mainly in the south-east. The cultivated form of *U. europaeus* called Irish Gorse was originally found wild in Co. Down.

Goshawk (Falconiformes: *Accipiter gentilis*). A large, broad-winged bird of prey, formerly breeding in the Scottish Highlands and some other areas, but latterly only a straggler from the Continent, occasionally nesting in the south. Much favoured by falconers, some of whose escaped birds may have bred.

Gosling. The young goose.

Gossamer. Web material spun chiefly by spiders of the family Linyphiidae, especially on sunny autumnal mornings, when it may float away in quantity on rising air currents.

Gout Fly (Cyclorrhapha: *Chlorops pumilionis*). A small two-winged fly, also called Ribbon-footed Corn Fly, which is a pest of cereal crops, related to frit fly.

Goutweed. ⇨ GROUND-ELDER.

Grain Beetle (Clavicornia: *Oryzaephilus*). Two species of stored-product pests found in warehouses.

Grampus. A seamen's name for several smallish whales, including Risso's Dolphin *Grampus griseus*, found off our south-western shores, and the Killer Whale.

Granite. A very hard rock composed of large crystals of quartz, felspar and mica or hornblende formed by the molten lava (magma) that was squeezed into the core of mountain ranges during the great earth-building movements. After millions of years during which the mountains were being worn down this core became exposed and resisted weathering. Granite mountains include Ben Nevis and the Cairngorms, which are the two highest mountain masses in Britain, as well as Dartmoor and the Mourne Mountains in Northern Ireland. Granite from Shap, Westmorland, with its large felspar crystals, is often seen in London buildings.

Grannom or Green-tail (Trichoptera: *Brachycentrus subnubilus*). A widespread caddis-fly of running water.

Grape Hyacinth (Liliaceae: *Muscari atlanticum*). A garden plant that is also a scarce native in East Anglia, the spike of attractive blue flowers being supposed to resemble a bunch of grapes.

Grass (Monocotyledones: Gramineae). Flowering plants with very narrow leaves and small greenish petal-less flowers in heads or spikes, providing the main green covering of most natural terrestrial habitats; ⇨ GRASSLAND. The cereals used in agriculture, such as wheat, barley and oats, are grasses. Many plants with 'grass' in their name, however, are not grasses at all, e.g. Grass of Parnassus, Arrow-grass, Carnation Grass.

Parts of a Grass Flower

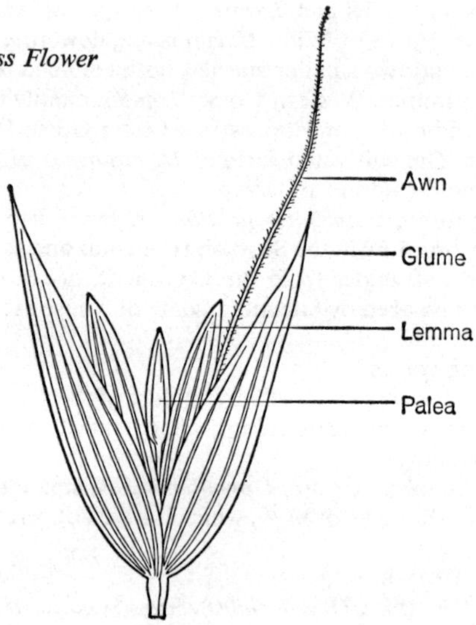

Grasshopper (Orthoptera: Acrididae). A group of jumping insects noted for their stridulating 'song' and distinguished from crickets by their short antennae. About a dozen species occur in Britain, perhaps the commonest being *Chorthippus parallelus* in grassland. The Great Green Grasshopper is a bush cricket. Also used in the wider sense to include the groundhoppers.

Grassland. More than half England and three-quarters of Wales and Ireland is grassland, most of it the result of man's activity (clearing forests) and the grazing of his domestic animals, combined with the favourable climate which gives the grasses the necessary rain throughout the growing season. Permanent grassland, i.e. excluding the leys which are sown, includes both chalk and limestone grassland such as the Downs and Cotswolds, kept short by grazing and with a distinctive flora; and acid siliceous grassland, such as the sheep pastures of the Southern Uplands of Scotland and the Welsh hills. Sub-maritime grassland on cliff tops and slopes facing the sea is probably natural in origin, where winds and spray have kept other plants out. Maritime grassland, usually grazed, close to the sea and affected by spray, includes some semi-halophytes in its flora. ⇨ GRAZING.

Grass Moth or Crambid (Pyraloidea: Crambinae). A large sub-family of small pyralid moths, which are often abundant in meadows and other grassy places.

Grass of Parnassus (Parnassiaceae: *Parnassia palustris*). A white-flowered relative of the saxifrages, frequent on wet moors and in fens and dunes in the north and west.

Grass Poly (Lythraceae: *Lythrum hyssopifolia*). A rare prostrate pink-flowered relative of purple loosestrife.

Grass-Wrack. ⇨ EEL-GRASS.

Gravel. A coarse deposit of sand mixed with small pebbles, and originally laid down by a river, well drained, usually dry and lacking in nutrients, and hence unfavourable to many plants.

Grayling. (1) (Salmonidae: *Thymallus thymallus*). A wholly freshwater relative of the trout, characterized by its long dorsal fin, preferring fast streams; common in England and Wales, introduced in Scotland. (2) (Papilionina: *Eumenis semele*). A large, strong-flying Brown Butterfly of moors, heaths and other open spaces.

Grazing, Effect of. The change from forest to grassland in Britain over the past 2,000 years has largely been produced by grazing animals, chiefly sheep and cattle, but also rabbits, often combined with tree felling. Under-grazing of pasture can lead to its gradual destruction: invasion by woody plants, e.g. heather, and the increase of un-nutritious grasses which are ignored by the grazing animals. Overgrazing, especially on light dry soils can lead to semi-desert, as e.g. in the Breckland; on wet soils, to the wet deserts of the Scottish Highlands.

Great Diving Beetle (Adephaga: *Dytiscus marginalis*). The largest and most familiar of the predatory water beetles, with the yellow borders to its elytra characteristic of its genus.

Grebe (Podicipitiformes: *Podiceps*). Highly specialized diving birds, rarely coming ashore even to breed, for their nests are floating piles of waterweed. Four species breed, all on fresh water, two widespread: the Great Crested Grebe *P. cristatus*, and the Dabchick or Little Grebe *P. ruficollis*; and two very local in Scotland, the Slavonian Grebe *P. auritus*, and the Black-necked Grebe *P. nigricollis*. The Red-necked Grebe *P. griseigena* is an uncommon winter visitor. The great crested grebe is noted for the remarkable courtship ceremonies it performs with the aid of the ruff and tippets both sexes assume on the head during the breeding season.

Green Algae (Chlorophyta or Isokontae). One of the largest phyla or divisions of algae, containing chlorophyll unmarked by other pigments and so mainly grass-green in colour; predominantly freshwater in habitat, though some species occur in damp soil and a few are marine: ⇨ GREEN SEAWEEDS. The ancestry of the higher plants is presumed to stem from this group. There are two classes, the Green Algae or Seaweeds (Chlorophyceae), and the Stoneworts (Charophyceae). Among the best known species of freshwater green algae are *Cladophora*, branched and dark green, in clumps on the bottom of lakes and slow-moving streams; *Enteromorpha* or blanket-weed, freshwater relative of several common green seaweeds, with long waving ribbons in slow-moving fresh and brackish water; *Spirogyra*, whose slimy, cotton-woolly masses are often abundant in stagnant ponds; and the free-moving *Volvox*, a minute spherical colonial plant, beloved of amateur microscopists. ⇨ DESMID, SIPHON ALGA.

Green-Bone. ⇨ GARFISH.

Greenbottle (Diptera: Cyclorrhapha). Various metallic-green, two-winged flies, similar to and sometimes closely related to the bluebottles, especially *Lucilia* spp. *L. bufonivora* has the singular habit of laying its eggs on the eye-rims and nostrils of the common toad, the larvae penetrating into the animal's body and usually causing its death. *L. sericata* attacks live sheep, causing 'strike'.

Greendrake. Angler's name for the duns of the mayflies *Ephemera danica* and *E. vulgata*. The spinners are Spent Gnats. ⇨ GREY DRAKE.

Greenfinch (Passeriformes: *Chloris chloris*). A green-plumaged finch, widespread and common in bushy places, including gardens; formerly known as Green Linnet.

Green-Fly. Winged green aphid, also applied in the wider sense to the wingless generations.

Green Oak Moth (Tortricoidea: *Tortrix viridana*). A small green 'micro'. whose larvae feed on oak and often defoliate the trees so severely that they are forced to put forth a second crop, known as Lammas leaves.

Greensand. A rock found below the Chalk in the geological succession. The Upper Greensand is often calcareous; the Lower is usually acid, and in some districts, e.g. south-west Surrey, produces extensive heathland. The name is due to the presence of green glauconite, though this is often bleached out. The two Greensands are usually separated by the Gault clay.

Green Seaweeds (Chlorophyceae). Marine algae, mainly grass-green in colour; the best known of the comparatively few species on our shores are Sea-Lettuce and various species of *Enteromorpha* and *Cladophora*. *E. compressa* is also common in brackish water. ⇨ HOG'S BRISTLE.

Greenshank (Charadriiformes: *Tringa nebularia*). Medium-sized wader with greenish legs, breeding locally on moors in the Scottish Highlands and frequent as a passage migrant further south.

Green-Tail. ⇨ GRANNOM.

Greenweed (Leguminosae: *Genista*). Three undershrubs, the two commonest being Dyer's Greenweed *G. tinctoria* with large yellow flowers like broom, locally frequent in grassland in the south, and Petty Whin.

Green-winged Orchid (Orchidaceae: *Orchis morio*). A close relative of the early purple orchid, more local and decreasing, mainly in grassland and differing in the green veins of the flower's 'hood'.

Grey Drake. Angler's name for the spinner of the mayfly *Ephemera danica* before oviposition. ⇨ SPENT GNAT; GREENDRAKE.

Grey Flag or Grey Sedge. (Trichoptera: *Hydropsyche*). Widespread caddis-flies of fast flowing streams.

Greyhen. The female black grouse.

Greylag. ⇨ Greylag GOOSE.

Grey Sedge. (1) ⇨ GREY FLAG. (2) *Carex divulsa* (Cyperaceae), a widespread and frequent sedge of hedge-banks and grassy places.

Grey Wether. ⇨ SARSEN STONE.

Gribble, The (Isopoda: *Limnoria limnorum*). A common small marine crustacean related to the woodlice, which burrows into wood and is often found in quantity in timber cast ashore.

Grike. A fissure in the surface of limestone rock. ⇨ LIMESTONE PAVEMENT.

Grilse or Peal. Salmon, after spending one or more years in the sea, return to the rivers to spawn as grilse.

Grim The Collier. ⇨ FOX AND CUBS.

Grisette (Agaricaceae: *Amanita vaginata*). An edible toadstool, found in beech and other broad-leaved woods, closely related to the highly poisonous death cap. Tawny Grisette *A. fulva* is commoner, especially in birch woods, and also edible.

Grit. A coarse-grained sandstone, the best known being Millstone Grit.

Gromwell (Boraginaceae: *Lithospermum*). Three flowering plants: Corn Gromwell *L. arvense*, an arable weed, and Common Gromwell *L. officinale*, a taller plant of bushy places, both with creamy white flowers and mainly southern in distribution; Purple Gromwell *L. purpurocaeruleum*, with purple flowers, very local on limestone in the south.

Grosbeak. An old name for the hawfinch, also applied to two rare finches: Pine Grosbeak *Pinicola enucleator*, a very rare straggler; and Scarlet Grosbeak *Carpodacus erythrinus*, an irregular visitor from northern Europe, mainly to the Scottish isles.

Ground Beetle or Carabid Beetle (Adephaga: Carabidae). A group comprising all members of this 344-strong family of carnivorous beetles, except the five tiger beetles; one of the commonest and most familiar is the Violet Ground Beetle *Carabus violaceus*. ⇨ BOMBARDIER BEETLE.

Groundbug (Heteroptera: Lygaeidae). A large family of rather elongated black or dark brown plant-bugs, some with red markings; over sixty British species, including the Nettle Groundbug *Heterogaster urticae*, common on stinging nettles in southern England, the Pine-cone Bug *Gastrodes grossipes*, widespread and common on Scots pine, and the Chinchbug.

Ground-Elder, Goutweed or Bishop's weed (Umbelliferae: *Aegopodium podagraria*). A common white-flowered umbellifer, with leaves somewhat like elder, whose far-spreading and tenacious root system makes it one of the most pestilential of garden weeds.

Ground-Hopper (Orthoptera: Tetrigidae). A group of small grasshopper-like insects, all three British species belonging to the genus *Tetrix*; also known as Grouse Locusts.

Ground-Ivy or Ale-hoof (Labiatae: *Glechoma hederacea*). Common purplish-blue flower of woods and hedge-banks, formerly used in the brewing of ale.

Ground-Pine (Labiatae: *Ajuga chamaepitys*). A relative of the bugle, with pink-purple flowers, and deeply cut narrow leaves, which make it look like a small pine seedling.

Groundsel (Compositae: *Senecio*). Close relative of ragwort, lacking, or apparently lacking ray-florets. Much the commonest and most widespread is the abundant Common Groundsel *S. vulgaris*; Heath Groundsel *S. sylvaticus* and Sticky Groundsel *S. viscosus* are progressively less common; all have brush-like yellow flowers and grow in bare ground.

Grouper. Skin-divers' name for two of the larger Mediterranean sea perches.

Grouse (Galliformes: Tetraonidae). Large to medium-sized game-birds, four species breeding: Capercaillie, Ptarmigan; Red Grouse *Lagopus lagopus*, formerly known as moorfowl or moorgame (and the cock as moorcock), widespread and common on heather moors in the north, less so in the west; and Black Grouse *Lyrurus tetrix* (male, blackcock; female, greyhen; collectively blackgame), much more local on the moorland edges of woodland. Blackgame are noted for their Leks. ⇨ SAND-GROUSE.

Grouse Wing. ⇨ Brown SILVERHORNS.

Grove. A group of trees isolated from other trees; sometimes used for patches of woodland between a half and five acres in extent, in which case smaller groups are called clumps.

Grub. General term for fairly large insect larvae found in the soil.

Gudgeon (Cyprinidae: *Gobio gobio*). A small freshwater fish with two barbels, frequent in rivers, streams and ponds, but absent from Scotland.

Guelder Rose (Caprifoliaceae: *Viburnum opulus*). A widespread deciduous shrub with white flowers and red berries, the origin of the garden snowball tree.

Guillemot (Charadriiformes: Alcidae). The Common Guillemot *Uria aalge* is an auk noted for its massed breeding colonies on cliff ledges all round the British Isles, but mainly in the north and west; winters at sea; lays a single pear-shaped egg. The Black Guillemot or Tystie *Cepphus grylle* is more local, with much smaller colonies using separate nesting crevices; wintering inshore. Brünnich's Guillemot *U. lomvia* is a rare straggler from the Arctic.

Gull (Charadriiformes: Laridae). Medium to large grey and white or black and white sea-birds, some species also seen on land. Six species breed: the abundant Black-headed Gull *Larus ribidundus*, the smallest and the commonest gull inland, especially in towns, with a dark brown hood in breeding plumage; Common Gull *L. canus* widespread and common in winter but breeding only in Scotland and Ireland; Herring Gull *L. argentatus*, the commonest round the coast at all seasons; Lesser Black-backed Gull *L. fuscus*, mainly a summer visitor; Great Black-backed Gull *L. marinus*, the largest, breeding only in the north and west; and the Kittiwake. Eight other gulls occur as visitors in varying degrees of frequency, the most often seen being the Glaucous Gull *L. hyperboreus*, winter visitor from northern seas, and the Little Gull *L. minutus*, passage migrant from the Continent.

Gully. A narrow, steep-sided miniature valley carved by water, usually torrential rain, but occasionally also overflowing flood-water, on a hillside. A series of gullies may lead into a ravine.

Gunnel. (Percomorphi: *Pholis gunnellus*). A small (usually 3–5 in.) eel-shaped inshore marine fish, allied to the blennies and characterized by a row of conspicuous white-ringed black spots along the back; widespread and common.

Gurnard (Scleroparei: *Trigla*). A group of odd-looking sea-fish, their large square heads and armour-plated cheeks giving them a wedge-shaped appearance, while the front three rays of their pectoral fins are free and look like feelers. The commonest of the six British species are the Grey Gurnard *T. gurnardus*, the mainly southern and western Red Gurnard or Elleck *T. cuculus*, and the Yellow Gurnard or Tub-fish *T. lucerna*; they are all coloured according to their names.

Gwyniad (Salmonidae: *Coregonus clupeoides pennantii*). The endemic race of the powan found in Lake Bala, Merioneth.

Gymnosperm (Spermatophyta: Gymnospermae). Primitive woody seed-plants bearing their seeds exposed, usually in cones, instead of protected in a fruit, and comprising the conifers and their allies.

Gynandromorphism. An abnormality, found especially in butterflies, moths and some other insects, in which the organism is partly male and partly female, e.g. a butterfly may be male on the one side and female on the other, always being sterile in such cases.

Gynoecium. ⇨ PISTIL.

H

Haar. A sea fog or mist driven by easterly winds from the North Sea on to the east coast of Great Britain, especially in Scotland and northern England.

Habitat. The environment of an animal or plant, comprising the whole complex of vegetation, soil and climatic factors to which it is adapted.

Habituation ⇨ LEARNING.

Haddock (Gadidae: *Gadus aeglifinus*). An abundant fish of British seas, with an important fishery, easily told by the large black spot on its side. The unrelated Norway Haddock (Scleroparei: *Sebastes marinus*), a northern deep-water fish, looks somewhat like a reddish freshwater perch, and is allied to the gurnards. Jerusalem Haddock ⇨ OPAH.

Haddy (Salmonidae: *Salvelinus alpinus killinensis*). A race of char inhabiting two lakes in Inverness-shire.

Hag-fish (Cyclostomata: *Myxine glutinosa*). A primitive eel-shaped deep-water marine fish, allied to the lampreys, uncommon off Britain.

Hail. Raindrops which get caught in the updraughts of thunderstorms get carried upwards to levels where the temperature is freezing, and may then fall only part of the way to the ground and be carried up several times, each time growing by the addition of another layer of ice. When they eventually fall in a hailstorm, they may therefore be quite large, though in Britain usually around the size of a pea and rarely if ever larger than a golf-ball. Most hail falls in the spring, especially in March, when the freezing layer of the atmosphere is much lower than in summer, when most of the thunderstorms occur.

Hair Grass (Gramineae). A name given to several species of grass with thin hair-like leaves, especially the small Early Hair Grass *Aira praecox*, and its close relative Silver Hair Grass *A. caryophyllea*, which both prefer acid soils; the four species of *Deschampsia*, notably the abundant Tufted Hair Grass *D. caespitosa* of damp grassland and the common Wavy Hair Grass *D. flexuosa* of dry acid grassland; Crested Hair Grass *Koeleria cristata*, common in calcareous grassland; and *Corynephorus canescens*, scarce on dunes.

Hair Moss (Musci: *Polytrichum*). A genus of mosses frequent in woods and on moors and heaths, usually on acid soils, the commonest species being *P. commune*.

Hairstreak Butterfly (Papilionina: Lycaenidae). A group of five smallish butterflies named from the white hair-like streak on their undersides: the two commoner species are the Green Hairstreak *Calliophrys rubi* and the Purple Hairstreak *Thecla betulae*.

Hair-tail or Cutlass Fish. (Percomorphi: *Trichiurus lepturus*). Eel-shaped oceanic tropical or sub-tropical fish, infrequent off our southern coasts.

Hairworm. ⇨ HORSEHAIR WORM.

Hairy Back. ⇨ BRISTLE BACK.

Hake (Anacanthini: *Merluccius merluccius*). A large deep-water marine fish, closely allied to the Cod Family (Gadidae), important economically, and found mainly off our western coasts. Forked Hake, Trifurcated Hake, ⇨ FORK-BEARD.

Halibut (Pleuronectidae: *Hippoglossus hippoglossus*). A large, right-sided, deepwater flatfish, highly esteemed for its flesh, northern in distribution.

Halophyte. A plant that can grow in soil with a high concentration of salt, such as a salt marsh, and withstand immersion in the sea at high tide, e.g. glasswort *Salicornia*.

Halteres. A pair of small organs, reduced from the hind wings, used as balancers in flight by two-winged flies (Diptera), and possibly also as flight activators; their use is not yet fully understood.

Hanger. A wood growing on the side of a steep hill, usually beech on chalk hills in the south of England. The most famous beech hanger is at Selborne, in Hampshire, home of the 18th-century naturalist Gilbert White.

Haploid. An organism with a single set of unpaired chromosomes in each cell nucleus. ⇨ POLYPLOIDY.

Hardback Beetle (Heteromera: *Tenebrio*). Two species of beetle whose larvae are known as Mealworms; *T. molitor* is the commoner.

Hard Fern (Blechnaceae: *Blechnum spicant*). A common fern of acid woods, heaths and moors, especially in the north and west, with distinct fertile fronds arising in the centre of a tuft of tough shiny dark green pinnate infertile fronds.

Hard Grass (Gramineae: *Parapholis*). Two small maritime grasses *P. strigosa* and *P. incurva*, both rather local, with the flowering spikes so narrow as to appear like a continuation of the stem.

Hardhead. The black knapweed, from its hard, knobby buds.

Hard Pan, or Iron Pan or Moor Pan. A hardened layer of soil below the surface through which water cannot drain easily. ⇨ PODSOL.

Hardwood. The timber of a broad-leaved tree. ⇨ TREE.

Hare (Lagomorpha: *Lepus*). Male, jack; female, doe; young, leveret. Hares resemble rabbits, but are larger and more ungainly, and always live above ground. The two native species are the widespread Brown Hare *L. europaeus* of the lowlands and the Blue or Varying Hare *L. timidus* of hill country, so called because in winter its bluish-grey fur changes to white. The blue hare is native in Ireland and the Scottish Highlands, and has been introduced, with varying degrees of success, into the hills of southern Scotland, northern England and North Wales. The phrase 'mad as a March hare' derives from the spectacular courtship and aggressive behaviour of brown hares in March, when they chase and 'box' each other and leap about.

Harebell (Campanulaceae: *Campanula rotundifolia*). Widespread and common blue bell-shaped flower of dry grassy and heathy places, named from the hair-like flower-stalks; called Bluebell in Scotland.

Hare's-ear (Umbelliferae: *Bupleurum*). Three atypical species of yellow-flowered umbellifer, the least uncommon being the grass-like Slender

Hare's-ear *B. tenuissimum* of dry grassy spots near the sea between the Tees and the Severn.

Harestail (Cyperaceae: *Eriophorum vaginatum*). A cotton-grass with a single white cottony fruiting head, resembling the tail of a rabbit or hare. The name is also given to a rare alien sand-dune grass, *Lagurus ovatus*.

Harrier (Falconiformes: *Circus*). Medium-sized birds of prey with long wings and tails, three of which breed in Britain, all rather uncommonly. The Hen-Harrier *C. cyaneus* breeds in the Orkneys and scattered over the Scottish Highlands, and is not uncommon in winter further south. Montagu's Harrier *C. pygargus* is an uncommon summer visitor to the south. The Marsh Harrier *C. aeruginosus*, also a summer visitor, breeds only in a few widely scattered marsh and fen districts, notably in East Anglia. The Pallid Harrier *C. macrourus* is a very rare straggler from eastern Europe.

Hart. The old term for a stag.

Hartstongue Fern (Aspleniaceae: *Phyllitis scolopendrium*). Our only true fern with undivided leaves, commonest in the damper climate of the west where it may be abundant in woods and on hedge-banks.

Hartwort (Umbelliferae: *Tordylium maximum*). A white-flowered umbellifer somewhat resembling hedge parsley, now very rare on grassy banks in Essex.

Harvestman, or Harvest Spider (Arachnida: Opiliones). An order of often very long-legged carnivorous arachnids, distinguished from the spiders by their undivided body and the absence of silk glands; one of the commonest of the twenty British species is the long-legged *Phalangium opilio*, widespread in low vegetation towards the end of summer.

Harvest Mite or Bracken Bug (Prostigmata: *Trombicula autumnalis*). A notorious pest, frequent on wild mammals and birds, and especially common in rabbits' ears, whose larvae are liable to produce an exceedingly irritating itch on the human skin, persisting for several days after the removal of the mite.

Harvest Moon. The September full moon.

Harvest Spider. ⇨ HARVESTMAN.

Hastate. ⇨ LEAF-SHAPES.

Hastings Rarities. A paper by E. M. Nicholson and I. J. Ferguson-Lees in *British Birds*, August 1962, threw doubt on the authenticity of some 600 records of rare birds obtained in the Hastings district of Sussex and Kent between 1892 and 1930. As a result six full species and thirteen subspecies were formally struck off the British List.

Haugh. The flat alluvial ground on the floor of a river valley, especially in Scotland.

Haw. The fruit of the hawthorn.

Hawfinch (Passeriformes: *Coccothraustes coccothraustes*). The largest British finch, formerly known as grosbeak from its disproportionately large and powerful beak; local in woods and other well timbered places in England, Wales and southern Scotland.

Hawk. A general name for the smaller diurnal birds of prey, more specifically applied to the genus *Accipiter*. ⇨ SPARROWHAWK, GOSHAWK. The Fish Hawk is the osprey.

Hawkbit (Compositae: *Leontodon*). Three common grassland plants with yellow dandelion-like flowers: Greater Hawkbit *L. hispidus*, Lesser Hawkbit *L. taraxacoides* and Autumn Hawkbit *L. autumnalis*, which actually starts to flower in July.

Hawker. ⇨ DRAGONFLY.

Hawking. ⇨ FALCONRY.

Hawk-Moth (Notodontoidea: Sphingidae). Large moths whose larvae may be as thick as a man's finger and usually carry a horn at one end. The seventeen British species include the Death's Head Hawk-moth *Acherontia atropos*, a rare immigrant named from the skull-like mark on its thorax, whose larvae feed on potato plants; the Elephant Hawk-moth *Deilephila elpenor*, one of the smaller species, feeding on willow-herbs and bedstraws; the Eyed Hawk-moth *Smerinthus ocellata*, one of the commoner large species, feeding on sallow, willow, and apple foliage; the Humming-bird Hawk-moth *Macroglossum stellatarum*, a frequent immigrant from the Mediterranean, which is often mistaken for a humming-bird as it hovers in front of flowers; the Lime Hawk-moth *Mimas tiliae*, a common species feeding on lime foliage; the Pine Hawk-moth *Hyloicus pinastri*, a local but increasing species feeding on pine; the Poplar Hawk-moth *Laothoe populi*, a common large species feeding on poplar, sallow and willow; and the Privet Hawk-moth *Sphinx ligustri*, another common large species, whose privet-feeding larva may be 3 ins. long.

Hawksbeard (Compositae: *Crepis*). Six plants with dandelion-like flowers, the commonest being Smooth Hawksbeard *C. capillaris*, with small flowers, Beaked Hawksbeard *C. vesicaria*, an established alien, mainly in the south, and Marsh Hawksbeard *C. paludosa*, a native of grassland, mainly in the north.

Hawkweed (Compositae: *Hieracium*). Plants with yellow dandelion-like flowers: Common Hawkweed *Hieracium vulgatum* agg., one of the most variable British plants, with 223 microspecies, common on walls, banks, heaths and mountains; and Mouse-ear Hawkweed *H. pilosella* with lemon-yellow flowers, common and widespread in grassland. They are apomicts, whence the numerous microspecies.

Hawthorn (Rosaceae: *Crataegus*). A common thorny shrub or small tree, much used in hedges as Quick or Quickthorn. Flowers white, known as 'May blossom'; fruit a dark red berry, known as a haw. Two closely allied species: the widespread Common Hawthorn *C. monogyna*, and the Midland or Woodland Hawthorn *C. oxyacanthoides*, mainly in the south.

Haze. Low density cloud produced not by suspended water particles but by particles of dust, smoke or other solid matter.

Hazel (Corylaceae: *Corylus avellana*). Deciduous nut-bearing monoecious shrub, whose yellow catkins, known as 'lambs' tails', expand in January and February. Common and widespread in woods, especially coppiced oakwoods, and scrub.

Head. A deposit of rocky rubble in valleys, especially in south-west England, caused by Solifluction during the Ice Age.

Heartsease. ⇨ PANSY.

Heart-Rot. The decay of the heart-wood of a tree by fungal action.

Heart-Urchin. ⇨ SEA-URCHIN.

Heart-Wood or Sap-wood. The inner wood of a tree trunk.

Heath. (1) An area of poor acid soil, usually sandy or gravelly, often dominated by low-growing ericaceous undershrubs, especially heather *Calluna*. Chalk heath is a vegetation type that develops when the lime has all been leached out of the upper layers of soil overlying chalk, so that the soil becomes acid and is able to support heather and other heath plants. (2) (Ericaceae: *Erica*). Six undershrubs with pink or purple flowers, only the pink-flowered Cross-leaved Heath *E. tetralix*, which prefers wet heaths, and Bell Heather being widespread and common. The rare or local ones are Dorset Heath *E. ciliaris* in south-west England and Connemara, Cornish Heath *E. vagans* on the Lizard Peninsula and in Co. Fermanagh, and Irish Heath *E. hibernica* and Mackay's Heath *E. mackaiana* in west Ireland. The related St Dabeoc's Heath *Daboecia cantabrica* is another western Irish speciality.

Heath Butterfly (Papilionina: Satyridae). Two of the smaller Brown Butterflies, the widespread and abundant Small Heath *Coenonympha pamphilus* of grassy places, and the local and northern Large Heath *C. tullia* of bogs and damp moorland. There are also two common Heath Moths, both day-flyers and geometrids: Common Heath *Ematurga atomaria* and Latticed Heath *Chiasmia clathrata*.

Heather or Ling (Ericaceae: *Calluna vulgaris*). The widespread and often abundant undershrub that makes heaths and moors purple in August; the favourite food of red grouse. The related Bell Heather *Erica cinerea* flowers earlier and prefers drier heaths and moors.

Heathpoult. Local name for the black grouse in Somerset, especially in the Quantocks.

Hedgehog (Insectivora: *Erinaceus europaeus*). A widespread small mammal, covered with spines, which rolls into a ball when disturbed.

Hedgehog Fungus (Basidiomycetes: *Hydnum erinaceus*). A curious fungus growing on beech, with long hanging spines, supposedly resembling a hedgehog.

Hedgesparrow (Passeriformes: *Prunella modularis*). Small thin-billed, insect-eating song bird, widespread and common, frequent in gardens; once known as hedge accentor, now often called by the revived folk name of dunnock. Not related to the seed-eating house and tree sparrows (*Passer*).

Heft. A term applied both to an area of rough grazing in Scotland and to the flock of sheep which graze it.

Heligoland Trap. A large trap consisting of two roofed wire fences converging into a narrow point at which is situated a small catching trap; the birds are walked into the open end. A type of trap especially used in areas where vegetation is sparse, such as coastal migration stations, and constructed around a few bushes, which attract any migrant birds in the vicinity. Named from the German North Sea island bird observatory where it was invented.

Hellebore (Ranunculaceae: *Helleborus*). Two widespread but local early flowering plants of calcareous woods: Green Hellebore *H. viridis* and Stinking Hellebore *H. foetidus.*

Helleborine (Orchidaceae: *Cephalanthera* and *Epipactis*). The woodland *Cephalanthera* has three species, the common White Helleborine *C. damasonium,* frequent in southern beechwoods, the scarce Sword-leaved Helleborine *C. longifolia* and the rare Red Helleborine *C. rubra.* The seven species of *Epipactis* are more varied in habitat, the commonest being the mainly woodland Common Helleborine *E. helleborine* and the damp-loving Marsh Helleborine *E. palustris;* Dark Red Helleborine *E. atrorubens* is local on limestone rocks in the north and west, and Violet Helleborine *E. purpurata* in woods in the south.

Helminth. A parasitic worm, especially flatworm or roundworm.

Helminthology. The study of parasitic worms.

Helm Wind. A strong wind that blows from Crossfell (2,930 ft) in Cumberland at certain times of year, especially late spring. It blows severely for three to four miles westwards into the Eden valley from the bottom of the fell, often for days at a time, damaging crops and haystacks. It is named from the helm or cloud cap that settles over the summit while it is blowing. The helm bar is a long roll of cloud that forms in the air above the point where the wind ceases its violence.

Helophyte. A marsh plant.

Hemichordata. A primitive sub-phylum of the Chordata containing the enteropneusts or acorn worms.

Hemicryptophyte. A perennial plant, almost always non-woody, with its overwintering buds at soil level, often covered by the surface litter.

Hemiparasite. A plant which derives its food supply partially from other plants, to which it is attached, usually by its roots fastening on to their roots, e.g. Eyebright, Yellow Rattle.

Hemiptera (Arthropoda: Insecta). An order containing a wide variety of often very different looking insects, broadly known as bugs, in two distinct sub-orders, the Homoptera (aphids, frog-hoppers, leaf-hoppers, white-flies, mealy-bugs, scale-insects) and the Heteroptera (plant bugs water bugs); more than 1,400 British species. Hemiptera are characterized by the adaptation of their mouth-parts to sucking, usually the sap of plants but in a few species blood. Many species exude conspicuous substances, such as honey-dew, foam (cuckoo-spit) and waxy wool (woolly aphis).

Hemlock (Umbelliferae: *Conium maculatum*). A tall white-flowered umbellifer with purple-spotted stems, highly poisonous and common in damp woods and by ditches.

Hemp-Nettle (Labiatae: *Galeopsis*). Plants with pinkish-purple or pale yellow flowers, and harmless nettle-like leaves, mostly weeds of cultivation. Common Hemp-nettle *G. tetrahit,* the most widespread and frequent, is believed to have arisen naturally from Large Hemp-nettle *G. speciosa* and the Continental *G. pubescens* by a process of hybridization and back-crossing.

Hen. Name for female bird, especially song birds and game birds. (Male, cock).

Henbane (Solanaceae: *Hyoscyamus niger*). An evil-smelling poisonous plant with dull creamy-buff bell-shaped flowers, widespread but local in disturbed ground.

Henbit (Labiatae: *Lamium amplexicaule*). A common weed with pinkish-purple flowers, closely related to red dead-nettle.

Hep. Outmoded spelling of hip, the rose fruit.

Hepatic (Hepaticae). ⇨ LIVERWORT.

Herald Moth (Caradrinoidea: *Scoliopteryx libatrix*). A common autumn-flying moth, related to the Silver Y, 'the herald' of winter, often found hibernating in houses and caves.

Herb. (1) A non-woody vascular plant; adj. herbaceous. (2) A plant whose leaves are used for food, for medicinal purposes, for their scent or for their flavour. A herb in sense 2 need not be a herb in sense 1, though it usually is. Herbaceous, in botany as distinct from horticulture, means having a soft green texture like a leaf.

Herb Bennet (Rosaceae: *Geum urbanum*). Common yellow-flowered plant of hedge-banks.

Herb Christopher. ⇨ BANEBERRY.

Herbicide. Chemical weed-killer.

Herbivore. A plant-eater. Deer, rabbits, hares, mice and voles are the principal herbivores among British wild mammals.

Herb Paris (Liliaceae: *Paris quadrifolia*). A curious-looking plant with a solitary greenish flower surmounting a whorl of four leaves; local in woods, usually on limy soils, in England.

Herb Robert (Geraniaceae: *Geranium robertianum*). A common pink-flowered hedge-bank plant, closely related to the cranesbills.

Hermaphrodite. A plant with both male and female organs (stamens and pistils) in the same individual, or an animal with both male and female sexual organs, e.g. an earthworm.

Heron (Ardeiformes: Ardeidae). A family of mostly large, long-legged marsh birds, including also the egrets and bitterns, feeding off fish, frogs and other aquatic life. The Common or Grey Heron *Ardea cinerea* nests in colonies known as heronries; about 6,000 pairs breed every year. The Bittern is the only other British breeding species, but three other species of heron and three egrets are irregular visitors or stragglers to this country.

Herpetology. The scientific study of reptiles and amphibians.

Herptile. An American term to cover reptiles and amphibians together, now coming into use in Britain.

Herring (Clupeidae: *Clupea harengus*). A highly gregarious marine fish, abundant in British seas, and the basis of an important fishery on the coasts of Scotland and eastern England. Kippers and bloaters are smoked herrings.

Hessian Fly (Nematocera: *Mayetiola destructor*). A cecidomyiid gall midge, which is a greater pest of wheat in America and southern Europe than it is in Britain.

Heterocera. Outmoded term for the moths.

Heterochrosis. Colour variation, especially in the plumage of birds. Dilution occurs when pigments are present in reduced quantities, so that the plumage or pelage appears pale. ⇨ ALBINISM, ERYTHRISM, LEUCISM, MELANISM, XANTHISM.

Heteromera (Coleoptera: Polyphaga). A superfamily of beetles containing 142 species in thirteen families, including the cardinal beetles, Meloidae (oil and blister beetles), and Tenebrionidae (churchyard and hardback beetles).

Heteroneura or Ditrysia (Insecta: Lepidoptera). Much the largest of the three sub-orders of the Lepidoptera, containing the superfamily of the Butterflies and eleven superfamilies of moths: Caradrinoidea (noctuid, tussock and tiger moths), Notodontoidea (geometers, waves, pugs, hawkmoths, prominents), Drepanoidea (hooktips), Pyraloidea (micros), Lasiocampoidea (Kentish Glory), Bombycoidea (lackeys and eggars), Psychoidea (micros, burnets and foresters), Cossoidea (goat and leopard moths), Tortricoidea (micros), Tinaeoidea (micros, clearwings,) and Stigmelloidea (micros).

Heteroptera (Insecta: Hemiptera). A sub-order of bugs, characterized by the wings being closed flat over the back, comprising the Plant Bugs, the Shore Bugs and the Water Bugs, and totalling 509 species in Britain.

Heterosomata (Vertebrata: Pisces). An order of bony sea fishes comprising the flatfish, in the families Bothidae, Pleuronectidae and Soleidae.

Heterostylous. Botanical adjective used of plants whose flowers have styles of differing sizes relative to the stamens and other organs.

Heterozygote. Animal or plant bearing dissimilar Alleles in relation to a particular character. ⇨ HOMOZYGOTE.

Hexaploid. ⇨ POLYPLOIDY.

Hibernation. A special adaptation to more or less complete dormancy in winter, whereby the animal becomes torpid and its metabolism is greatly slowed down, the temperature of 'warm-blooded' animals (⇨ HOMOIOTHERMIC) falling to that of their surroundings. In Britain many invertebrates, e.g. snails, wasps, and small tortoiseshell and brimstone butterflies, hibernate, and so do all reptiles and amphibians, but only a few mammals. The dormouse is our only truly hibernating mammal, for hedgehogs and bats are liable to wake up in warm intervals in winter and feed a little. Squirrels, contrary to popular belief, do not hibernate.

Hide (Blind in North America). An artificial structure within which an observer of wild animals or birds conceals himself in order to watch or photograph them.

Hide Beetle. ⇨ LEATHER BEETLE.

Hill Creep. ⇨ SOIL CREEP.

Hind. The female in red and Japanese sika deer. A yeld or eild hind is a barren one.

Hip or Hep. The fruit of the rose.

Hirsel. The area of Scottish moorland which will carry about 600 ewes and should be manageable by a single shepherd.

Hirudinea. ⇨ LEECH.

Hispid. Botanical adjective used of plants with numerous rather large, coarse hairs.

Hive Bee. ⇨ HONEYBEE.

Hoar Frost. Frozen dew. ⇨ FROST.

Hobby (Falconiformes: *Falco subbuteo*). A small falcon with long wings and short tail like a peregrine, feeding especially on swifts, swallows and martins. A rather local summer visitor to southern England.

Hog Bristle (Chlorophyta: *Chaetomorpha melagonium*). A widespread but local green seaweed of pools on the lower shore, characterized by stiff bristle-like fronds.

Hoggin. Gravel which has been sifted.

Hog Slater. Water-louse.

Hogweed or Cow Parsnip. (Umbelliferae: *Heracleum sphondylium*). The commonest white umbellifer of roadsides and grassy places in summer and autumn, a large stout plant up to 5 ft high. Giant Hogweed or Cartwheel Flower *H. mantegazzianum* is a garden escape locally established by fresh water in various places, and growing to 10 ft high with umbels up to 18 ins. across.

Holly (Aquifoliaceae: *Ilex aquifolium*). A common evergreen tree or shrub, noted for its prickly leaves and red berries, which are highly prized as Christmas decorations.

Holly Fern (Aspidiaceae: *Polystichum lonchitis*). A scarce mountain fern, allied to the shield ferns, with simply pinnate, markedly spine-toothed fronds.

Holm. A Yorkshire term for a water meadow.

Holocene Period. The Geological Period of the present day, starting at the end of the Ice Age.

Holothuroidea. ⇨ SEA CUCUMBER.

Holotype. In taxonomy the individual specimen chosen to describe the whole species.

Holt. The otter's lair or sleeping place.

Holy Grass (Gramineae: *Hierochloë odorata*). A rare hay-scented grass of Scotland and northern Ireland, so named because it was formerly strewn at church doors on festival days in parts of the Continent.

Homeless Bee. ⇨ PARASITIC BEE.

Homoiothermic. Warm-blooded, with a temperature maintained constantly higher than the surroundings, as in birds and mammals, as opposed to Poikilothermic, cold-blooded.

Homologous. Similar in structure, development or origin.

Homoneura or Monotrysia (Insecta: Lepidoptera). The primitive sub-order of moths containing the Swift Moths (Hepialidae).

Homonym. In taxonomy, a name that has been applied by various authors to two quite different organisms; e.g. the name *Melilotus officinalis* has been applied by several authors to the melilot now known as *M. altissima* as well as to the melilot so named by Linnaeus.

Homoptera (Insecta: Hemiptera). A sub-order of four-winged or wingless insects containing the frog-hoppers (Cercopidae), thorn insects (Membracidae), leaf-hoppers (Jassidae or Cicadellidae), aphids or plant-lice (Aphididae), white-flies (Aleurodidae) and mealy-bugs and scale-insects (Coccidae), characterized by the wings sloping at an angle over the body.

Homozygote. Animal or plant bearing similar Alleles in relation to a particular character. ⇨ HETEROZYGOTE.

Honesty (Cruciferae: *Lunaria annua*). A purple-flowered plant with flat rounded seed-pods, a favourite cottage garden plant, often escaping.

Honewort (Umbelliferae: *Trinia glauca*). A low-growing glaucous white-flowered umbellifer, very local in limestone grassland in the south-west.

Honey. A sweet substance prepared in the hive by worker honeybees, which manipulate it with their mouthparts, from nectar collected by foraging bees, and stored in cells in the hive as food for the bee larvae. Nectar from different species of plant produces honeys of different and often characteristic flavours, as with the heather honey prepared from ling heather. The honeycomb is made of beeswax inside the hive to store surplus quantities of honey.

Honeybee or Hive Bee (Aculeata: *Apis mellifera*). The most important honey-gathering bee, and the only one which it has been worth while to induce to nest in specially constructed beehives, the honeybee cannot be said to be domesticated like dogs or cattle, but far more now nest in beehives than in natural wild situations. The characteristic original British form of the honeybee (var. *lehzeni*) is dark brownish-black, but in fact the present-day honeybee stocks of Britain are mongrel to a high degree. Honeybees are highly social insects, their colonies being founded by fertile females called queens, which first produce many infertile females called workers and only late in the season produce fertile females and males (drones). They feed on honey and pollen.

Honeycomb Moth (Pyraloidea: *Galleria mellonella*). A pyralid moth remarkably large for a 'micro', whose larvae feed on honeycomb and were formerly regarded as a serious beehive pest.

Honey-Dew. A sweet-tasting substance exuded by various plant-bugs (Hemiptera), especially aphids, suckers, scale insects and white-flies, consisting of the unwanted constituents, mainly sugars, of the plant juices on which they live. It is often abundant on plant leaves, especially those of the lime tree, in summer, and is much relished by ants, bees, flies and other insects. Ants 'farm' aphids for their honey-dew, which they milk from them.

Honey Fungus or Bootlace Fungus (Basidiomycetes: *Armillaria mellea*). A common yellow edible toadstool, related to beech tuft, parasitic on the roots of trees and shrubs, which it may kill by 'root rot', often appearing in quantity on the roots of dead trees or tree stumps; the mycelium looks like black bootlaces.

Honey Moth (Pyraloidea: *Achroia grisella*). A fairly common pyralid moth whose larvae feed on the wax in beehives, sometimes seen hovering around the entrances of beehives at dusk.

Honeysuckle or Woodbine (Caprifoliaceae: *Lonicera periclymenum*). Our only twining woody climber, with familiar fragrant flowers, red berries, leaves appearing in midwinter, and stems twining clockwise; widespread and common.

Hook-Tip Moth (Drepanoidea: Drepanidae). A group of five, mostly brown, moths characterized by having the tip of the forewings pointed and curved backwards.

Hoopoe (Coraciiformes: *Upupa epops*). Strikingly coloured, conspicuously crested, long-billed bird, breeding in western Europe and a regular straggler to Britain in small numbers; has bred several times. Named from its call-note.

Hop (Cannabiaceae: *Humulus lupulus*). Our only native member of the Hemp Family, also occurring widely as an escape from cultivation; a climber twining clockwise, it is dioecious; the cones resulting from the female flowers are used to flavour beer.

Hop-Flea. ⇨ Turnip FLEA BEETLE.

Horehound (Labiatae). The name of two aromatic plants, the disagreeably smelling Black Horehound *Ballota nigra*, a common wayside weed, and the pleasantly scented whitely downy White Horehound *Marrubium vulgare*, which is much less common.

Horizon. A layer in a soil profile.

Hornbeam (Corylaceae: *Carpinus betulus*). A smallish deciduous catkin-bearing tree, confined as a native to the south-eastern quarter of England, where it occurs commonly in some oakwoods, but planted elsewhere. Pollarded for firewood in Epping Forest and elsewhere.

Hornet (Aculeata: *Vespa crabro*). Our largest wasp, a tawny yellow social wasp more than an inch long, which uses its fearsome sting on man only if attacked; nesting in hollow trees.

Hornet Moth (Tinaeoidea: *Sesia apiformis*). One of the larger clearwing moths, which looks remarkably like a hornet; mainly in eastern England. The more widespread Lunar Hornet Moth *Sphecia bembeciformis* is similar but smaller.

Horn Fly (Muscidae: *Haematobia stimulans*). A close relative of the stable fly, with the same blood-sucking habits, though it does not attack humans; named from its habit of settling in groups on the heads, and especially horns, of cattle.

Horn of Plenty (Basidiomycetes; *Craterellus cornucopioides*). A dark brown, trumpet-shaped edible fungus related to the Chanterelle, growing in beech and other broad-leaved woods.

Horns. Goats are the only wild British mammals which have these bony growths from the skull covered with thick and hardened skin (horn). Unlike the antlers of deer they are never shed. In Scotland 'horns' may mean antlers.

Horn-Tail. ⇨ WOOD WASP.

Hornwort (Ceratophyllaceae: *Ceratophyllum*). Two species of free-floating submerged waterweed, with obscure green flowers at the base of whorls of repeatedly forked leaves; *C. demersum* is the commoner.

Hornwrack (Polyzoa: *Flustra foliacea*). An offshore sea-mat, resembling a pale yellowish seaweed, often washed ashore after a storm.

Horse. ⇨ PONY.

Horse Fly. A general name for the blood-sucking tabanid flies.

Horsehair Fungus or Horsehair Toadstool (Basidiomycetes: *Marasmius androsaceus*). A small toadstool, closely related to the fairy ring champignon, with thin black wiry stems resembling horsehair, common on dead pine needles and often carpeting the ground in pinewoods.

Horsehair Worm, Hairworm or Threadworm (Aschelminthes: Nematomorpha). A small class of freshwater worms, allied to the roundworms, and named from the resemblance of the adults, some of which may grow to more than a foot long, to animated horsehair. Indeed in former days this was what they were believed to be. The commonest British genus *Gordius* is named from the habit of the adults of getting themselves tangled into knots of Gordian complexity. The larvae are largely parasitic, passing through both aquatic and terrestrial insect hosts in the course of their development.

Horseradish (Cruciferae: *Armoracia rusticana*). White-flowered plant cultivated for the use of its roots as a condiment; widely established on roadsides and in waste places in England.

Horse-Stinger. An old country name for dragonflies, which were erroneously believed to sting horses.

Horsetail (Sphenopsida: *Equisetum*). Leafless flowerless plants whose spores are borne in cones at the tip of the sometimes unbranched stems. Of the nine British species, three, including Common Horsetail *E. arvense* and Giant Horsetail *E. telmateia*, have separate short-lived unbranched brown fertile stems and long-lived sterile branched green stems. Other common species are Water Horsetail *E. fluviatile*, Marsh Horsetail *E. palustre*, and Wood Horsetail *E. sylvaticum*, whose names describe their habitats. The name is often erroneously applied to the aquatic flowering plant marestail. ⇨⇨ DUTCH RUSH.

Hot-Bed Bug (Heteroptera: *Xylocoris galactinus*). A predatory bug allied to the bedbug and inhabiting manure-heaps and similar warm places.

Hottentot Fig. ⇨ MESEMBRYANTHEMUM.

Hound. Name given to several species of dogfish.

Houndstongue (Boraginaceae: *Cynoglossum officinale*). Maroon-coloured wild flower, smelling of mice and with greyish leaves shaped like a dog's tongue, locally frequent in the south; Green Houndstongue *C. germanicum* is now rare.

House Fly (Muscidae: *Musca domestica*). This universal pest is both the eponymous fly and the archetypal two-winged fly. Smaller flies seen with them are not 'young flies', as many people suppose, but the Lesser House Fly *Fannia canicularis*. An orange-yellow house fly is another species, *M. autumnalis*. Biting House Fly, ⇨ STABLE FLY.

House Moth (Tinaeoidea: Oecophoridae). Two members of this family of 'micros' are household pests: the Brown House Moth or False Clothes Moth *Borkenhausia pseudopretella* is even commoner and more destructive than the common clothes moth in many houses, its larvae attacking many kinds of fabric and dried animal and vegetable matter; the White-shouldered House Moth *Endrosis sarcitrella* is easily recognized by its white head and 'shoulders'.

House Spider (Araneae: *Tegenaria*). Three large dark blackish-brown spiders, common in houses, and especially liable to arouse human alarm or distaste; ⇨ CARDINAL SPIDER.

Houting (Salmonidae: *Coregonus oxyrhynchus*). Our only anadromous whitefish, frequent in the North Sea but rarely caught in British rivers.

Hover. A floating mass of living vegetation in the Norfolk Broads.

Hover-Fly or Flower Fly (Cyclorrhapha: Syrphidae). A large family of nearly 250 species of two-winged flies, including some economic pests, noted for their habit of hovering apparently motionless in the air, and darting away so swiftly as to seem to disappear by magic; some of their larvae are known as rat-tailed maggots; some adults are remarkably bee-like. ⇨⇨ BULB FLY, DRONE FLY.

Howe. A hillock or mound in the north of England.

Humblebee. ⇨ BUMBLEBEE.

Hummel. A stag whose antlers do not grow owing to some physiological defect. A nott in Devon and Somerset.

Humming-Bird. There are no humming-birds native to Britain, though at least one has escaped and been seen in a garden many miles from its cage; ⇨ HAWK-MOTH, Humming-Bird.

Humpback (Cetacea: *Megaptera novaeangliae*). A large, rather stout whale, allied to the rorquals, that migrates regularly through the Atlantic off our western coasts.

Humus. Decaying organic material in the soil, derived mainly from dead plants. The decay is mainly the work of numerous soil bacteria – a gramme of soil may contain hundreds of millions – together with earthworms, which drag dead leaves down and pass them through their bodies, and other soil animals and fungi. The soil bacteria reduce plant and animal remains to the materials from which the plants derived their original nutriment – carbon dioxide, water, soluble salts. In favourable conditions, where the decay is rapid, the resulting humus is called Mild Humus, or Mull; in unfavourable soil with no earthworms, the result is Raw Humus, or Mor.

Hunter's Moon. The November full moon.

Hurt or Whort. Southern name for fruits of the bilberry or whortleberry. The Hurt Wood near Leith Hill in Surrey is named from them.

Hyaline. Botanical adjective used of thin, transparent or shining sepals or other organs.

Hybrid. The progeny of animal or plant parents of different genera, species, subspecies or even varieties. Hybrid animals are rarely fertile (⇨ ALLOTETRAPLOID), so that this is not a common method of speciation among animals. Several plants in the British flora, however, are fertile hybrids, notably Russian comfrey, Townsend's cord-grass and probably also common hemp-nettle. Occasionally, as with the yellow water-lilies, an infertile or partially infertile hybrid may persist in the absence of both parents.

A hybrid swarm is a large population consisting of various hybrids, including back-crosses, together with one or both of their parents.

Hydra (Hydrozoa: *Hydra* and *Chlorohydra*). Three species of small carnivorous freshwater coelenterate animals, related to the marine sea-firs of the order Athecata, the commonest being the Green Hydra *C. viridissima*, coloured bright green by a tiny green alga *Zoochlorella*. Hydras have only a single polyp and no medusa stage, spending their lives settled on water plants, and reproduce themselves either by budding off new individuals from their sides or by a sexual process.

Hydroid. ⇨ SEA-FIR.

Hydrology. The scientific study of water, especially fresh water in a natural state, in lakes, rivers, etc.

Hydrophyte. A plant that grows in water or very wet places.

Hydrosere. An aquatic plant community.

Hydrotropism. A tropism stimulated by water.

Hydrozoa (Cnidaria). Aquatic, mainly marine, mostly colonial animals, some of which, such as the sea-firs or hydroids and the hydras, spend all or most of their lives as sedentary polyps or zooids, while others, such as the siphonophores, the freshwater Limnomedusae and the jellyfish-like Trachymedusae and Narcomedusae of the marine plankton spend much or most of their lives as free-swimming medusae. The first freshwater medusa known to science was *Craspedacusta sowerbii*, found in a hothouse tank in Regent's Park, London, in 1880.

Hygrophilous. Inhabiting or growing in wet or damp places.

Hymenoptera (Arthropoda: Insecta). The large order (over 6,000 British species) of insects containing the bees, wasps, ants (Aculeata) and ichneumon flies (Parasitica) in the sub-order Apocrita, and the saw-flies and wood-wasps in the sub-order Symphyta. The winged members of the order have four transparent wings.

Hypha. One of the tubular threads which make up a fungal mycelium.

Hypochile. The rear part of the lip or labellum of certain orchid flowers, as opposed to Epichile.

Hypogeal. Botanical adjective meaning underground.

Hypolimnion. The cold lower layer of water in a lake. ⇨ SEICHE.

Hypotremata (Vertebrata: Selachii). The class of marine fishes containing the rays and skates.

I

Ibis, Glossy (Ardeiformes: *Plegadis falcinellus*). A medium-sized, long-legged marsh bird with a curved bill; an irregular visitor to Britain.

Ice. Frozen water, or water in the solid state, forming on the surface of puddles as soon as the temperature falls below 0° C. (32° F.) and on larger pools, ponds, lakes and even rivers as it falls still further. In extremely hard weather even shallow arms of the sea and estuaries may freeze over.

Ice Age. ⇨ GLACIAL ACTION.

Iceland Moss (Lichenes: *Cetraria islandica*). An edible moorland lichen growing especially on mountains in the Highlands, along with reindeer moss.

Ice Sheet or Ice Cap. An extensive mass or sheet of ice, such as now covers Greenland and Antarctica and formerly covered much of Great Britain during the Ice Ages.

Ichneumon (Parasitica: Ichneumonoidea). A large superfamily of parasitic wasps, with some 2,825 species, which lay their eggs in the eggs or larvae of other insects, the ichneumon larvae completing this stage in their life history inside their host which dies when they emerge to pupate; many species are highly specific in their hosts. The principal families are the Braconidae and the Ichneumonidae. The largest British ichneumon is *Rhyssa persuasoria*, which parasitizes the greater horntail. *Ophion luteus*, another large species, often comes to light in houses. The braconid *Apanteles glomeratus* is a well known parasite of cabbage white butterflies. Some ichneumons parasitize spiders.

Ichthyology. The scientific study of fishes.

Igneous Rocks. *Ignis:* Latin for fire. Rocks which have solidified from molten magma, either below the surface (plutonic rocks) or on reaching the surface (volcanic rocks or lavas).

Imago. The fourth and final stage of the metamorphosis of various insects, also known as the perfect insect, especially in butterflies and moths, in some of which it is not a feeding stage but has mature sex organs.

Imbricate. Botanical adjective used of organs whose edges overlap.

Immature Bird. ⇨ YOUNG BIRD.

Imparipinnate. Botanical adjective used of pinnate leaves terminating in a single leaflet.

Imperfect Fungi. A miscellaneous group of fungi which cannot be classified in any of the other three main groups as no sexual stage is known.

Imprinting. In animal behaviour, a rapid and usually very stable form of learning, which takes place in the early life of social species, whereby, often apparently without immediate reinforcement, broad supra-individual characteristics of the species come to be recognized as the species pattern and subsequently used as releasers (W. H. Thorpe), e.g. a duckling learns what its mother looks like and thenceforward follows it. It is possible for young animals to become accidentally or experimentally imprinted on the wrong objects, e.g. a duckling can follow a man instead of its mother.

Indian Summer. A short period of warm, sunny weather in October or early November; if the latter, it is St Martin's Summer, named from Martinmas, 11 November; 2–6 November is one of the most frequent periods for an Indian summer, and before the reform of the calendar in 1751 Martinmas fell eleven days earlier, on what is now 31 October.

Indumentum. The hairs or down on a plant.

Indusium. The covering of the heaps of spores on the underside of a fern frond.

Industrial Melanism. A phenomenon whereby melanic forms of certain moths have replaced the normal forms by natural selection, largely because their darkness affords them better protection from enemies while resting on soot-blackened walls, fences or trees in industrial towns, against which the normal paler forms would show up more distinctly. The phenomenon was first noticed in the peppered moth around Manchester in the latter half of the nineteenth century, but has since become known in other species both in Britain and Germany. Recently it has also occurred in spiders in northern England.

Inflorescence. The whole flowering part of a plant, comprising the stems, peduncles (stalks) and bracts as well as the actual flowers.

Types of Inflorescence

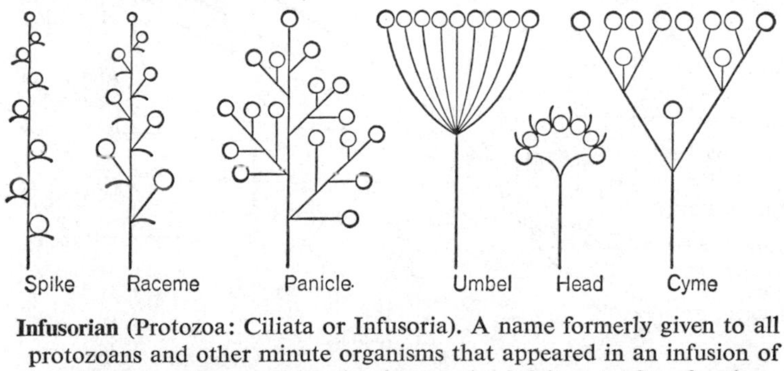

Infusorian (Protozoa: Ciliata or Infusoria). A name formerly given to all protozoans and other minute organisms that appeared in an infusion of hay, i.e. a wisp of hay or other dried vegetation left in water for a few days, but now confined to a large class of minute freshwater protozoan animals, a few also occurring on the shore and in the marine plankton. They move by means of cilia, and so are also called ciliates. The best known are the Slipper Animalcules *Paramecium* and Bell Animalcules *Vorticella*, both named from their shape. Some, including the bell animalcules, are attached to algae and other objects by stems. Others, such as *Carchesium* and *Epistylis*, are both colonial and attached.

Ings. Damp grass fields in Yorkshire.

Ink Cap (Agaricaceae: *Coprinus*). A group of toadstools, often with conical caps, whose gills deliquesce into an inky black fluid. Two edible species are the Common Ink Cap *C. atramentarius*, which however contains the same substance as makes antabuse a method of curing alcoholism by creating nausea, and so should not be eaten with strong drinks; and Shaggy Ink Cap, Shaggy Cap or Lawyer's Wig *C. comatus*, whose cap is almost cylindrical, and looks wig-like with its shaggy scales.

Inquiline. An animal which cohabits with another animal, e.g. in a gall, without actually parasitizing it but sometimes getting a share of its food.

Insect (Arthropoda: Insecta). The most abundant class of arthropods on land, the adults characterized by having their bodies divided into head, thorax and abdomen, the head bearing a pair of feelers or antennae, and the thorax *three pairs of legs* and usually also wings. Insects are the only winged arthropods. The life history of insects usually passes through the three stages of egg, larva and pupa before attaining adulthood. The largest and best known of the twenty-seven British classes of insects are the Odonata (dragonflies), Orthoptera (grasshoppers), Phthiraptera (lice), Hemiptera (bugs, aphids), Coleoptera (beetles), Lepidoptera (butterflies and moths), Diptera (two-winged flies), and Hymenoptera (bees, wasps, ants). Approximately 21,500 species of insect are known to occur in the British Isles.

Parts of an Insect

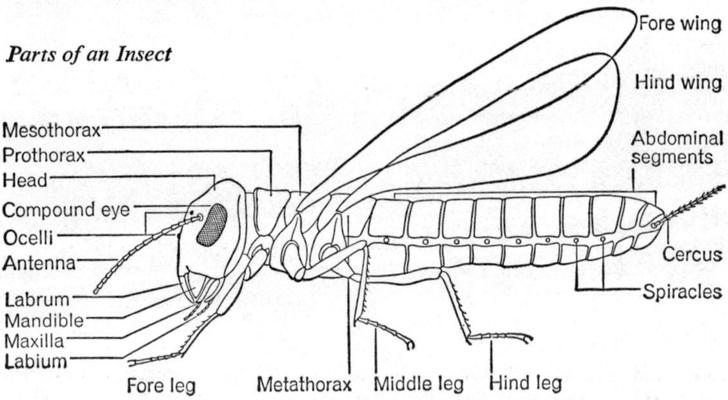

Insectivora (Vertebrata: Mammalia). The most primitive order of British mammals, characterized by long snouts and numerous sharp teeth adapted to their diet of invertebrates. The six British species are the hedgehog, mole and four shrews.

Insectivore. An animal or plant that feeds on insects. The specialized insectivores among British mammals are the Insectivora and the bats, among the many birds the warblers are the most specialized. Insectivorous plants include the sundews and lords and ladies.

Insight. In animal behaviour, the apprehension of relations. (W. H. Thorpe.)

Instar. A stage between moults of a nymph or larva.

Instinct. Innate behaviour of animals, as distinct from learned behaviour, resulting from internal and/or external stimuli. More precisely, an inherited and adapted system of coordination within the animal's central nervous system as a whole, which when activated finds expression in behaviour culminating in a fixed action pattern. (W. H. Thorpe).

Intelligence. A term now regarded as being too vague for use in animal behaviour studies.

Intercalary. Botanical adjective used of the leaves between the upper branches on the stem and the lowest bracts, especially in the yellow rattles.

International Council for Bird Preservation. The international body for protecting bird life throughout the world, with fifty-five national sections, founded in 1922. The British Section, which has fourteen member bodies, was founded in 1923 and publishes an *Annual Report*.

Internode. The stem between two nodes or leaf-junctions.

Intertidal. Occurring between high and low water marks on the shore.

Intravaginal. Botanical adjective meaning within a sheath.

Introduced Animals and Plants. Animals deliberately set free, or plants deliberately sown or planted, with the intent that they should form a feral breeding stock. There are, for instance, about a dozen naturalized mammals and ten each of birds and freshwater fishes in the British fauna, nearly all of which have been introduced. ⇨ ADVENTIVE.

Introgression. The acquisition by one species or population of genetic characters originating in another closely related one, usually by means of hybridization and back-crossing.

Introrse. Botanical adjective used of anthers which open inwards, towards the centre of a flower.

Invertebrate. Animals without a backbone, i.e. animals belonging to all phyla except the Chordata, and including the primitive chordates – cephalochordates, tunicates, etc.

Involucel. The epicalyx in the Teasel Family (Dipsacaceae).

Involucre. The leaf-like bracts which form a calyx-like structure at the base of a capitulum or compound flower-head, as in the composites.

Involute. Botanical adjective used of leaves or other organs with their margins rolled upwards.

Iris. ⇨ GLADDON, YELLOW FLAG.

Irish Moss. ⇨ CARRAGHEEN MOSS.

Iron Bacteria. Certain chemophytic bacteria are able to oxidize soluble ferrous iron compounds to produce insoluble ferric iron, and these produce the reddish-yellow deposit sometimes seen in springs and other small watercourses. They are also responsible for the rusting up of water pipes.

Irruption. A migratory movement undertaken in some strength at irregular intervals; in Britain usually by birds from northern Europe, especially waxwings and crossbills.

Isle of Wight Disease, or Acarine Disease. A disease of honeybees caused by a tiny parasitic mite, *Acarapis woodi* (Prostigmata), which invades their tracheae.

Isokontae. Alternative name for Chlorophyta, green algae.

Isomerous. Botanical adjective used when the different parts of a flower are the same in number, e.g. five petals and five sepals, or five stamens and five carpels.

Isopoda (Crustacea: Malacostraca). An unusually catholic order of small crustaceans, with flattened bodies, having some terrestrial species such as the well-known woodlice or skaters; some freshwater species, the water-lice or water skaters; some marine or brackish species, including the gribble and the sea-slater: and even a few parasitic species.

Isospondyli (Vertebrata: Pisces). The order of bony fishes including the Herring (Clupeidae) and Salmon (Salmonidae) Families.

Itch Mite (Astigmata: *Sarcoptes scabei*). A parasitic mite producing the itch or scabies in man and sarcoptic mange in domestic animals. The Hay Itch Mite is *Pyemotes ventricosus* (Prostigmata), a parasite of beetles and moths.

Ivy (Araliaceae: *Hedera helix*). Our only evergreen climber, and the only one to climb by means of adventitious rootlets. Leaves on non-flowering shoots of the familiar ivy shape, those on flowering shoots shaped like an ace of spades; flowers not appearing till September; widespread and common in woods and hedges and on rocks.

J

Jack. A small pike.

Jack by the Hedge. ⇨ Hedge GARLIC.

Jackdaw (Passeriformes: *Corvus monedula*). The smallest of the British true crows, all black in plumage with a grey nape patch. Widespread, common and highly gregarious, often associating with rooks. Named from its call-note, 'chack'.

Jack-Go-to-Bed-at-Noon. ⇨ GOATSBEARD.

Jack o'Lantern. ⇨ MARSH GAS.

Jack-Sail-by-the-Wind. ⇨ BY-THE-WIND-SAILOR.

Jacob's Ladder (Polemoniaceae: *Polemonium caeruleum*). A handsome plant with spikes of large blue-violet flowers, local on limestone in the Peak District and the Craven Pennines.

Jaeger. North American name for Skua.

Jay (Passeriformes: *Garrulus glandarius*). A brightly coloured relative of the crows with a harsh voice, noted for its habit of burying acorns in autumn as a store of food against the winter. Widespread and common in woods.

Jellyfish (Cnidaria: Scyphozoa). Marine coelenterate animals, umbrella-shaped with tentacles hanging down. Our half-dozen larger species are planktonic, spending much the greater part of their lives in the medusa stage, and may attain sizes of well over a foot. Our commonest species, often blown inshore or cast in quantity on the beach by summer and autumn storms, is *Aurelia aurita*, transparent except for four pale purple crescents. *Chrysaora isosceles* is the largest jellyfish in the world, attaining 6 ft across in Arctic seas, but rarely as much as 2½ ft in British waters. The balloon-like *Rhizostoma octopus* with its many mouths is the largest jellyfish regularly seen in British seas, often up to 2 ft. The bioluminescent *Pelagia noctiluca* is a not infrequent invader from the warmer parts of the Atlantic. Our four species of Stalked Jellyfish (Lucernaridae) are much smaller; they do not swim but attach themselves to seaweeds and stones by a sucker, and move only by motions somewhat resembling those of a looper caterpillar. None are both widespread and common, but *Haliclystus auricula* is frequent in the south-west.

Jelly Fungus (Basidiomycetes: Tremellales). An order of waxy or gelatinous, sometimes slimy, cushion-shaped, saprophytic fungi, mostly found on dead wood, such as the bright yellow *Tremella mesenterica*.

Jet. A semi-precious black stone which, like coal, is fossilized wood and burns. The principal British deposits of jet are at Whitby, on the north Yorkshire coast, where fragments may still be found on the beach and are drifted by the tides to beaches further south on the east coast.

Jew's Ear (Basidiomycetes: *Auricularia auricula*). A brown, somewhat ear-like edible fungus, growing almost entirely on elder trees.

John Dory (Pisces: *Zeus faber*). Strikingly shaped (almost circular) marine fish, often tinged yellow (whence its name, from the French *doré* gilded), not uncommon off our southern coasts; moderately good to eat.

Jumping Spider (Araneae: Salticidae). A family of hunting spiders named from their ability to jump anything from twice to twenty times their body length, the champion jumper, according to Dr Bristowe, being *Attulus saltator*.

Juniper (Coniferae: *Juniperus communis*). An evergreen coniferous shrub, undershrub or small tree, bearing berries instead of cones, which are used in the manufacture of gin. Local, especially on chalk and limestone in the south; more frequent in hill districts and on sea cliffs in the north.

Jurassic. ⇨ GEOLOGICAL PERIODS, LIMESTONE.

Juvenile Bird. ⇨ YOUNG BIRD.

K

Kaffir Fig. ⇨ MESEMBRYANTHEMUM.

Karst. Limestone formation consisting in Britain largely of Limestone Pavements, with only thin or no overlying soils and much underground water; named after Karst in Yugoslavia.

Keck, or Kex. A general term for the taller white-flowered umbellifers, most often applied to cow parsley and hogweed.

Ked (Diptera: Hippoboscidae). Wingless flies parasitic on hoofed animals, notably the Deer Ked or Deer Fly *Liptoptena cervi* and the Sheep Ked or Sheep Louse *Melophagus ovinus,* sometimes miscalled Sheep Tick.

Kelp. ⇨ OARWEED.

Kelp Fly (Cyclorrhapha: Coelopidae). A family of seven species of small two-winged flies, inhabiting the kelp and other seaweed thrown up on the shore at or above high-tide mark, on which their larvae feed; they sometimes occur in swarms, to the annoyance of holidaymakers, and even invade buildings abutting on the shore.

Kelt. The spent salmon after spawning. Kelts return to the sea and may either die or return to spawn again.

Kentish Glory (Lasiocampoidea: *Endromis versicolora*). A large handsome moth, now no longer found in Kent, but locally plentiful in other parts of southern England and in Scotland.

Kentish Rag. ⇨ RAGSTONE.

Keratinophilic. Growing on feathers, e.g. of certain fungi.

Kestrel or Windhover (Falconiformes: *Falco tinnunculus*). A small falcon with long wings and tail, generally the commonest British diurnal bird of prey, though now rather local in eastern England owing to secondary poisoning from toxic farm pesticides. Noted for its habit of hovering in search of its prey, mostly mice and voles. The Lesser Kestrel *F. naumanni* is a rare straggler from southern Europe.

Kew. Vernacular abbreviation for the Royal Botanic Gardens at Kew, Surrey, founded in 1759 and now a centre of botanical research of international importance, especially in the fields of taxonomy, phytogeography and economic botany.

Kex. ⇨ KECK.

Key. A winged fruit (achene), e.g. of ash or sycamore, technically a samara.

Kid. The young of goats and roe deer.

Killarney Fern. ⇨ FILMY FERN.

King Alfred's Cakes. ⇨ CRAMP BALL.

Kingcup. ⇨ Marsh MARIGOLD.

King-Fish. ⇨ OPAH.

Kingfisher (Coraciiformes: *Alcedo atthis*). A brilliantly coloured small bird with long red bill, noted for diving after fish either from the air or from a perch. Widespread and frequent.

Kinglet. Name given to North American relatives of the goldcrest and firecrest.

King of the Herrings. Name given to several large fishes supposed to accompany herring shoals, including the Rabbit-fish and the Ribbon-fish.

Kite (Falconiformes: *Milvus*). Large buzzard-like birds of prey with forked tails. The Red Kite *M. milvus* is a scarce resident, about a dozen pairs breeding in mid Wales. The Black Kite *M. migrans* is a very rare straggler from the Continent.

Kitten Moth (Notodontoidea: *Harpyia*). Three of the smaller Prominents, closely allied to the Puss Moth and with the same curious larvae: the scarce Alder Kitten Moth *H. bicuspis*, the common Poplar Kitten Moth *H. bifida* and Sallow Kitten Moth *H. furcula*.

Kittiwake (Charadriiformes: *Rissa tridactyla*). A rather small gull, named from the cries, often deafening, at its breeding colonies off our northern and western coasts; spends the winter at sea.

Knapweed (Compositae: *Centaurea*). Two plants with attractive, somewhat thistle-like purple flowers, common in grassy places: the widespread Black Knapweed or Hardhead *C. nigra*, with or without rayed flowerheads, and the more local Greater Knapweed *C. scabiosa*, which always has rayed flowerheads and prefers limy soils.

Knawel (Caryophyllaceae: *Scleranthus*). Small weedy green-flowered plants, *S. annuus* fairly common but *S. perennis* rare.

Knot (Charadriiformes: *Calidris canutus*). Small wader, often abundant on shore in winter; named from its call-note.

Knotgrass (Polygonaceae: *Polygonum*). Name given to those small polygonums whose small pink or white flowers form 'knots' along the stem together with a leaf and a papery sheath. The commonest, *P. aviculare* agg., is a widespread and abundant weed.

Krill (Malacostraca: Euphausiacea). The Norwegian whalers' name for the shrimp-like planktonic crustaceans on which the Antarctic whales feed; several species, including the large *Meganyctiphanes norvegica*, *Nyctiphanes couchi* and *Thysanoessa rachii* occur in British inshore waters.

Kyle. A sea loch or inlet of the sea in the north of Scotland.

L

Labiate (Labiatae). Family of mostly aromatic flowering plants with two-lipped flowers and square stems, including the mints, thymes, woundworts and dead-nettles.

Lacebug (Heteroptera: Tingidae). Plant-bugs with a lace-like network of veins on the forewings; twenty-two British species, including the Gorse Lacebug *Dictyonota strichnocera*, common on gorse and broom, and two Thistle Lacebugs, *Tingis ampliata* on creeping thistle and *T. cardui* on spear thistle. ⇨ RHODODENDRON BUG.

Lace-Wing (Neuroptera: Planipennia). Gauzy-winged insects in three families, fourteen Green Lace-wings (Chrysopidae), two of which are actually brown; twenty-eight Brown Lace-wings (Hemerobiidae); and the handsome Giant Lace-wing *Osmyius fulvicephalus*, which is semi-aquatic. A common green lace-wing *Chrysopa carnea* often enters houses to hibernate in autumn and then undergoes a striking colour change to red, becoming green again in the spring. The larvae of both green and brown lace-wings feed on aphids and other small insects and mites.

Laciniate. Botanical adjective used of leaves and other organs raggedly edged with very narrow lobes.

Lackey Moth (Bombycoidea: *Malacosoma*). Two brown moths, whose strikingly coloured furry caterpillars live in colonies, retiring to a communal web when they are small; *M. neustria* is much the commoner and more widespread.

Ladybird (Clavicornia: Coccinellidae). A distinct family of forty-five small beetles, often coloured red or yellow with black spots, well known as friends of the gardener because both adults and larvae (known as 'niggers') feed on scale insects and aphids. One of them must have almost the record scientific name for length: *Subcoccinella vigintiquatuorpunctata*.

Lady Fern (Athyriaceae: *Athyrum filix-femina*). A large, tufted fern of damp woods, rocks, banks and hillsides, preferring acid soils and mainly in the north and west, with more delicate and graceful fronds than the male fern. Its mountain relative, Alpine Lady Fern *A. alpestre* is very local in the Highlands.

Lady Orchid (Orchidaceae: *Orchis purpurea*). A handsome tall orchid with flowers of two shades of pinkish-purple and a manikin lip like the man orchid; almost confined to woods and scrub on the Kentish chalk.

Lady's Fingers. ⇨ Kidney VETCH.

Lady's Mantle (Rosaceae: *Alchemilla*). Low-growing, somewhat tufted, green-flowered plants with palmate leaves: Common Lady's Mantle *A. vulgaris* agg., variable, in all kinds of grassland but scarce in the south, and Alpine Lady's Mantle *A. alpina*, with leaves silvery beneath, on mountains, common in the Highlands, less so elsewhere.

Lady's Slipper (Orchidaceae: *Cypripedium calceolus*). One of our handsomest and rarest orchids, with slipper-shaped lower lip, now confined to a very few limestone woods in Yorkshire.

Lady's Smock or Cuckoo Flower (Cruciferae: *Cardamine pratensis*). A common white or lilac spring flower of damp meadows and marshes.

Lady's Tresses (Orchidaceae). Four species of white-flowered orchid; the local southern grassland Autumn Lady's Tresses *Spiranthes autumnalis*, the local northern woodland Creeping Lady's Tresses *Goodyera repens*, the scarce western marshland American Lady's Tresses *S. romanzoffiana*, and the possibly extinct Summer Lady's Tresses *S. aestivalis*, formerly only in the New Forest.

Lagg. The vegetation, especially alderwood, on the edge of a raised bog.

Lagomorpha (Vertebrata; Mammalia). An order of ground-living, herbivorous mammals, formerly included with the rodents, and comprising the hares, of which two species are native, and the introduced rabbit.

Lagoon. A sheet of fresh or brackish water close to the sea and often separated from it only by a sand or shingle bar.

Lair or Ligging. The place where a deer lies. A red deer 'harbours' and is 'unharboured'; a fallow deer 'lodges' and is 'roused'; a roe deer 'beds' and is 'dislodged'.

Lake. A large sheet of inland fresh water, also sometimes applied to broad, almost landlocked arms of the sea. There is no quantitative distinction between a lake and a pond.

Lamarckism. The theory that physical characters acquired during the lifetime of an organism can be inherited by its descendants, commonly attributed to the French scientist Lamarck (1744–1829), who, however, did not hold the theory strictly in this form, believing rather that the habits an animal developed during its lifetime might be so inherited. The inheritance of acquired characteristics is therefore more correctly described as Neo-Lamarckism.

Lambs' Tails. Folk name for hazel catkins.

Lamellibranchia. ⇨ BIVALVE.

Lamellicornia (Coleoptera: Polyphaga). A superfamily of beetles containing ninety-two species in two families, the Scarabaeidae (dung beetles, chafers) and the stag beetles.

Lamina. Any thin piece of plant tissue, but especially a leaf-blade or petal.

Lammas Leaves. The second crop of leaves put forth around Lammas Day (2 August) by oaks which have been severely defoliated by the larvae of various moths, especially the green oak moth.

Lampern. ⇨ River LAMPREY.

Lamprey (Cyclostomata: Petromyzontidae). Primitive eel-shaped fishes with circular suctorial lips; three British species: the River Lamprey or Lampern *Lampetra fluviatilis*, beloved of medieval gourmets, which migrates from the sea to spawn in rivers; the smaller Brook Lamprey *L. planeri*, confined to fresh water; and the larger Sea Lamprey or Stone-sucker *Petromyzon marinus*, which comes into estuaries to spawn.

Lancelet (Cephalochordata: *Branchiostoma*, formerly *Amphioxus*). A primitive fish-like jawless chordate, about 2 ins. long, found on sandy sea bottoms, only rarely uncovered by the tide.

Lanceolate. ⇨ LEAF-SHAPES.

Lappet Moth (Bombycoidea: *Gastropacha quercifolia*). A large reddish-brown eggar moth, named from the fleshy lappets on the flanks of it furry caterpillars; widespread in England.

Lapwing, Peewit or Green Plover (Charadriiformes: *Vanellus vanellus*). A widespread and common wader noted for its crest and rounded wings, breeding on farmland and moors, and so useful to the farmer that it once had a special Act of Parliament to protect it. The three names derive respectively from the lapping sound of its wings in display flight, its 'song' and most frequent call-note, and its dark green plumage. Its eggs are the plovers' eggs prized by gourmets.

Larch (Coniferae: *Larix*). A tall deciduous coniferous tree. The European Larch *L. decidua*, the Japanese Larch *L. leptolepis* and their hybrid *L. x eurolepis* are all widely planted, and the first-named readily regenerates and is more or less naturalized in many places.

Lard Beetle or Larder Beetle. ⇨ BACON BEETLE.

Lari-Limicolae. ⇨ CHARADRIIFORMES.

Lark (Passeriformes: Alaudidae). Small song birds, mostly brown in colour. The two British breeding species, both with songs of very high quality delivered on the wing, are the Skylark *Alauda arvensis*, widespread and common on downs, moors and farmland, and the Woodlark *Lullula arborea*, mainly in the south. The Shore Lark *Eremophila alpestris* is a local winter visitor on the east coast. Three other species are rare stragglers. Titlark, ⇨ MEADOW PIPIT.

Larva. The active second stage of the development of various animals, especially invertebrates; in insects coming between the ovum and the pupa. ⇨ CATERPILLAR, PLANKTON, TADPOLE.

Lasher. An artificial waterfall that takes that part of a river or stream which does not pass through a lock.

Lasiocampoidea (Lepidoptera: Heteroneura). The superfamily whose sole British representative is the Kentish Glory moth.

Latex. The milky juice of certain plants, such as poppies and dandelions.

Launce. ⇨ SAND EEL.

Laver. A seaweed name. Green Laver is the sea-lettuce. Laverbread is the red seaweed *Porphyra* made into cakes with oatmeal and fried with bacon; it is a delicacy in South Wales, especially around Swansea and Llanelly.

Lawyer's Wig. ⇨ INK CAP.

Layer. In any plant community different plants grow to different heights, and the levels to which they grow are called layers; e.g. a deciduous woodland commonly has four layers: tree, shrub, herb or field, and moss or ground.

Lazy-Bed. A cultivated plot in the Hebrides, fertilized by tangle and other seaweed cast up on the shore and carried or carted inland by the crofter.

Leaching. The process by which rain water percolating through the soil washes down the soluble and finer insoluble substances, such as mineral salts, to lower levels, e.g. the clay-with-flints capping much of the Chilterns is a deposit from which the chalk has been leached out. The surface soils are leached and become coarse and poor in nutritional salts, especially in a wet climate.

Leadwort. A vernacular name in the Peak District for the local Spring Sandwort (Caryophyllaceae: *Minuartia verna*), which grows there on old lead workings.

Leaf Beetle (Phytophaga: Chrysomelidae). Our second largest family of beetles, containing some 264 species, many of them brightly coloured, often with a metallic sheen, including the bloody-nosed, Colorado, flea, mustard and tortoise beetles.

Leaf-cutting Bee (Aculeata: *Megachile*). A genus of carpenter bees which make the walls of the egg-cells in their nests of pieces cut from leaves; the work of *M. centuncularis* on garden roses is familiar to gardeners.

Leaf-Hopper or Sucker (Homoptera: Jassidae or Cicadellidae). A family of numerous small hopping and flying bugs, some of which are pests because they spread virus diseases while sucking plant juices.

Leaf-Miner. A larva, e.g. of many Microlepidoptera and the two-winged fly family Agromyzidae, which tunnels within the tissues of a leaf, often making elaborate arabesques.

Leaf Shapes. ⇨ p. 151.

Learning. In animal behaviour, the process which produces adaptive change in individual behaviour as the result of experience. There are five types of learning: Habituation, the waning of a pre-existing response as a result of repeated stimulation that is not followed by any kind of reward or penalty (such reward or penalty being known as Reinforcement); ⇨ CONDITIONING; Trial-and-Error Learning; Latent Learning, which is independent of rewards and penalties, but may be brought into play by them; and Insight Learning, which is the sudden adaptive re-organization of experience, ⇨⇨ INSIGHT (based on W. H. Thorpe).

Leather Beetle (Diversicornia: *Dermestes maculatus*). A pest of hides and other stored animal products, also known as Hide Beetle.

Leather Jacket. The larva of certain crane flies, which feeds on the roots and lower stems of grasses and other plants and can be a destructive pest of turf and grassland.

Lectotype. In taxonomy a specimen chosen as a substitute for the holotype when this is missing and no paratype is available.

Lee or Leeward. The side of an object away from the direction from which the wind blows.

Leech (Annelida: Hirudinea). Worm-like aquatic animals, with a sucker at each end, enabling them both to grip their victims and to anchor themselves in swift currents. They are mostly bloodsuckers and secrete hirudin, which prevents the coagulation of blood, so that a leech wound takes a long time to heal; a leech is almost impossible to detach from its prey until it is fully fed. Eleven species of freshwater leech occur in Britain, including the Duck Leech *Protoclepsis tesselata*, sometimes found in the nostrils of ducks; the common Horse Leech *Haemopis sanguisuba*, which

Leaf Shapes

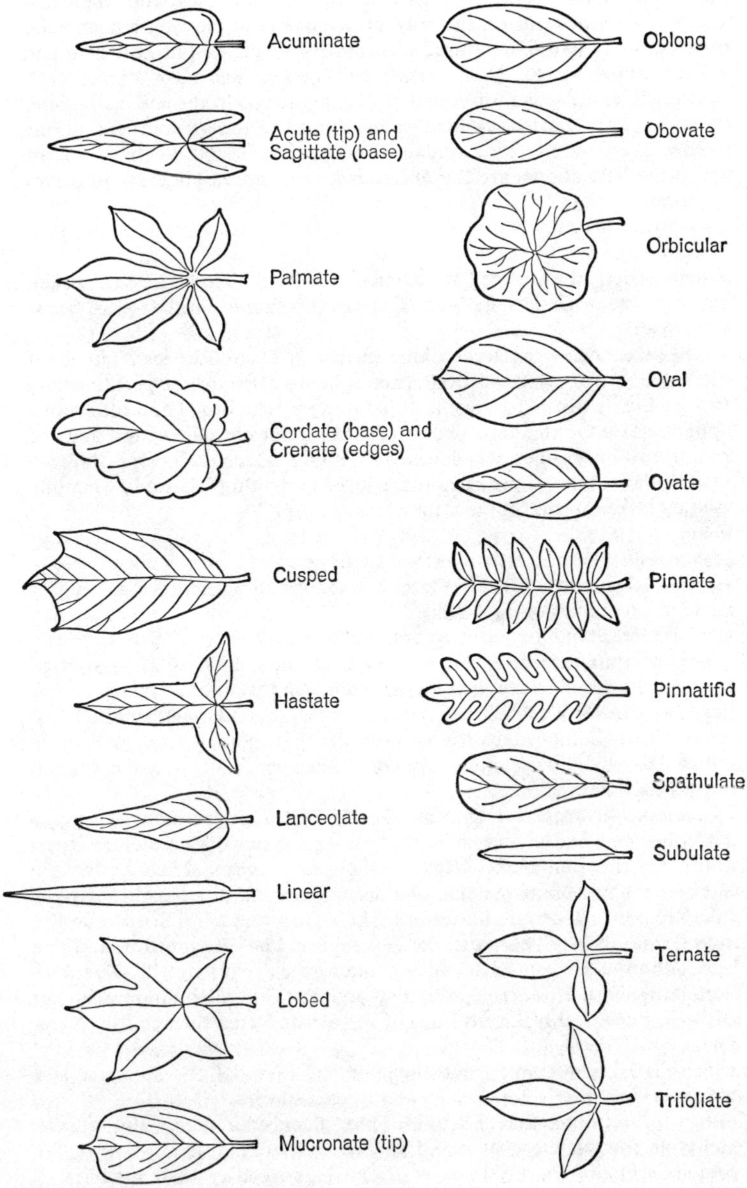
Acuminate
Oblong
Acute (tip) and
Sagittate (base)
Obovate
Orbicular
Palmate
Oval
Cordate (base) and
Crenate (edges)
Ovate
Cusped
Pinnate
Hastate
Pinnatifid
Spathulate
Lanceolate
Subulate
Linear
Ternate
Lobed
Trifoliate
Mucronate (tip)

despite its name feeds on frogs and fishes, not horses; the Medicinal Leech *Hirudo medicinalis*, the only leech capable of piercing human skin, once common and widely used in primitive medicine but now rare and almost confined to the New Forest; and *Trocheta subviridis*, a large leech whose adults leave the water and hunt earthworms in the soil, sometimes being dug up in gardens close to water. There are also a number of marine leeches of the family Piscicolidae, none of them found on the shore, of which the best known are the *Pontobdella*s often seen adhering to skates and rays.

Leek. ⇨ GARLIC.

Leeward. ⇨ LEE.

Legume. Strictly the seed pod of a member of the Pea Family (Leguminosae), but used by farmers generally for crops of this family, such as peas, beans and clovers.

Leguminosae or Papilionaceae (Angiospermae: Dicotyledones). The great Pea Family, characterized by the seeds being carried in a pod (legume) and the highly distinctive five-petalled flowers. The broad and often erect upper petal is the standard, the two narrower ones are the wings, and the two central lower ones, which cover the stamens, are the keel; p. 193.

Lek. Display ground, for ceremonial courtship fighting followed by mating, used by certain birds, notably the black grouse.

Lemma. ⇨ GRASS.

Lemon-scented Fern. ⇨ MOUNTAIN FERN.

Lenticular. Botanical adjective meaning shaped like a lentil, i.e. roughly circular but convex on each face.

Lentic Water. Standing water, as opposed to Lotic (flowing) water.

Leopard Moth (Cossoidea: *Zeuzera pyrina*). A medium-sized black-spotted grey moth, related to the goat moth; mainly in the south. The very local Reed Leopard Moth *Phragmataecoa castaneae* is found mainly in the Fens.

Leopardsbane (Compositae: *Doronicum*). Two naturalized plants with large yellow daisy-like flowers, much the commoner, especially in east Scotland, being *D. pardalianches*.

Lepidoptera (Arthropoda: Insecta). The order of insects containing the butterflies and moths and characterized by the powdery scales on their wings, which often make brightly coloured patterns. Their larvae are caterpillars, and the name chrysalis derives originally from the brightly coloured pupae of certain butterflies. There are about 2,190 British species, in two small primitive suborders, Zeugloptera and Homoneura, and one large suborder Heteroneura which contains the great majority. The division between the Macrolepidoptera and the Microlepidoptera is not scientific but a highly illogical one of custom and convenience for collectors.

Leptocephalid. The transparent and elliptical larva of the common and conger eels, formerly believed to be separate species.

Lettuce, Wild (Compositae: *Lactuca*). Three plants with small yellow dandelion-type flowers, closely related to the Garden Lettuce *L. sativa*, the commonest being Prickly Lettuce *L. serriola*, a weed of waste ground and known as Compass Plant because its prickly leaves, when in full sun, are held vertically in the north-south plane. Wall Lettuce *Mycelis muralis* is locally frequent on walls and dry banks and in beechwoods. For the unrelated Lamb's Lettuce, ⇨ CORNSALAD.

Leucism (Adj. Leucistic). Properly a form of albinism in which pigments are lost from fur or feathers, but not from the rest of the body, but sometimes wrongly applied to dilute forms; ⇨ HETEROCHROSIS.

Leveret. The young hare.

Ley. Grass sown as a crop.

Lias. The lowest strata of the Jurassic rocks, consisting of clays, marls and shales, with a few limestone bands and some sand and sandstone. Its greatest area is in the Midlands.

Liberty Cap (Agaricaceae: *Psilocybe semilanceata*). A small toadstool, with a conical shape resembling the Cap of Liberty of the French Revolution, fairly common in richer grassland.

Lice. Plural of Louse.

Lichen (pronounced ly-ken) (Lichenes). A group of lower plants consisting of a fungus (usually an ascomycete, sometimes a basidiomycete), which enfolds an alga (either green or blue-green) the two living together symbiotically, with fruit-bodies like those of the fungi. They grow on bare surfaces, such as the ground, especially on mountains, rocks, walls and tree trunks, and the great majority will not grow in smoke-polluted air in or to the leeward of the prevailing wind of large towns. Lichens take three forms: crustose (forming a thin crust), foliose (somewhat leaf-like) and fruticose (with strap-shaped or cylindrical 'stems'). Among the few species with vernacular names are Common Cup-lichen *Cladonia pyxidata*, with orange-red 'cups', which is related to Reindeer Moss, and Cudbear Lichen *Ochroelchia tartarea*, growing especially on birch trees and once used to make crimson dyes. ⇨⇨ ICELAND MOSS, OLD MAN'S BEARD, ROCK TRIPE, Tree LUNGWORT.

Lichenology. The scientific study of lichens.

Life Sciences. The subjects which comprise biology: including pure sciences such as botany, zoology, biophysics and biochemistry, and applied ones such as forestry, agriculture, medicine.

Ligging. ⇨ LAIR.

Ligulate. Botanical adjective meaning strap-shaped, not to be confused with Lingulate.

Lightning. Very high-voltage electricity generated in thunder clouds and discharging itself into the ground, often through some prominent object, such as a tree or building, which may be burned in the process. Summer lightning is the reflection on the horizon of lightning from a distant thunderstorm.

Lignicolous. Growing on or inhabiting wood.

Ligule. A small straplike piece of tissue at the base of a leaf, especially in the grasses.

Liliaceae (Angiospermae: Monocotyledones). A family containing many attractive bulbous plants, including the bluebell, lily of the valley, Solomon's seal, spring squill and fritillary.

Lily (Liliaceae). No true lilies *Lilium* are native in Britain, except perhaps the widely naturalized pink-flowered Turkscap Lily *L. martagon*, which may be native in a few woods in the south. The closely related yellow *L. pyrenaicum* is also naturalized in hedge-banks in north Devon and elsewhere. The fragrant Lily of the Valley *Convallaria majalis* is widespread in woods. May Lily *Maianthemum bifolium* is very rare in woods in northern England. Lent Lily, ⇨ DAFFODIL; Loddon Lily, ⇨ Summer SNOWFLAKE; Snowdon Lily, ⇨ SPIDERWORT.

Lime. A term strictly applicable to calcium oxide or quicklime, CaO, but often loosely used for the presence of calcium itself, most often as calcium carbonate, $CaCO_3$, in chalk or limestone rock.

Lime or Linden (Tiliaceae: *Tilia*). Tall deciduous tree with sweet-scented flowers, much affected by honey-dew. Two native species, both rather local: Small-leaved Lime *T. cordata* and Large-leaved Lime *T. platyphyllos*. Their hybrid the Common Lime *T. x europaea* is widely planted, especially in avenues, and the commonest lime generally.

Limestone. A rock with a high content of calcium carbonate; one form of it, chalk, may have up to 99 per cent. Because limestone is soluble in rain water that has absorbed carbon dioxide from the air, rain water seeping into the ground gets into the fissures in the rock and slowly dissolves it, thus forming the underground streams and caves that are a feature of limestone hills, e.g. Cheddar and Wookey in the Mendips, and river gorges such as that of the Dove in Derbyshire. The Carboniferous Limestone, found in the Mendips, Derbyshire and the Craven Pennines, is the oldest and being harder produces a more craggy landscape. The Magnesian Limestone is a comparatively soft rock found in a narrow strip northwards through Yorkshire. The Jurassic Limestones include the Oolite of the Cotswolds and Cleveland Hills. Sugar Limestone is Carboniferous Limestone that has been metamorphosed into sugar-like crystals. ⇨ CHALK.

Limestone Pavement. A rock formation found on the flat tops or terraces of bare limestone hills. The rock becomes fissured as a result of the lime being dissolved by rain water acting with carbon dioxide in the air. The fissures are called clints or grikes. Soil is blown into them, enabling plants to root, and the narrowness of the fissures provides excellent shelter for the plants in a wind-swept place. Found mainly in western Ireland (Co. Clare), south Westmorland and north-west Yorkshire (the Craven Pennines).

Limicolae. An old term for the waders, now included in the Charadriiformes. Adjective: Limicoline.

Limnology. The scientific study of fresh water, including its flora and fauna.

Limpet (Prosobranchia: Archaeogastropoda). Marine molluscs with shells shaped like a flat cone, lampshade or Chinaman's hat, and adhering tenaciously to intertidal rock surfaces. Much the commonest is the Common Limpet *Patella vulgaris;* other widespread or common species include the pretty Blue-veined or Blue-eyed Limpet *Patina pellucida*, the handsome Tortoiseshell Limpet *Acmaea tessulata*, with its paler relative the White Tortoiseshell Limpet *A. virginea*, the Slit Limpet *Emarginula* *reticulata* with a slit in its shell, and the Keyhole Limpet *Diodora apertura* with a hole at the apex of its shell. The name is also given to two of the Mesogastropoda: the Slipper Limpet *Crepidula fornicata*, a serious alien pest of oyster beds, which lives in chains of up to a dozen individuals each piled on another's back; and the Bonnet Limpet *Capulus ungaricus*, whose apex curves over like an old-fashioned bonnet. There are also two species of Freshwater Limpet with hooded elliptical shells, the Common River Limpet *Ancylastrum fluviatile* and the more local Lake Limpet *Ancylus lacustris*.

Linden. An old name for the lime tree.

Ling. (1) (Pisces: Gadidae). Name of several fishes of the Cod Family, of which the commonest in British seas is *Molva molva*, a fish of some economic importance. (2) ⇨ HEATHER.

Lingulate. Botanical adjective meaning tongue-shaped.

Linnaea (Caprifoliaceae: *Linnaea borealis*). A charming little pink-flowered undershrub, named by the great Linnaeus after himself because it was his favourite flower; now rather local in Highland pinewoods.

Linnet (Passeriformes: *Carduelis cannabina*). A small finch, the cock having a red breast in spring and summer; widespread and common in bushy places. Green Linnet ⇨ GREENFINCH. Mountain Linnet ⇨ TWITE.

Linnean. Pertaining to, or in honour of, Carl Linnaeus, the eminent eighteenth-century Swedish naturalist, who established the binominal system of naming animals and plants.

Linnean Society of London. The premier natural history society of Britain, founded in 1788; housed by the Government in Burlington House which is the repository of Linnaeus' natural history MSS. and collections, acquired by the Society from Sir J. E. Smith in 1828; publishes *Transactions*, *Proceedings*, *Journal* and *Synopses of the British Fauna*.

Lithophyte. A plant which grows on rocks, stones or other surfaces virtually devoid of soil, especially lichens and mosses.

Litter. (1) Dead leaves, twigs and other vegetable detritus lying on the surface of the soil, and so forming its topmost horizon, and destined to be transformed into humus. (2) The brood of young of certain mammals.

Little Robin (Geraniaceae: *Geranium purpureum*). A local coastal cranesbill like a small-flowered Herb Robert.

Livelong. ⇨ ORPINE.

Liver-Fluke. ⇨ FLUKE.

Liverwort (Bryophyta: Hepaticae). A class of green lower plants, with or without distinctly differentiated leaves and stems, growing on damp surfaces or in water; nearly 300 British species, none except crystalwort with vernacular names, though the *Marchantia* type are sometimes known as ribbon hepatics. Common liverworts include *Lunularia cruciata*, frequent on damp ground and brickwork in gardens; *Marchantia polymorpha*, on damp ground in gardens and by streams; *Conocephalum conicum*, whose broad, dark green, ribbon-like stems are familiar on rocks and banks by fresh water; and *Ricciocarpus natans*, one of our only two free-floating aquatic liverworts, frequent in still or slow-moving lowland eutrophic waters.

Lizard (Squamata: Sauria). Small scaly reptile; three species are native: the legless slow-worm (Anguidae) and the quadrupedal common and sand lizards (Lacertidae). The Common or Viviparous Lizard *Lacerta vivipara* is widespread and common, but rather local in Ireland, where it is the only native reptile. The Sand Lizard *L. agilis* is confined to heaths between Kent and Dorset and to sand dunes on the Lancashire coast; males in the breeding season are bright green and often confused with the European Green Lizard *L. viridis*, which has from time to time been unsuccessfully introduced into England. There is a naturalized colony of Wall Lizards *L. muralis*, also from Europe, in Surrey.

Lizard Orchid (Orchidaceae: *Himantoglossum hircinum*). A scarce but increasing orchid with an unpleasant goat-like smell, named from the supposed likeness to a lizard of its very long-lipped flowers.

Loach (Ostariophysi: Cobitidae). Small freshwater fish, closely related to the Carp Family, with elongated bodies and barbels, preferring small clear streams with gravelly bottoms. Two species: the widespread Stone Loach *Nemacheilus barbatula* and the more local Spined Loach *Cobitis taenia*, confined to England.

Loam. A soil with a good mixture (from the point of view of plant growth) of the different-sized particles of which all natural mineral soils consist, and containing all essential mineral salts.

Lobelia (Lobeliaceae: *Lobelia*). Plants with milky juice and two-lipped flowers; Heath Lobelia *L. urens*, very local on damp heaths in the south; Water Lobelia *L. dortmanna*, not infrequent in acid lakes and tarns in the the north.

Lobster (Malacostraca: Decapoda Reptantia). Large marine crustacean with two large claws. The true lobster *Homarus vulgaris*, which is blue not red when alive, is widespread but local on rocky coasts. There are five species of Squat Lobsters *Galathea*, which carry their tails well tucked under the body; the greenish-brown *G. squamifera* is the commonest, the red and blue *G. strigosa* the most strikingly coloured. Rock or Spiny Lobster, ⇨ CRAWFISH.

Lobster Moth (Notodontoidea: *Stauropus fagi*). A large Prominent, named from the curious shape of its caterpillar; locally frequent in beechwoods.

Lobworm. ⇨ LUGWORM.

Loch. A lake in Scotland. A Sea Loch is a narrow arm of the sea penetrating well inland.

Lochan. A small loch. A Dubh Lochan is a peaty lochan, with dark water.

Locust (Orthoptera: *Locusta migratoria*). A large highly gregarious grasshopper, a great plague in tropical and sub-tropical countries where it destroys vegetables and crops over wide areas; migrating in vast swarms, odd individuals of which occasionally reach Britain, but fortunately cannot breed. Grouse Locust, ⇨ GROUND-HOPPER.

Locust Tree. ⇨ ACACIA, FALSE.

Lode. A broad ditch or dyke in eastern England.

Log-Fish. ⇨ BARREL-FISH.

Loggerhead. ⇨ TURTLE.

London Pride. A hybrid between St Patrick's Cabbage *Saxifraga spathularis* and Pyrenean Saxifrage *S. umbrosa*, commonly grown in gardens, and often escaping.

Longhorn Beetle or Longicorn (Phytophaga: Cerambycidae). A family of wood-eating beetles noted for their long antennae, including the House Longhorn *Hylotrupes bajulus* and the musk beetle. The sub-family Laminae have the longest antennae, and are known as timbermen.

Longicorn. ⇨ LONGHORN BEETLE.

Longleaf (Umbelliferae: *Falcaria vulgaris*). A white-flowered umbellifer whose greyish leaves have long strap-like lobes, established here and there in waste and grassy places.

Longspur, Lapland. ⇨ Lapland BUNTING.

Looper. The caterpillar of a geometrid moth, characterized by having only two pairs of claspers or false legs at the rear. It progresses by stretching to its full length, grasping hold with its true legs in the front, and then looping up its body to bring the claspers up close to the true legs. In North America they are known as spanworms or measuring worms.

Loosestrife. Name of two common but quite unrelated plants, both growing among tall vegetation in marshes and by fresh water. Purple Loosestrife (Lythraceae: *Lythrum salicaria*) is often unaccountably confused with the quite different Rosebay Willowherb. Yellow Loosestrife (Primulaceae: *Lysimachia vulgaris*) is like several garden plants which sometimes escape. Tufted Loosestrife *L. thyrsiflora* is now very local in the north.

Lords and Ladies or Cuckoo Pint (Araceae: *Arum maculatum*). Two of the many names of a singular spring-flowering plant, with a strangely shaped inflorescence, consisting of a dull purple spike of tiny individual flowers inside a large, pointed green hood, together with often black-spotted arrow-shaped leaves, which have earned it more than sixty folk-names; widespread and common in woods and hedge-banks. Large Cuckoo Pint *A. italicum* is very local near the sea along the south coast.

Loricata. ⇨ AMPHINEURA.

Lotic Water. Flowing water, as opposed to Lentic (standing) water.

Lough. A lake in Ireland. Pronounced like the Scottish loch.

Louse (pl. Lice) (Insecta: Phthiraptera). An order of small parasitic insects, divided into the Sucking Lice (Anoplura), found only on mammals, sucking their blood, and the Biting or Feather Lice (Mallophaga), found mainly on birds with a few on mammals and living on fragments of skin, hair or feathers; 286 British species. The Human Louse *Pediculus humanus* is a sucking louse, existing in two races, the head louse *capitis* and the body louse *corporis;* ⇨ NIT. All lice are highly adapted to clinging to their hosts, being difficult to dislodge, because of their flattened forms, tough skins and big claws. Bee-lice, Book-lice, and Plant-lice (⇨ APHID and SUCKER) are unrelated. Sheep Louse, ⇨ KED.

Lousewort (Scrophulariaceae: *Pedicularis sylvatica*). Hemiparasite with pink flowers, closely related to red rattle and named from its supposed insecticidal properties; widespread and common on damp moors and heaths.

Lousy Watchman. ⇨ DOR BEETLE.

Lovage (Umbelliferae: *Ligusticum scoticum*). An umbellifer with white or greenish flowers, local on sea cliffs in Scotland and northern Ireland.

Lucerne or Alfalfa (Leguminosae: *Medicago sativa*). A large purple-flowered medick widely grown as a fodder crop, and widespread on waste ground.

Lugworm or Lobworm (Polychaeta: Arenicolidae). The tube-dwelling marine greenish- or yellowish-black bristle worms, whose casts on small mounds are so conspicuous on tidal sands at low tide; much dug for bait, especially *Arenicola marina.* Their tubes are mucus-lined burrows in the sand.

Luminescence. ⇨ BIOLUMINESCENCE.

Lump-Sucker or Lump-fish (Scleroparei: *Cyclopterus lumpus*). A dark-coloured sea fish, attaining 2 ft, characterized by an adhesive disc beneath; named Sea-Hen because, as in the stickleback, the male guards the egg-masses, which are laid on the shore above low-water mark; mainly in the north.

Lungwort (Boraginaceae: *Pulmonaria*). Blue-flowered plants with pale-spotted leaves; one scarce native, *P. longifolia,* in clayey woods in the New Forest, Dorset and Isle of Wight; one frequent garden escape, *P. officinalis.* Tree Lungwort *Lobaria pulmonaria* is a large lichen, up to 6 ins. high, growing on old trees and rocks in the north and west.

Lupin (Leguminosae: *Lupinus*). Two introduced species: the shrubby Tree Lupin *L. arboreus,* with yellow flowers, naturalized in sandy places, mainly on the coast; and the Scottish Lupin *L. nootkatensis,* with blue-purple flowers, locally naturalized on moors and river shingle in north Scotland.

Lurg. ⇨ RAGWORM.

Lusitanian Fauna and Flora. A term, derived from the Roman name for Portugal, applied to animals and plants found in the British Isles only in the warm wet climate of the west and south-west coasts of Great Britain and Ireland, and again on the west coasts of France, Spain and Portugal; e.g. the strawberry tree, the Kerry slug *Geomalacus maculosus,* pale butterwort and Mediterranean heath.

Luth. ⇨ Leathery TURTLE.

Lycopsida (Pteridophyta). A class of fern allies containing the Clubmosses (Lycopodiales and Selaginellales) and the Quillworts (Isoetales).

Lyme Grass (Gramineae: *Elymus arenarius*). A frequent grass of coastal sand dunes, with greyish leaves.

Lynchet. A low grassy bank bounding the so-called 'Celtic fields' of the prehistoric and Roman periods, the remains of the squarish arable fields ploughed both along and across hillsides by the primitive ard plough.

M

Machair. Natural grassland on calcareous (shell) sand behind dunes on the west coasts of Scotland and the Hebrides. Colonized by many lime-loving plants and abounding in snails, it is preserved by controlled grazing of the crofters' animals, which also add organic matter. In many coastal settlements on the islands it is the basis of human subsistence.

Mackerel (Percomorphi: *Scomber scombrus*). A gregarious surface-living marine fish, common off our western coasts in summer and early autumn. The related Spanish Mackerel *Pneumatophorus colias* is much less common, mainly in Cornish seas. The unrelated Horse Mackerel or Scad *Trachurus trachurus* is also mainly a western fish.

Mackerel Sky. Cirro-cumulus cloud. ⇨ CLOUD.

Macrolepidoptera. An unscientific division of the Lepidoptera much favoured by collectors, and consisting of the butterflies, the larger moths and a highly arbitrary selection of the smaller moths: the Homoneura and all or part of nine of the twelve superfamilies of the Heteroneura. ⇨⇨ MICROLEPIDOPTERA.

Madder, Wild (Rubiaceae: *Rubia peregrina*). A bedstraw-like plant with yellow-green flowers, straggling over other vegetation and found near the coast from Kent to Cornwall and thence to North Wales; madder dye comes from the closely related *R. tinctorum*, no longer grown as a crop in Britain.

Madrepore (Anthozoa: Scleractinia). A hard or true coral, in which the zooid is encased in a hard calcareous exoskeleton but may expand beyond it to look like a sea-anemone. Two very handsome species occur, the pink and white Devonshire Cup Coral *Caryophyllia smithii* and the rare Scarlet and Gold Star Coral *Balanophyllia regia*.

Maggot. Larva of many two-winged flies, especially bluebottles and house flies; it is legless, progressing by wriggling or squirming, and its head is reduced to minute proportions; breathes through pores at the tail end. Rat-tailed Maggots are the larvae of certain hover-flies, living in very stagnant water or semi-liquid rotting vegetation, and named from their long respiratory siphons.

Magpie (Passeriformes: *Pica pica*). A strikingly piebald, long-tailed relative of the crows, making a large domed nest; widespread and common.

Magpie Moth (Notodontoidea: *Abraxas grossulariata*). A handsome black, white and yellow geometrid moth, whose looper caterpillars of the same colour feed on currant and gooseberry bushes, among other plants, and may even assume pest proportions. The Small Magpie Moth (Pyraloidea: *Eurrhypara hortulata*) is one of our larger pyralids, whose larvae spin up in nettle leaves.

Maidenhair Fern (Adiantaceae: *Adiantum capillus-veneris*). A scarce fern of damp rocks in the west, with fan-shaped leaflets like those of the allied species often seen as pot plants.

Malacology. The study of molluscs.

Malacostraca (Arthropoda: Crustacea). The sub-class of crustaceans which contains the Mysidacea (opossum shrimps), Cumacea, Isopoda (woodlice), Amphipoda (freshwater shrimps and sandhoppers), Euphausiacea (krill), and Decapoda (crabs, lobsters, shrimps, prawns and crayfish).

Male Fern (Aspidiaceae: *Dryopteris filix-mas*). The commonest of our larger tufted ferns, with coarser and less graceful fronds than lady fern; widespread in woods, on hedge-banks and among rocks. *D. borreri* is a form with numerous golden-brown scales on the stalks of its fronds.

Mallard (Anatidae: *Anas platyrhynchos*). A large surface-feeding or dabbling duck, the commonest British duck and the origin of most domestic ducks, widespread both in and out of the breeding season; the handsome drake has a bottle-green head.

Mallophaga (Insecta: Phthiraptera). The sub-order containing the biting lice.

Mallow (Malvaceae). Rather showy wild flowers, Common Mallow *Malva sylvestris*, named Cheesecakes from its flat fruits, being common on roadsides, and Musk Mallow *M. moschata* in woods and scrub. Marsh Mallow *Althaea officinalis*, whose roots provide the sweetmeat, and the sub-shrubby Tree Mallow *Lavatera arborea* are both local near the sea in the south.

Mammal (Vertebrata: Mammalia). The most advanced group of backboned animals, characterized by their superior intelligence, the hair on their bodies, and the nourishment of their young on milk secreted by the females in the mammary glands which give them their name. Nine orders are represented in the British fauna: Insectivora (mole, hedgehog, shrews); Chiroptera (bats); Primates (man); Lagomorpha (rabbit, hares); Rodentia (rats, mice, voles, squirrels); Cetacea (whales, porpoises, dolphins); Carnivora (fox, badger, otter, wild cat, weasel, stoat, pine marten, polecat); Pinnipedia (seals); and Artiodactyla (deer). There are fifty-three resident land mammals in the British Isles, of which thirteen are established aliens, and in British seas some twenty-nine mammals, mostly cetaceans, have been recorded.

Mammalogy. The scientific study of mammals.

Mammal Society. The national society for the study of mammals, especially those of the British Isles and British seas, founded in 1954 as the Mammal Society of the British Isles; publishes a *Bulletin*.

Mange. Disease of mammals, often caused by mites, as red mange of dogs by *Demodex folliculorum* (Prostigmata), sarcoptic mange in domestic animals produced by the itch mite, mange in cats by *Notoedres cati*, and psoroptic mange in domestic animals, known as sheep scab in sheep, by *Psoroptes equi* (Astigmata).

Mangold Fly. ⇨ ONION FLY.

Man Orchid (Orchidaceae: *Aceras anthropophorum*). A local orchid of calcareous turf in the south, named from the manikin-like lip of its yellow

Typical manikin lip of an orchid

flowers, which is four-lobed to make 'arms and legs'; ⇨ LADY, MILITARY and MONKEY ORCHIDS.

Maple or Field Maple (Aceraceae: *Acer campestre*). A usually small deciduous tree or hedgerow shrub, with palmate leaves; common in England and Wales.

March Brown. Anglers' name for both duns and spinners of the mayfly *Rhithrogena haarupi*, appearing in late March. The dun of *Ecdyonurus venosus* is the Late March Brown, appearing in April.

Marestail (Hippuridaceae: *Hippuris vulgaris*). An aquatic flowering plant, with leafy spikes projecting from the water; often confused with the non-flowering horsetails.

Marestail Cloud. Cirrus; ⇨ CLOUD.

Marigold. A name for yellow daisy-like plants, including Bur Marigolds; Corn Marigold *Chrysanthemum segetum*, a decreasing but still locally common cornfield weed; and the orange-flowered Pot Marigold *Calendula officinalis*, a frequent garden escape. Marsh Marigold or Kingcup (Ranunculaceae: *Caltha palustris*), is a marsh plant resembling a large stout buttercup, with similar bright shiny yellow flowers.

Marine Biological Association of the United Kingdom. The national body for the study of marine biology, founded in 1884; runs a Laboratory at Plymouth, Devon, with an aquarium that is open to the public; publishes a *Journal*.

Marine Bug (Heteroptera: *Aepophilus bonnairei*). A shorebug that has become adapted to life around low water mark, where it lives in tiny rock crevices.

Marjoram or Sweet Marjoram (Labiatae: *Origanum vulgare*). Thyme-scented aromatic plant with pinkish-purple flowers, common in calcareous grassland, used as a herb.

Marking Schemes. The marking of individual animals with small light metal rings or tags, serially numbered, is an increasingly valuable tool for biological field research. The best known scheme in Britain is the one for Bird Ringing, but there are also schemes for marking bats, grey seals and various fish of commercial value. Many research schemes also involve the marking of animals, including invertebrates, with dabs of colour, or even of radioactive matter.

Marl. A clay soil with a high lime content.

Marram (Gramineae: *Ammophila arenaria*). A common xeromorphic grass of sand dunes all round the coast, with far-creeping rhizomes and sharply pointed leaves; frequently planted to fix the dunes and prevent them blowing away.

Marsh. Ground that is waterlogged, the summer water level being normally at or near the surface. The soil has an inorganic (i.e. mineral) basis, usually silt or clay, in contrast to Fen which is organic (peat).

Marsh Fern (Thelypteridaceae: *Thelypteris palustris*). A local fern of marshes and fens, not tufted but producing its pale green fronds at intervals along the rhizome.

Marsh Fly (Cyclorrhapha: Sciomyzidae). A family of smallish two-winged flies, frequenting damp or marshy places and mostly with aquatic larvae.

Marsh Gas. Gases, mainly methane (CH_4), produced during the decomposition of organic matter by anaerobic bacteria in muddy places, and rising to the surface in bubbles, which may become spontaneously ignited and give rise to the folklore of 'will o' the wisp' and 'jack o' lantern'.

Marsh Orchid (Orchidaceae: *Dactylorhiza*). A complex group of pinkish-purple orchids, related to the spotted orchids, and growing in damp meadows, marshes and fens: the three commonest are Early Marsh Orchid *D. incarnata*, Southern Marsh Orchid *D. praetermissa* and Northern Marsh Orchid *D. purpurella*.

Marsh Snail (Pulmonata: *Limnaea palustris*). A widespread and common pond snail of marshes, freshwater margins and shallow fresh water.

Marsh Worm or Red Worm (Oligochaeta: *Lumbricus rubellus*). A reddish-brown earthworm with a pale belly, very common in damp, humus-rich soil, in such places as gardens, parks and agricultural grassland. The chestnut worm is also called marsh worm by anglers.

Marshwort (Umbelliferae: *Apium inundatum*). A small white-flowered aquatic umbellifer related to celery and fools' watercress.

Marsipobranchii (Chordata: Vertebrata). A class of primitive fishes whose only living representatives are the Cyclostomes (lampreys, etc.).

Marten, Pine (Carnivora: *Martes martes*). The rarest of our eight native carnivores, formerly widespread but now confined to remote areas of the north and west, especially in the Highlands. It somewhat resembles a large stoat.

Martin (Passeriformes: Hirundinidae). Small swallow-like birds with short forked tails, both British species being summer visitors, and feeding, like swallows, on the wing. The House Martin *Delichon urbica* has blue-black upperparts with a white rump and nests under leaves outside buildings. The Sand Martin *Riparia riparia* has sandy brown upperparts and breast-band, nests in tunnels which it excavates in sandy and gravelly banks, and is one of the earliest summer migrants to arrive, often in late March.

Martin Bug (Heteroptera: *Oeciacus hirundinis*). A predatory bug, closely related to the bedbug, whose principal host is the house martin.

Masked Bug. ⇨ FLYBUG.

Mason Wasp (Aculeata: Vespidae). A group of small black and yellow solitary wasps named from the habit of some of them, e.g. *Odyneius* *spinipes* and *O. melanocephalus*, of building around the entrance to its burrow a small hollow tower of pellets excavated from inside; some make their nests in hollow stems and other existing holes.

Masterwort (Umbelliferae: *Peucedanum ostruthium*). A stout white-flowered umbellifer, formerly used as a herb, and still established on roadsides in the north.

Mastigophora. ⇨ FLAGELLATE.

Mat Grass (Gramineae: *Nardus stricta*). A common grass of infertile moors and heaths, of little nutritional value to sheep.

Mavis. Old name for the song thrush.

May Blossom. Country name for the flowers of the hawthorn, which blooms in May.

Maybug. ⇨ COCKCHAFER.

Mayfly (Insecta: Ephemeroptera). An order of aquatic winged insects, much eaten by fish and imitated by flyfishers. After hatching, they become first nymphs, then duns and finally spinners; nymphs are aquatic, the other two aerial in habit. The spinners cannot eat, so rarely live for more than a day or two. Though mayflies start to appear in May, they can be found throughout the summer; about forty-seven species, living in and by both still and moving fresh water. In the north of England large stone-flies are called mayflies.

Mayweed (Compositae). Plants with a white and yellow daisy-like flower: Scentless Mayweed *Tripleurospermum maritimum*, a common weed of cultivated ground and shingle by the sea; and Stinking Mayweed *Anthemis cotula*, a less common aromatic weed with a sickly sweet scent.

Meadow. Strictly a field of permanent grass used for hay, but also generally used for rich waterside grass fields that are grazed. In the nursery rhyme Little Boy Blue was at fault for allowing the sheep to stray into the meadow, which was being reserved for hay. A Water Meadow is regularly inundated with water from a river, either naturally or through sluices; found especially in Hampshire and other parts of southern England.

Meadow Grass (Gramineae: *Poa*). A genus containing some of our commonest grasses, including the ubiquitous weed Annual Meadow Grass *P. annua*, the common *P. pratensis* and *P. trivialis*, and in woods the widespread *P. nemoralis*. The name is also sometimes given to some allied grasses of the genus *Puccinellia*, which grow by the sea.

Meadow-Rue (Ranunculaceae: *Thalictrum*). Three locally frequent species with much divided leaves and clusters of small flowers consisting mainly of yellow stamens; Common Meadow-rue *T. flavum* in fens and damp grassland; Lesser Meadow-rue *T. minus*, very variable, on mountains, limestone rocks, dunes and freshwater shingle and in sandy and calcareous grassland; Alpine Meadow-rue *T. alpinum* on mountains in the Highlands, scarce elsewhere.

Meadow Saffron (Liliaceae: *Colchicum autumnale*). A pale mauve crocus-like wild flower growing in damp woods and meadows, mainly in the Cotswolds and the west of England; the leaves appear in spring and die before the flowers emerge in late August on long thin white tubes that have the country name of Naked Ladies. It yields colchicine, a drug of increasing importance in genetic studies because it induces chromosome doubling. Named from its resemblance to the cultivated Saffron Crocus *Crocus sativus*, which produces the saffron used for colouring and flavouring in cookery.

Meadowsweet (Rosaceae: *Filipendula ulmaria*). A fairly tall, handsome plant with fragrant creamy-white flowers, widespread and common in marshes and damp woods and grassland.

Meal Moth (Pyraloidea: *Pyralis farinalis*). A small pyralid moth whose larvae feed on stored grain, chaff etc., and occur in mills, barns, stables and warehouses, especially heated ones. The larvae of the Indian Meal Moth *Plodia interpunctella*, another small but not very closely related pyralid, feed on stored products of many kinds, and are often abundant in warehouses, also especially heated ones.

Mealworm. The larva of one of the hardback beetles, pests of flour; widely bred and used for feeding insectivorous birds and mammals.

Mealy-Bug. ⇨ SCALE INSECT.

Mecoptera. ⇨ SCORPION-FLY.

Medick (Leguminosae: *Medicago*). Small peaflowers, with trefoil leaves and heads of mostly yellow flowers, the commonest and most widespread being Black Medick or Nonsuch *M. lupulina*, which closely resembles Common Yellow Trefoil, and Spotted Medick or Calvary Clover *M. arabica*, with a spot on each leaflet.

Medlar (Rosaceae: *Mespilus germanica*). Small deciduous fruit tree, occasional in hedgerows bird-sown from gardens.

Medusa. The free-swimming stage of the marine animals formerly included in the Coelenterata, sometimes, as with the jellyfish, occupying the main part of the life history. The only freshwater medusae are the Limnomedusae; ⇨ HYDROZOA.

Megaloptera (Insecta: Neuroptera). The sub-order containing the alder flies (Sialidae) and snake flies (Raphididae).

Megaphanerophyte. ⇨ PHANEROPHYTE.

Megrim or Sail-fluke (Bothidae: *Lepidorhombus*). Two not uncommon left-sided deep-water flatfish, somewhat resembling the sole in shape, though not closely related to it. *L. whiff-iagonis* is much the commoner.

Melanism (adj. Melanic). A colour variation of animals due to the excessive presence of melanin pigments, making the animal look very dark, sometimes black. ⇨ INDUSTRIAL MELANISM.

Melick (Gramineae: *Melica*). Two graceful grasses, with roundish individual flower spikelets, the common woodland *M. uniflora* and the more local *M. nutans* of limestone districts.

Melilot (Leguminosae: *Melilotus*). Four species of peaflower with long spike-like heads of small yellow flowers (white in the frequent weed White Melilot or Bokhara Clover *M. alba*); the two commonest are Common Melilot *M. officinalis* and Field or Golden Melilot *M. altissima*, both mainly on waste ground.

Mendelism. An early name for genetics, which originated in the researches on the inheritance of characters in sweet peas of the Abbé Gregor Mendel of Brno in Moravia, published in 1865 but not discovered by the scientific world till 1900, sixteen years after his death.

Mercury. ⇨ DOG'S MERCURY.

Mere. A shallow lake or pond, especially in Cheshire and Shropshire, where the meres are a noted haunt of wildfowl.

Merganser (Anatidae: *Mergus*). Diving ducks with narrow serrated bills, known as saw-bills. Four British species; the Goosander; the Red-breasted Merganser *M. serrator*, breeding in Scotland and extreme northern England, and widespread round our coasts in winter; the Smew; and the Hooded Merganser *M. cucullatus*, a rare straggler from North America.

Merlin (Falconiformes: *Falco columbarius*). A small falcon, the smallest British diurnal bird of prey. Widespread and frequent in hill districts in the north and west, moving to coastal marshes in winter. Known as Pigeon Hawk in North America.

Merse. A coastal marsh in south-west Scotland.

Mesembryanthemum (Ficoidaceae: *Carpobrotus edulis*). A pink or yellow flowered succulent well established on cliffs in the south-west; also known as Hottentot Fig or Kaffir Fig.

Mesophanerophyte. ⇨ PHANEROPHYTE.

Mesostigmata (Arachnida: Acari). An order or sub-order of mites, mostly free-living in the soil or decaying organic matter, but including some parasites of both vertebrates and invertebrates. ⇨ FOWL MITE, RAT MITE, SOIL MITE.

Mesozoic Era. ⇨ GEOLOGICAL PERIODS.

Metamorphic Rock. Igneous or sedimentary rock which has been changed by the action of heat, pressure or water into a quite different type of rock, as granite is changed to gneiss and shale into slate.

Metamorphosis. The change in both form and structure which occurs when the larval form of an animal changes into the adult one, e.g. a tadpole into a frog or toad, or a caterpillar into a chrysalis and a perfect insect.

Metazoa. The sub-kingdom containing all multicellular animals; i.e. the whole Animal Kingdom except the uni- or non-cellular Protozoa; the sponges, which though multicellular are very primitive, are also usually placed in a separate sub-kingdom.

Meteorology. The scientific study of the weather.

Methane. ⇨ MARSH GAS.

Mezereon (Thymelaeaceae: *Daphne mezereum*). Deciduous undershrub with highly fragrant pink flowers appearing before the leaves in early spring; now very rare in calcareous woods in the south, due to being often dug up for cottage gardens.

Mice. Plural of mouse.

Micro. Vernacular abbreviation of the Microlepidoptera.

Microclimate. The special climate of a strictly limited habitat where some particular factor or factors differentiate it from its surroundings. The area may be extremely small, e.g. the crevices of bark on a tree, under an overhanging rock (sheltered from rain, frost, sun), or in the angle of a wall giving shelter from cold winds); or quite large, e.g. a valley which acts as a frost trap.

Microlepidoptera. An unscientific division of the Lepidoptera used by collectors, and consisting of an arbitrary selection of the smaller moths; the Zeugloptera and all or part of five superfamilies of the Heteroneura.

Microphanerophyte. ⇨ PHANEROPHYTE.

Microspecies. A species separated from other, very closely related species or microspecies by rather small differences, but not a geographical race or subspecies.

Microscopy. The study of minute animals and plants by means of a microscope; normal medium-power microscopes magnify up to about 400 times, but the process of oil immersion produces magnifications of 1,000 times or more, and the electron microscope up to about 100,000.

Midge (Diptera: Nematocera). Several families of small two-winged flies: Chironomid or Non-biting Midges (Chironomidae), like small gnats, the males often dancing in swarms, the aquatic larvae known as bloodworms, 370 species; Biting Midges (Ceratopogonidae), over 130 species, of which the genus *Culicoides* includes the human bloodsuckers, which are such a pest on summer evenings, especially in Scotland; Gall Midges (Ceciomyiidae), over 600 species, whose larvae create Galls on their food-plants, including three pests of wheat, the Hessian Fly and the Wheat Midges *Contarinia tritici* and *Sitodiplosis mosellana*; and Owl Midges or Hairy Moth Flies (Psychodidae), about seventy species.

Mignonette, Wild (Resedaceae: *Reseda lutea*). Green-flowered plant somewhat resembling garden mignonette, common in bare and grassy places on calcareous soils in the south and east; two other species are casuals. ⇨⇨ WELD.

Migration. Regular movements of animals, usually between breeding places and winter quarters, occurring cyclically or annually through the animal's life, as with swallows or cuckoos between their British breeding places and their winter quarters in Africa, or with salmon and sea trout between their spawning places in fresh water and their recuperating grounds in the sea. Some of these journeys may be performed once only, as with eels migrating from the Sargasso Sea, coming to maturity in our rivers and ponds and returning to the ocean to spawn (probably) and die; or with painted lady butterflies migrating here from the Mediterranean in spring, some of which appear to return in autumn. Migration in some form or other is found in many groups of animals, though especially in mammals, birds, fishes and insects, but needs to be distinguished from the once-and-for-all dispersal movement of such insects as aphids. Much of our recent information about the geography of bird migration has been acquired as a result of ringing; ⇨ BIRD RINGING, IRRUPTION.

Mildew. Two groups of small fungi mostly parasitic on the higher plants. The Downy Mildews (Phycomycetes: Peronosporaceae) penetrate deeply into their plant hosts, and include the Onion Mildew *Peronospora destructor*. The True or Powdery Mildews (Ascomycetes: Erysiphyceae) are more like moulds, living on the surface of their hosts, and often forming a kind of white meal, such as the well-known Oak Mildew *Microsphaera quercina*, abundant on oak seedlings in late summer and autumn.

Military or Soldier Orchid (Orchidaceae: *Orchis militaris*). A very rare orchid whose pinkish-purple flowers have a manikin lip like the man orchid (⇨ p. 161), found only in two places on the chalk.

Milk Cap (Basidiomycetes: *Lactarius*). A group of toadstools named from the milky juice they exude when broken, especially the reddish edible Saffron Milk Cap *L. deliciosus*, with orange milk, growing in coniferous woods, and the pinkish inedible Woolly Milk Cap *L. torminosus*, with white milk and a woolly surface on the cap.

Milkweed Butterfly (Papilionina: *Danaus plexippus*). A large, handsome and uncommon vagrant from North America, where it is known as the Monarch Butterfly and noted for its long-distance migrations. Many of the British specimens, however, have probably crossed the Atlantic on ships. The food-plant is the American milkweed *Asclepias*, not the European *Polygala*. Also called Black-veined Brown.

Milkwort (Polygalaceae: *Polygala*). Four low-growing plants with blue, pinkish-purple or white two-lipped flowers, the two commonest and most widespread being Common Milkwort *P. vulgaris* in most kinds of drier grassland, and Heath Milkwort *P. serpyllifolia* on heaths, moors and grassland on acid soils.

Miller. (1) (Basidiomycetes: *Clitopilus prunulus*). A common edible white woodland fungus, named from its strong smell of new meal. (2) A local name for noctuid moths, and also applied specifically to the noctuid *Apatele leporina*.

Miller's Thumb. The only freshwater Bullhead.

Millet (Gramineae). The name of several grasses used in birdseed mixture and often seen on waste ground, especially Common Millet *Panicum miliaceum* and Foxtail Millet *Setaria italica*. Wood Millet *Milium effusum* is a locally common native grass of woodlands.

Millipede (Arthropoda: Diplopoda). A class of myriapods, living in the soil, with two pairs of legs on most segments of the body: forty-four British species, including the Black Millipede *Tachypodoiulus niger*, common in calcareous soils; the Pill Millipede *Glomeris marginata*, looking like an elongated woodlouse; and the Spotted Snake Millipede *Blaniulus guttulatus*, an agricultural pest.

Mill Moth. ⇨ FLOUR MOTH.

Millstone Grit. A coarse tough sandstone, found especially in the West Riding of Yorkshire, that made good grindstones for flour mills and for the cutlery industry, hence its name.

Mimetic. Adjective from Mimicry.

Mimicry. (1) Superficial similarity of one species to another, especially in insects. In Batesian Mimicry a relatively defenceless species is protected from its predators by mimicking another species which is poisonous or distasteful or deters predators in some other way, and is often warningly coloured, e.g. the bee hawk-moths and clearwing moths resemble various aggressive hymenopterous insects. Müllerian Mimicry exists when both species are equipped with some form of deterrent to predators, but gain double protection by resembling each other. (2) Imitation of one bird by another; the songs of some birds such as the marsh, reed and sedge warblers, are largely mimetic; jays and starlings are also highly mimetic.

Mimic Wasp. ⇨ BLACK WASP.

Mind Your Own Business or Mother of Thousands (Urticaceae: *Helxine soleirolii*). A tender mat-forming plant with numerous small flowers and leaves, a garden escape, mainly in the south-west.

Mining Bee (Aculeata: Apidae). A group of solitary bees, which burrow in dry earth to make their nests, often in loose aggregations of varying size, especially the genera *Andrena* and *Halictus* and the Flower Bees *Anthophora*. *Andrena* bees are parasitized by homeless bees of the genus *Nomada*. A few *Halictus* bees have developed sub-social habits, with a distinct form of female, which never mates, doing most of the work.

Mink (Carnivora: *Mustela vison*). A North American fur-bearing relative of the stoat that escaped from fur-farms and has established itself on some rivers in Devon, Hampshire and various other parts of Britain since 1953. It is a serious threat to stocks of wildfowl, and perhaps also to fish stocks.

Minnow (Cyprinidae: *Phoxinus phoxinus*). The smallest member of the Carp Family, widespread and common in brooks and streams; a favourite bait for anglers.

Mint (Labiatae: *Mentha*). Sweetly scented aromatic plants with whorls, heads or spikes of small lilac flowers: four native and three introduced species, and numerous hybrids, the two commonest being Water Mint *M. aquatica* by fresh water and *M. arvensis* on bare ground. ⇨ PENNYROYAL, PEPPERMINT, SPEARMINT.

Miocene Period. ⇨ GEOLOGICAL PERIODS.

Mire. A marsh, especially in the uplands and in Yorkshire.

Mist. Low-density fog.

Mistletoe (Loranthaceae: *Viscum album*). Our only woody plant that grows parasitically on trees, producing glutinous white berries much relished by birds, especially the mistle thrush; it has chlorophyll and is therefore only a hemiparasite. Often grown in orchards for sale on the Christmas market.

Mite (Arachnida: Acari). Those members of the sub-class or order Acari, which are not called Ticks; Mesostigmata, Cryptostigmata, Astigmata and Prostigmata. They occupy a very wide range of habitats, including the soil, decaying organic matter, debris on the sea shore and stored products, and can be both internal and external parasites of both vertebrates and invertebrates. ⇨⇨ SOIL MITE.

Mole (Insectivora: *Talpa europaea*). A widespread small mammal, which burrows underground in search of its prey, earthworms, throwing up 'molehills' at intervals; has poor eyesight, but is not blind.

Mollusc (Mollusca). The second largest phylum of higher animals, after the arthropods, generally characterized by the possession of a hard Shell and a tough muscular organ, the foot, but having no standard shape, varying in appearance as widely as an octopus, a razor shell and a garden snail.

They occupy almost every possible habitat in the sea, and a wide range of fresh and brackish water and terrestrial habitats as well. Five of the six living classes are represented in Britain; the Amphineura (chitons or coat-of-mail shells), Gastropoda (land snails, slugs, sea slugs, and univalve marine shells) Scaphopoda (tusk shells), Lamellibranchia (bivalves) and Cephalopoda (squids, octopuses). The British species total about 760.

Molluscicide. A pesticide for killing molluscs.

Monarch Butterfly. ⇨ MILKWEED BUTTERFLY.

Money Spider (Araneae: Liniphiidae). A family of some 250 tiny spiders, existing in enormous quantities – over 1,000,000 per acre have been estimated in a grass field in Sussex in autumn – and particularly conspicuous in autumn, when their innumerable webs catch the dew and glisten in the early morning sun, and when they themselves take to the air and drift considerable distances on their silken ropes of gossamer.

Moneywort. ⇨ CORNISH MONEYWORT.

Monkey Flower (Scrophulariaceae: *Mimulus*). Waterside plants with large yellow flowers, related to musk, now widespread as garden escapes; *M. guttatus* is the commonest species.

Monkey Orchid (Orchidaceae: *Orchis simia*). A miniature version of the military orchid, found in a very few places on the chalk in Kent and the Chilterns.

Monkfish or Angel Fish (Pleurotremata: *Squatina squatina*). A curious primitive fish, intermediate between the sharks and rays, with shark-like tail, flattened back-fins like those of a ray and a squarish head with the hooded or cowled appearance of a monk, or, to the French, who call it *l'ange de mer*, an angel; not infrequent offshore, especially in spring. Monk fish is also a name of the Angler.

Monkshood or Wolfsbane (Ranunculaceae: *Aconitum anglicum*). Handsome blue wild flower, named from the hooded shape of the flowers, closely resembling the garden plant, local by streams in south-west England and South Wales; contains a deadly poison, aconitin.

Monocotyledon (Angiospermae: Monocotyledones). The division of the flowering plants whose seedlings have only one seed-leaf (cotyledon), and flower-parts nearly always in multiples of three; leaves usually narrow and unstalked, often parallel-sided and nearly always parallel-veined. Our only woody species is butcher's broom. The principal monocotyledonous families are the Liliaceae (lilies), Juncaceae (rushes), Cyperaceae (sedges), Orchidaceae (orchids), Potamogetonaceae (pondweeds) and Gramineae (grasses).

Monoecious. A plant that has separate male and female organs on the same individual; e.g. hazel, with dangling yellow male catkins and erect bud-like red female flowers. In animals, a Hermaphrodite.

Monomorphic. An organism that is not polymorphic.

Monophagous. Of an animal which eats only one kind of food.

Monotrysia. Alternative name for the Homoneura.

Montane Habitats. Habitats on mountains, i.e. above 2,000 ft.

Montbretia (Iridaceae: *Crocosmia x crocosmiiflora*). A well known garden plant with orange flowers, widely established on cliffs and banks, especially in the west.

Moon-Face Observation. A technique for watching bird migration; birds flying at night can be watched through binoculars and counted passing across the face of the moon.

Moon-Fish. ⇨ OPAH.

Moonwort (Pteridophyta: *Botrychium lunaria*). A curious aberrant little fern, allied to adderstongue, in which the spores are borne on a branched spike arising from the middle of the pinnate frond, whose segments are moon- or fan-shaped; widespread but increasingly local in dry grassland.

Moor. Open, usually high ground, with acid peaty soil, typically dominated by heather (*Calluna* and *Erica*) and allied undershrubs, or by such unnutritious grasses or sedges as purple moor grass *Molinia*, mat grass *Nardus*, and deer sedge *Scirpus caespitosus*. ⇨ WHITE MOOR. The red grouse is the typical bird of heather moors. At one time the word was also used for bogs and swamps, e.g. Moorfields in London. The moorhen is named from this sense of moor.

Moorcock, Moorfowl, Moorgame. ⇨ Red GROUSE.

Moor Grass, Purple Moor Grass or Flying Bent (Gramineae: *Molinia caerulea*). A tussock-forming grass, with purplish flower spikes, common on damp moors and heaths, also in fens and marshes. Blue Moor Grass or Blue Mountain Grass *Sesleria caerulea* is an early flowering grass of limestone districts in the north and in western Ireland.

Moorhen or Waterhen (Ralliformes: *Gallinula chloropus*). Widespread and common water bird with red forehead and conspicuous white under tail coverts, often frequenting quite small ponds.

Mor or Raw Humus. The humus found in poor soils unfavourable to animal and plant life, where organic material decays slowly and sometimes incompletely, e.g. a heather moor. There are no earthworms, unlike Mull. ⇨⇨ HUMUS.

Moraine. A deposit of ice-worn pieces of rock brought down by the glaciers of the Ice Ages and left behind when the ice melted.

Moray. ⇨ MURRY.

Morel (Ascomycetes: *Morchella*). A group of edible fungi with oval caps reticulated like a honeycomb, especially *M. esculenta*, common in the south in spring, on banks and open ground, and *M. vulgaris*, in woods. False Morels *Helvella*, with concave caps, can also be eaten by most people, but give some individuals indigestion.

Morphology. The study of the shape, form or appearance of animals and plants, used especially in the more elementary systems of classification.

Moschatel or Town-hall Clock (Adoxaceae: *Adoxa moschatellina*). A widespread and common small woodland plant, named from its alleged musky smell and the arrangement of its five small greenish flowers at right angles to each other, like a four-sided public clock with the fifth on top.

Mosquito (Nematocera: Culicinae). Large gnats, with aquatic larvae and pupae, both of which are active; some species bite and suck the blood of mammals, including man. The commonest, which fortunately rarely bites man, is the Common Gnat *Culex pipiens*; it often hibernates in houses and its larvae are commonly seen in stagnant pools, rainwater butts, etc. It gets a bad name from the activities both of a very similar species, *Culex molestus*, and of the larger, dark-spotted *Theobaldia annulata*, which may also hibernate in houses. Another biter is *Aëdes punctor*, which frequents acid pools. A brackish-water form of the biting *Anopheles maculipennis* used to carry ague, our now extinct native form of malaria, and is still capable of transmitting tropical malaria if it encounters a human carrier.

Moss. (1) The northern name for bog. (2) (Bryophyta: Musci). A class of green lower plants, always with distinctly differentiated leaves and stems, growing mainly on damp surfaces; over 600 British species. Few, apart from the Bog Mosses and Hair Mosses, have vernacular names, and most plant names containing 'moss' refer either to flowering plants, e.g. Dovedale Moss, or to other lower plants, e.g. Reindeer Moss, which is a lichen, and Irish and Carragheen Mosses, which are seaweeds. Such names as Bristle Moss for *Orthotrichum*, Fork Moss for *Dicranum* and Thread Moss for *Bryum*, are little more than book names. Mosses occupy a fairly wide range of habitats, though the majority grow on moist or damp surfaces. Some of the commoner species include *Rhacomitrium lanuginosum*, often dominant on mountain tops; the widespread and variable *Hypnum cupressiforme*; the woodland carpeter *Mnium hornum*; the aquatic *Fontinalis antipyretica*; *Bryum argenteum*, which makes silvery cushions at the edges of town pavements; *Tortula muralis*, a common wall moss even in cities; *Leucobryum glaucum*, whose pale grey-green tufts sometimes become balls of moss on the woodland floor; and *Grimmia maritima*, which grows on rocks by the sea.

Mother Carey's Chicken. Seamen's name for small petrels of the storm-petrel group.

Mother-of-Pearl. ⇨ SHELL.

Mother of Pearl Moth (Pyraloidea: *Sylepta ruralis*). One of our largest pyralids, bigger than many 'macros', widespread and common among stinging nettles, in the leaves of which its wriggling larvae wrap themselves.

Mother of Thousands. A name given to several plants with numerous flowers, including ivy-leaved toadflax and mind your own business.

Mother Shipton (Caradrinoidea: *Euclidimera mi*). One of our few day-flying moths, related to the red underwings, named from the supposed resemblance of the pattern on its forewings to the legendary Mother Shipton.

Motherwort (Labiatae: *Leonurus cardiaca*). A former medicinal herb with pink-purple flowers, still surviving near buildings in a few country districts.

Moth Fly. ⇨ Owl MIDGE.

Moth (Insecta: Lepidoptera). All the Lepidoptera not included in the Papilionina (Butterflies) are moths, the great majority of which are night-flying, and are characterized by not having club-tipped antennae but possessing a mechanism for hooking the fore and hind wings together. There are some 2,120 British species, formerly classed as Heterocera as distinct from the Rhopalocera (Butterflies), but this classification is now outmoded. For the distinctions between the Macrolepidoptera and the Microlepidoptera, see those entries. Nearly all 'macros' and even a great many 'micros' have been given English names, but most of these are the veriest book names and are far too numerous to be listed here.

Mottled Umber (Notodontoidea: *Erannis defoliaria*). A geometrid moth, related to the spring usher, flying in autumn and winter, whose looper caterpillars often abound in the spring and may defoliate trees.

Mould. (1) Soil that is fairly rich in humus. (2) A general term for tiny fungi that collectively produce a moss-like felt on the surfaces where they grow, usually dead or decaying animal or vegetable matter, including food and clothing in damp houses, and stored fruits and bulbs. Most belong to the fungal classes Phycomycetes and Ascomycetes. *Penicillium notatum*, the origin of the famous drug penicillin, is a common blue mould and an ascomycete, while *Mucor* is a white mould of bread and stored fruits.

Mountain Everlasting, Catsear or Catsfoot (Compositae: *Antennaria dioica*). A locally common plant of hill districts in the north and west, with dioecious pink or white flowers, daisy-like in the males, brush-like in the females.

Mountain Fern or Mountain Shield Fern (Thelypteridaceae: *Thelypteris limbosperma*). A common fern of acid soils in hill districts, somewhat resembling a yellow-green male fern, also known as Lemon-scented Fern from the fragrance of its leaves when crushed.

Mournful Wasp. ⇨ BLACK WASP.

Mouse. (1) (Rodentia: Muridae). The four species in Britain are the alien House Mouse *Mus musculus*, the common pest which may have arrived with our Neolithic ancestors more than 3,000 years ago; the abundant Wood Mouse or Long-tailed Field Mouse *Apodemus sylvaticus*, the widespread but local Yellow-necked Mouse *A. flavicollis*, and the tiny Harvest Mouse *Micromys minutus*, less than 2½ ins. long, which lives in cornfields and rough herbage. Both species of *Apodemus* have numerous island races on the Scottish islands. (2) (Caradrinoidea: *Amphipyra tragopogonis*). A grey-brown noctuid moth, related to the Old Lady, which often, when disturbed by day, runs mouse-like instead of flying.

Mouse-Ear or Mouse-ear Chickweed (Caryophyllaceae: *Cerastium*). Small plants with five deeply cleft white petals; half with small flowers, of which the commonest is the abundant Common Mouse-ear *C. fontanum*, in grassland everywhere; half with large flowers, of which the most frequent is Field Mouse-ear *C. arvense*. ⇨ SNOW-IN-SUMMER.

Mouse-hunt. Local name for weasel.

Mousetail (Ranunculaceae: *Myosurus minimus*). One of the most curious looking members of the Buttercup Family, its long (1–2½ ins.) seed-head, the 'mouse's tail' mimicking the flower-spike of a plantain; local and decreasing on rather bare ground in the south.

Mucro. A tiny point on a leaf or sepal; *mucronate*, ⇨ LEAF-SHAPES.

Mud Snail (Pulmonata: *Limnaea glabra*). A mainly northern pond snail with a distinctively narrow and almost cylindrical shell.

Mudwort (Scrophulariaceae: *Limosella*). Waterside plants with tiny pink or white flowers growing on mud; Common Mudwort *L. aquatica* widespread but local; Welsh Mudwort *L. subulata* is rare.

Muirburn. The practice of burning moorland periodically to encourage the vegetation, and especially ling heather to produce fresh shoots in the interests of sheep and/or grouse; on Exmoor it is called swaling. Modern conservationists look askance at burning, unless very carefully planned and controlled, on the grounds that in the long run it impoverishes the soil. Muirburn is responsible for the maintenance of many moors under heather, instead of grass, but too frequent burning can lead to the conquest of the heather by bracken.

Mule. A hybrid; used most frequently of the offspring of an ass and a mare, and of hybrids between canaries, goldfinches and other finches commonly kept as cage birds.

Mull or Mild Humus. The humus found in good soils favourable to animal and plant life, where organic material (dead plants, animals) decays rapidly. Earthworms are found in mull, in contrast to Mor.

Mullein (Scrophulariaceae: *Verbascum*). Seven tall plants, all but one with yellow flowers, much the commonest and most widespread being the whitely woolly Common Mullein or Aaron's Rod *V. thapsus*; Dark Mullein *V. nigrum* is frequent on limy soils in the south.

Mullet, Grey (Percomorphi: *Mugil*). Gregarious sea fish, ranging up to 3 ft long, and entering inshore and even brackish waters in summer and autumn. The three British species are the Thin-lipped *M. capito*, the Thick-lipped *M. chelo*, and the Golden *M. auratus* Grey Mullets.

Mullet, Red or Surmullet (Percomorphi: *Mullus surmuletus*). Red, or red and yellow, marine fish, much esteemed by Roman gourmets; regular off our south and south-west coasts.

Munro. A 3,000-ft hill top in Scotland, separated from other summits by a dip of at least 500 ft, named from the Scottish mountaineer H. T. Munro, who climbed all 276 of them.

Muntjac or Barking Deer (Artiodactyla: *Muntiacus muntjak*). A small Asiatic deer, the size of a large dog, now widespread in the woodlands of south central England, having originally escaped from Woburn Park, Bedfordshire. Male, buck; female, doe; young, fawn.

Muraena. ⇨ MURRY.

Murmuration. A fanciful collective term for a chattering flock of starlings.

Murragh or Great Red Sedge (Trichoptera: *Phryganea*). The largest British caddis-flies, found in lakes and rivers; the larvae are predatory and will even eat small fish, but the adults are eaten by larger fish.

Murry, Moray or Muraena (Apodes: *Muraena helena*). A large Mediterranean marine eel, much esteemed by Roman gourmets; very rare in British waters.

Musci. ⇨ MOSS.

Muscid Flies (Cyclorrhapha: Muscidae). Economically one of the most important families of two-winged flies, containing the house flies, stable flies, horse bot-flies, wheat bulb fly, cabbage root fly and onion fly.

Museum Beetle (Diversicornia: *Anthrenus museorum*). A small scavenging beetle noted for the destruction it can cause to museum collections.

Mushroom (Agaricaceae: *Agaricus* or *Psalliota*). A group of edible toadstools, the best known being the Field Mushroom *A. bisporus* (*P. campestris*), which together with its whiter cultivated form is much the most widely eaten edible fungus in the British Isles. Horse Mushroom *A. arvensis* and Wood Mushroom *A. silvicola* are also edible, but the Yellow-staining Mushroom or Yellow-stainer *A. xanthodermus* is liable to produce unpleasant, though not fatal, symptoms when eaten. The name is also given to two other edible toadstools: St George's Mushroom *Tricholoma gambosum*, related to blewits and so named because it appears about St George's Day, 23 April, and the graceful Parasol Mushroom *Lepiota procera*. Several species, including St George's and parasol mushrooms make Fairy Rings, while others, such as the common and horse mushrooms grow in outward-spreading circles without making conspicuous rings. Fool's Mushroom *A. verna* is highly poisonous. Oyster Mushroom, ⇨ OYSTER FUNGUS.

Musk (Scrophulariaceae: *Mimulus moschatus*). A frequent garden escape, like a small monkey flower; no longer muskily scented.

Musk Beetle (Phytophaga: *Aromia moschata*). A handsome bright green longhorn beetle, with a strong, rather musky odour.

Muskrat (Rodentia: *Ondatra zibethica*). A large North American rodent prized for its fur (musquash), which escaped from fur farms between the wars and established itself in the Severn valley and elsewhere; it was so destructive to river banks that it had to be expensively exterminated.

Mussel (Lamellibranchia: Mytilidae). Marine bivalves with oblong or oval shells, notably the edible and highly gregarious Common Mussel *Mytilus edulis*, widespread and often abundant on rocky, stony and muddy shores, including estuaries, the local south-western Mediterranean Mussel *M. galloprovincialis*, and the three Horse Mussels *Modiolus*, of which *M. modiolus*, up to 6 ins. long, is our largest marine mussel and the commonest in the north, and the Bearded Horse Mussel *M. barbatus* is commonest in the south; the Bean Horse Mussel *M. phaseolinus*, shaped as it is named, is also widespread and common. The Fan Mussel *Pinna fragilis*, a deep-water mollusc occasional on the lower shore, is our largest bivalve, with fan-shaped shells up to a foot both long and broad; it is more closely related to the oysters. ⇨⇨ FRESHWATER MUSSEL.

Mustard (Cruciferae). Name of several, mostly yellow-flowered, annual or biennial weeds of the Cabbage Family. Wild Mustard, Charlock or Runch *Sinapis arvensis* is common in waste places and on arable land untreated with modern chemicals. White Mustard *S. alba*, the cultivated mustard that goes with cress, grows mainly on chalky soils in the south. Black Mustard *Brassica nigra* is frequent on disturbed ground, sea cliffs and river banks. Hoary Mustard *Hirschfeldia incana* is an alien, locally established in the south. Hedge Mustard *Sisymbrium officinale*, a common wayside and waste ground plant. Tumbling Mustard *S. altissimum* is a spreading alien. Garlic Mustard, ⇨ Hedge GARLIC. Tower Mustard

Turritis glabra is a scarce and Treacle Mustard *Erysimum cheiranthoides* a common weed of light soils.

Mustard Beetle (Phytophaga: *Phaedon*). Two species of leaf beetle which are pests of mustard grown for seed.

Mustelid. A mammal belonging to the Mustelidae, a family of the Carnivora, whose native representatives are the stoat, weasel, polecat, pine marten, badger and otter, together with the alien mink.

Mutant. An organism one or more of whose genes has changed so that it carries some abnormal, often physical, characteristic. Many so-called mutants, however, are the result of a recombination of genes rather than of actual mutation.

Mutation. A change in one of the genes of a chromosome.

Mycelium. The feeding, vegetative or non-fruiting part of a fungus, consisting of a web of minute tubular threads called hyphae, which are often invisible to the naked eye, but may be joined together in a white thread known as a rhizomorph.

Mycetophagy. The eating of fungi.

Mycology. The study of fungi.

Mycorrhiza. An association between the mycelium of a fungus with the roots of a tree or other higher plant, such as yellow birds'-nest *Monotropa*, which may be either symbiotic (probably suppressed parasitism), in that the host plant derives at least some benefit from it, e.g. in the establishment of its seedlings; or parasitic, only the fungus benefiting. With trees the mycorrhiza forms a kind of whitish cloak around the smaller roots (ectotrophic), but with some other plants, notably the orchids and heathers, the fungus may be actually inside the root (endotrophic).

Myiasis. The temporary parasitism involved when the larvae of two-winged flies develop in the superficial layers of other animals' tissue, as the greenbottle *Lucilia sericata* does in sheep, and bot flies in various ungulates.

Myriapod. An obsolescent general term for many-legged terrestrial arthropods, which are neither insects nor arachnids nor crustaceans, and including the centipedes, millipedes, Pauropoda and Symphyla. There are 104 known British species in these groups.

Myrmecophilous. Ant-loving, of mites and other small arthropods which inhabit ants' nests.

Mysid. ⇨ OPOSSUM SHRIMP.

Myxomatosis. A virus disease of rabbits, introduced into south-east England, probably from France, by unknown causes in the autumn of 1953, which spread almost throughout Great Britain during the next two years, almost completely eliminating the rabbits. The consequent effects on vegetation, released from heavy grazing pressure, and predators, temporarily deprived of their main prey, were striking. From 1956 onwards there has been a slow recovery of rabbit populations, interspersed with fresh outbreaks of myxomatosis, which has become endemic, at the same time lessening in the intensity of its virulence so that many more rabbits are recovering after an attack than in 1954–55, when the kill in many areas was virtually 100 per cent.

Myxomycetes. A class of the Myxomycophyta.
Myxomycophyta. ⇨ SLIME FUNGUS.
Myxophyceae. Cyanophyceae. ⇨ BLUE-GREEN ALGAE.
Myzostomid (Annelida: Myzostomaria). A small, highly specialized class of annelid worms, parasitic on or in echinoderms, especially the feather stars, such as *Antedon bifida*.

N

Nab. A steep, flat-topped headland of a range of sea-cliffs or inland hills, especially in Yorkshire.

Nacre. Mother-of-pearl. ⇨ SHELL.

Naiad. (1) (Najadaceae: *Najas*). Short-leaved monocotyledonous water-weeds; *N. flexilis* very local, *N. marina*, with toothed leaves, confined to the Norfolk Broads, where it is called Stagshorn Weed. (2) Alternative name for the dragonfly larva or nymph.

Naked Ladies. ⇨ MEADOW SAFFRON.

Nanophanerophyte. ⇨ PHANEROPHYTE.

Narcissus Fly. A large Bulb Fly.

Narwhal (Cetacea: *Monodon monoceros*). An Arctic dolphin with a single long tusk. Specimens have been stranded on our shores about half-a-dozen times.

Nastic Movement. Plant response to a stimulus; not directional or growth, e.g. sleep positions of leaves of wood sorrel. Photonasty is when the stimulus is light.

National Park. A natural geographical area, designated by the Government through the National Parks Commission in order to control its development so as to maintain a well-balanced rural economy while preserving some wild country and wildlife, and also providing access and recreation for visitors. Ten parks have so far been designated in England and Wales. National Park status does not involve any change of land-ownership or give outsiders the right to wander over private land.

National Trust. A national body, independent of the Government, which exists to acquire land and buildings worth preserving for their aesthetic, amenity or scientific value. Founded in 1895, it now owns over 1,000 properties in England, Wales and Northern Ireland, Scotland having its own National Trust for Scotland. Many important nature reserves belong to the National Trust, most notably Blakeney Point in Norfolk, Box Hill in Surrey, and Wicken Fen in Cambridgeshire.

Natterjack (Salientia: *Bufo calamita*). A small loud-voiced toad, with a yellow stripe on the back, distributed locally on sand dunes and heaths, mainly in south and east England from Yorkshire to Dorset, but also in the north-west from Dee to Solway, and in south-west Ireland.

Natural Environment Research Council. A body parallel with the Agricultural, Medical and Science Research Councils, set up in 1965 to oversee the work of the Nature Conservancy, as well as that of the official bodies concerned with research in meteorology, geology, oceanography, and other aspects of the natural environment.

Natural History. The study of natural phenomena, especially in the field, properly including all inanimate phenomena, such as rocks, soils and climate, but now commonly confined to living things, animals and plants in the wild.

Natural History Society or Field Club. A voluntary organization, consisting mainly of amateur naturalists and nature lovers, which holds both indoor and field meetings and usually publishes a bulletin or printed journal. Some societies are organized on a national, others on a local or regional basis. The great majority are in membership of the Council for Nature, including more than 264 local societies, sixty-four national societies and thirty-two naturalists' trusts. The oldest extant national natural history society in Great Britain is the Linnean Society of London, founded in 1788, and the oldest extant local natural history society the Ashmolean Natural History Society of Oxfordshire, founded in 1828 and named after the seventeenth-century botanist Elias Ashmole, who founded the Ashmolean Museum at Oxford.

Naturalists' Trust. A body of naturalists and conservationists, incorporated as a company limited by guarantee, who seek to conserve the natural habitats in a county or group of counties, by acquiring or leasing land as nature reserves, and in any other appropriate ways, such as by agreements with landowners for the management of land so as to preserve its natural communities unimpaired. The first such trust was founded in Norfolk in 1926, followed by Yorkshire in 1946 and Lincolnshire in 1948. Between 1956 and 1964 every other county in England, except Rutland, acquired its trust, while Wales was covered also, partly by the old-established West Wales Field Society (1938) turning itself into a trust. The Scottish Wildlife Trust covers Scotland as a whole. The English and Welsh trusts are serviced by the Society for the Promotion of Nature Reserves. By 1964 the English and Welsh trusts had acquired, leased or secured management agreements for 145 reserves.

Naturalized Animals and Plants. Alien animals and plants that have become completely established ferally and are breeding and holding their own in competition with the native fauna or flora. ⇨ ACCLIMATIZATION.

Natural Order. A term formerly applied to the Family in flowering plants.

Natural Selection. The normal mechanism of evolutionary change, first scientifically established by Charles Darwin and put forward in *The Origin of Species* in 1859. The mechanism involves the selection of the breeding stock of a species through elimination of those members not satisfactorily adapted to their environment, summed up in the phrases 'Struggle for Existence' and 'Survival of the Fittest'.

Natural Vegetation. Vegetation completely unaffected by the activities of man or his domestic animals, past or present, currently found in the British Isles only on the tops of the highest mountains and perhaps on the remotest parts of the sea-shore.

Nature Conservancy. The Government body set up by Royal Charter in 1949 responsible for conservation of wild life in Great Britain. It runs six research stations and has declared 111 nature reserves. In 1965 it became a Committee of the Natural Environment Research Council.

Nature Reserve. An area set aside for the conservation of its fauna, flora, geological or physiographical interest, to be controlled and managed so as to preserve and/or study it. A National Nature Reserve is designated by the Nature Conservancy; an official Local Nature Reserve by the council of a county or county borough.

Nature Trail. A trail designed to explain a piece of countryside, the type of soil, flora, fauna, etc., usually by setting up explanatory notices at intervals; or numbered boards referring to a printed sheet; or by a demonstrator.

Nauplius Larva. The earliest stage of the simpler and more primitive crustaceans, always free-swimming and often quite different from the adults, especially with the barnacles.

Navelwort. ⇨ Wall PENNYWORT.

Necklace Shell (Prosobranchia: *Natica*). Carnivorous marine mesogastropod operculate molluscs, with winkle-like shells, living in sand and preying on bivalves; the two commonest and most widespread are the Common Necklace Shell *N. alderi* and Large Necklace Shell *N. catena.*

Necrophagy. The eating of carrion or other dead matter.

Nectar. A sugary substance secreted by flowers, in glands known as nectaries, to attract bees and other insects which help the process of pollination. The sugars are mainly glucose, fructose and especially sucrose, and are the basis of Honey.

Needle. The long and very narrow leaf of a coniferous tree.

Needle-Fish. ⇨ Greater PIPE-FISH.

Needle-Fly (Plecoptera: *Leuctra fusciventris* and *L. hippopus*). Widespread small stone-flies, especially common in streams on limy soils, emerging in late summer and autumn.

Needle Shell (Prosobranchia: *Bittium reticulatum*). A small relative of the winkles, with a narrowly spired shell, widespread and common on sea shores.

Negro Bug (Heteroptera: *Thyreocoris scarabeoides*). A black shieldbug found on the ground on sand dunes and chalk downs in the south.

Nekton. Animals which swim purposively in the water, as distinct from the Plankton, which just floats.

Nemathelminthes. ⇨ ASCHELMINTHES.

Nematocera or Thread-horns (Insecta: Diptera). The most primitive of the three sub-orders of the Two-winged Flies, containing the crane flies (Tipulidae), winter gnats (Trichoceridae), mosquitoes and gnats (Culicidae), midges (Chironomidae), fungus gnats (Mycetophilidae), fever flies and St Mark's flies (Bibionidae), biting midges (Ceratopogonidae), black flies (Simuliidae), gall midges (Cecidomyiidae) and owl midges and moth flies (Psychodidae).

Nematoda. ⇨ ROUNDWORM.

Nematomorpha. ⇨ HORSEHAIR WORM.

Nemertina. ⇨ RIBBON WORM.

Neo-Lamarckism. ⇨ LAMARCKISM.

Neonate. New-born.

Neoteny. The persistence of early stages in the development of an animal into later periods of growth, either temporary, because of climatic or other factors, or permanent, in which case Paedogenesis occurs. The American amphibian axolotl is the classic example.

Neotype. In taxonomy, a specimen chosen as a substitute for the holotype, when this is missing and no paratype or lectotype is available.

Nerite (Prosobranchia: *Theodoxus fluviatilis*). A small operculate freshwater snail with a roundish shell, very variable in colour, of streams and other running water in England.

Ness. A headland, often quite low-lying, on the east coast, also applied in Yorkshire to a promontory of a range of hills inland.

Nest. Structure made by an animal for the care and safety of its eggs and/or young. Birds' nests are the most familiar, but nests are also made by mammals (mice, squirrels), fishes (sticklebacks) and invertebrates (spiders, bees, wasps, ants). Birds' nests are commonly cup- or saucer-shaped and made of dried grass, moss and similar materials, but some are domed and others consist of no more than a few scraps of material in a hollow in the ground. The nests of ants and other social Hymenoptera are homes for the whole insect colony as well as for the young.

Nest-Box. An artificial nest cavity, erected for hole-nesting or ledge-nesting birds: two main types, a cylindrical or rectangular box with a small entrance hole, for tits, redstarts, sparrows and pied flycatchers, and a ledge or tray, with or without a roof, for robins, spotted flycatchers, and pied wagtails. More specialized nest-boxes are made for owls, treecreepers, swallows and other birds.

Nest Fly (Cyclorrhapha: *Neottophilum praeustum*). A reddish two-winged fly, whose larvae inhabit the nests of song-birds and suck the blood of the nestlings.

Nestling. ⇨ YOUNG BIRD.

Nettle (Urticaceae: *Urtica*). Two green-flowered plants with stinging hairs on their stems and leaves. Stinging Nettle *U. dioica*, forming large patches on nitrogenous soils in fields, woods, and fens and on waste ground, is much the commoner. Small Nettle *U. urens*, a frequent weed of cultivation, has milder stinging hairs. Dead-nettles and Hemp-nettles are labiates with nettle-like leaves but no stinging hairs.

Neuroptera (Arthropoda: Insecta). An order of slender, weakly flying insects, with relatively large, net-veined wings closing over the back like a roof; sixty species in two sub-orders, the Megaloptera, including the alder flies and snake flies, and the Planipennia, including the lacewings and sponge flies.

Newt (Caudata: *Triturus*). Small lizard-like amphibians, formerly known as Efts. Three British species: Crested or Warty Newt *T. cristatus*, the largest European species, widespread and frequent; Palmate Newt *T. helveticus*, the smallest European species, widespread but local; Smooth or Common Newt *T. vulgaris*, widespread and common, the only newt in Ireland.

Nidicolous. Of birds with young which stay in the nest until they are fledged.

Nidifugous. Of ground-nesting birds with young which leave the nest shortly after they hatch, as with waders, game birds and ducks.

Nigger. (1) (Holothuroidea: *Holothuria forskali*). A sea-cucumber, uncommon off our south and west coasts, named from its dark colour; from its habit of ejecting sticky white threads to counter aggressors, it is also called Cotton Spinner. (2) Ladybird larva.

Nightingale (Passeriformes: *Luscinia megarhynchos*). Perhaps the best known British song bird, allied to the thrushes, and widely esteemed as having the highest quality song; named from its habit of singing at night, though it also sings regularly by day, and many other birds, notably the woodlark, also sometimes sing at night. A summer migrant, frequent in wooded areas in England, except the north and south-west; very retiring, and rarely seen in the open. The Thrush-Nightingale or Sprosser *L. luscinia* is a very rare straggler from eastern Europe, perhaps sometimes overlooked because of its extreme similarity to the nightingale.

Nightjar or Goatsucker (Caprimulgiformes: *Caprimulgus europaeus*). A medium-sized, long-winged, long-tailed nocturnal bird, feeding on moths and other night-flying insects, and noted for its churring voice at dusk. A now rather local summer visitor, on heaths and moors. Two other species are rare stragglers.

Nightshade (Solanaceae). Poisonous plants related to the potato and tomato, including Black Nightshade *Solanum nigrum*, a common white-flowered weed with black berries, Deadly Nightshade or Belladonna, and Woody Nightshade or Bittersweet. Enchanter's Nightshade is unrelated.

Nimbostratus. Type of rain cloud. ⇨ CLOUD.

Nipplewort (Compositae: *Lapsana communis*). A common weed of gardens and shady waste places, with small yellow dandelion-like flowers.

Nit. The egg of the human louse, which is firmly attached to a human hair.

Nit Grass (Gramineae: *Gastridium ventricosum*). An uncommon grass whose shiny individual flowers are supposed to resemble nits.

Noah's Ark Shell (Lamellibranchia: *Arca tetragona*). A marine bivalve with boat- or ark-shaped shells, widespread but scarce on the lower shore.

Nobody Crab. ⇨ SEA SPIDER.

Noctuid Moth (Caradrinoidea: Caradrinidae, formerly Noctuidae). A large family of stoutly built, mostly rather drably coloured moths, including such familiar species as the yellow underwings, the cabbage, antler and turnip moths.

Noctule (Chiroptera: *Nyctalus noctula*). The commonest and most widespread large British bat, often flying by day.

Node. The junction of the stem and one or more leaves in flowering plants, especially grasses and sedges.

Nomenclature. The present scientific system of naming animals and plants is Linnaeus's binominal system, based on the Latin language. Each organism is given two names, a generic name, which has a capital initial, followed by a specific or trivial name, which nowadays always has a small initial e.g. *Passer domesticus*, the house sparrow. Generic names may be abbreviated to their initial after the first mention, so that the tree sparrow may now be rendered *P. montanus*. Strictly the trivial name should be followed by that of the author who first gave the organism its present trivial name, in the case of the house sparrow Linnaeus, who is commonly abbreviated to L., making *Passer domesticus* L. If an animal was later removed by another author into a different genus, the original namer is put in brackets; thus the house-martin is *Delichon urbica* (L.) because Linnaeus originally called it *Hirundo urbica*. Botanical nomenclature differs here in that the name of the remover is also given, e.g. the yellow waterlily is called *Nuphar lutea* (L.) Smith, because Linnaeus having named it *Nymphaea lutea*, Sir J. E. Smith later removed it to the genus *Nuphar*. There is no need, however, to cite authors' names for common animals and plants in ordinary usage. A hybrid is indicated by 'x' between the generic and trivial name, or if it is an inter-generic hybrid, before the generic name, e.g. *Symphytum x uplandicum*, Russian comfrey, the hybrid between common *S. officinale* and rough *S. asperum* comfreys; *X Festulolium loliaceum*, the hybrid between meadow fescue *Festuca pratensis* and rye-grass *Lolium perenne*. Subspecies get a third name, e.g. *Motacilla alba yarrellii* the pied wagtail; varieties get a third name with 'var.', e.g. *Geranium sanguineum* var *lancastriense*, the pink variety of bloody cranesbill; forms get a third name with 'forma', e.g. *Cirsium acaulon* f. *caulescens*, the stemmed form of the stemless thistle.

Non-Passerine. A bird not belonging to the great order of Passeriformes or song-birds.

Nonsuch. Black medick grown as a fodder crop.

Northern Lights. ⇨ AURORA BOREALIS.

Nostoc (Cyanophyta: Cyanophyceae). A Blue-green Alga, capable of living on carbon dioxide, water and nitrogen, simpler constituents than any other organism; in the soil it helps to fix atmospheric nitrogen.

Notodontoidea (Lepidoptera: Heteroneura). The large superfamily of moths containing the hawk-moths (Sphingidae), prominents (Notodontidae), emperor moth (Saturniidae), and the assemblage of families loosely known as geometers.

Nott. ⇨ HUMMEL.

Nudibranch (Opisthobranchia: Nudibranchiata). Sea slug with no shell, internal or external. The Common Grey Sea Slug *Aeolidia papillosa*, which preys on sea-anemones, is the largest species (3 ins.) and the Sea Lemon the second largest.

Nullipore (Rhodophyta: Corallinaceae). A group of small red seaweeds forming calcareous encrustations on rocks, including *Lithophyllum* and *Lithothamnion*.

Nunatak. A mountain or plateau, which has escaped glaciation or other drastic physical or climatic change, and so becomes a refugium for relict animals and plants.

Nut. A dry fruit with a woody wall, like an achene but having a hard pericarp.

Nutcracker (Passeriformes: *Nucifraga caryocatactes*). A relative of the crows, which occasionally straggles to Britain from the coniferous forests of the Alps or northern and eastern Europe.

Nuthatch (Passeriformes: *Sitta europaea*). Small song bird, noted for its habits of creeping up *and down* trees and of hammering nuts which it wedges in cracks of bark; widespread and frequent in England and Wales, mainly in the south.

Nut Shell (Lamellibranchia: *Nucula*). Small yellow or brown offshore bivalve.

Nymph. A young insect in one of the more primitive groups, such as the dragonflies, mayflies or stone-flies, resembling small wingless adults.

O

Oak (Fagaceae: *Quercus*). Generally the commonest British tree, dominant in the majority of broad-leaved woodlands, especially on heavier soils, with fruits known as acorns. Two native species: the widespread *Q. robur* with long-stalked acorns and the mainly northern and western Durmast Oak *Q. petraea* with short or unstalked acorns. Two frequently naturalized species: Turkey Oak *Q. cerris* and the quite different-looking Evergreen or Holm Oak *Q. ilex*.

Oak-Apple. A spongy, apple-shaped, red-tinged plant gall made on oaks by the gall-wasp *Biorhiza pallida*.

Oak Fern (Thelypteridaceae: *Thelypteris dryopteris*). A graceful bright green fern with a creeping rootstock, widespread but local on acid soils in the north and west. Its lime-loving counterpart is Limestone Polypody *T. robertiana*.

Oar-Fish. ⇨ RIBBON-FISH.

Oarweed (Phaeophyta: Laminariaceae). Large brown seaweeds of the genera *Laminaria*, *Saccorhiza* and *Alaria*, with broad, sometimes crinkled, fronds up to 6 ft long, attached to rocks by a holdfast resembling a rootstock; common and sometimes abundant below high water mark, and often thrown ashore by storms. In the west of Scotland they are collectively known as Tangle; when thrown ashore they are called Kelp, and used either directly as manure or burnt to produce a chemically valuable ash, rich in sodium, potassium and magnesium salts. *A. esculenta* is edible.

Oat (Gramineae). Grasses with characteristic bent awns; the cultivated Oat *Avena sativa* is a frequent relic of cultivation, closely resembling the pestilential cornfield weed Wild Oat *A. fatua*, one of the problems created by the use of chemical weed-killers which have eliminated its less resistant weed rivals. Meadow Oat *Helictotrichon pratense* and Downy Oat *H. pubescens* are both frequent in calcareous grassland. Yellow Oat *Trisetum flavescens* and False Oat *Arrhenatherum elatius* are both widespread and common in grassy places.

Oat Fly. ⇨ FRIT FLY.

Ocellus. An eye-like spot or marking, e.g. on the wing of a peacock butterfly.

Oceanography. The study of the oceans or great seas of the world, and especially of their water, their bed and their flora and fauna.

Octoploid. ⇨ POLYPLOIDY.

Octopus (Cephalopoda: Octopoda). Marine molluscs with four pairs of long tentacles: the Common Octopus *Octopus vulgaris* with two rows of suckers is found among rocks on the south coast, and may be quite common in some years; the Curled Octopus *Eledone cirrhosa* with one row of suckers is widespread but mainly northern.

Odonata. ⇨ DRAGONFLY.

Oil Beetle (Heteromera: *Meloë*). A group of seven wingless beetles, related to the Blister Beetle, named from their habit of discharging their evil-smelling, oily, yellow blood, as a defence mechanism; only two species are at all common. Their larvae, which feed on the stored products in the nests of certain bees, look so like lice that they were originally classified as such, as *Triungulinus,* and are still called triungulin larvae.

Old Lady (Caradrinoidea: *Mormo maura*). A large, drably grey noctuid moth, often disturbed from its daytime hiding places behind curtains, etc., indoors.

Old Man's Beard. (1) ⇨ TRAVELLERS' JOY. (2) A group of lichens (*Usnea* spp.) growing in long branched tufts on trees and rocks.

Old Wife. ⇨ Black BREAM.

Oligocene Period. ⇨ GEOLOGICAL PERIODS.

Oligochaete Worms (Annelida: Oligochaeta). A class of mainly freshwater, segmented worms, with or without bristles, also including the terrestrial Earthworms and White Worms. There are eight families of freshwater species; many inhabit the mud at the bottom of lakes and ponds, but other habitats, such as rotting vegetation, algae, and snail shells are also used; among the commonest of the mud-dwellers are the River Worms.

Oligophagous. Of animals with a restricted range of diet.

Oligotrophic. Poor in the basic salts that are plant nutrients; used of lakes, peat, etc.; the opposite of Eutrophic.

Olive. Anglers' name for many mayflies.

Ombrogenous, Ombrotrophic. Growing in, nourished by, wet conditions.

Omnivore. An animal which eats both animal and vegetable food, e.g. a badger.

Onion Fly, Beet Fly or Mangold Fly (Muscidae: *Delia cepetorum*). A two-winged fly whose larvae are an economic pest of onions, and in the variety *betae* also of beetroot and mangolds.

Ontogeny. The course of development of the life history of an individual animal or plant.

Oolite. A type of limestone mainly found in the Cotswolds and named from the small egg-like particles it contains.

Opah (Pisces: *Lampris luna*). A large and strikingly coloured and shaped (oval) deep-sea fish of the open Atlantic that occurs in British waters from time to time; its many names include king-fish, sun-fish, moon-fish and Jerusalem haddock.

Operculate (Gastropoda: Prosobranchia). The sub-class containing the great majority of the shelled marine gastropod molluscs, with shells either spirally twisted (winkles, top shells, whelks) or conically hat-shaped (limpets), and normally large enough for the animal to withdraw right inside, protected by the horny or limy plug called the operculum. Three orders: Archaeogastropoda with two gills (limpets, ormer, pheasant shells and top shells), and the one-gilled Mesogastropoda (winkles, periwinkles, cowries, spire shells) and Stenoglossa (whelks).

Ophiuroidea. ⇨ BRITTLE-STAR.

Opiliones. ⇨ HARVESTMAN.

Opisthobranchia. ⇨ SEA SLUG.

Opossum Shrimp or Mysid (Malacostraca: Mysidacea). An order of small shrimp-like marine crustaceans, so-called because they carry their young in a pouch, like the marsupial opossums. Some species are found in rock pools and on the shore, including the Midge Shrimp *Hemimysis lamornae*, the Chameleon Shrimps *Praunus* spp., and the transparent Ghost Shrimp *Schistomysis spiritus;* some species are bioluminescent.

Orache (Chenopodiaceae: *Atriplex*). A genus of weedy-looking plants allied to and resembling the goosefoots. Common Orache *A. patula* and Spear-leaved Orache *A. hastata* are common weeds of cultivation; the three other species are found only on or close to the sea shore. *A. hortensis*, the orache of gardens, sometimes escapes.

Orange-Peel Fungus. ⇨ ELF-CUP.

Orange-Tip Butterfly (Papilionina: *Anthocharis cardamines*). A handsome relative of the White Butterflies, named from the bright orange tips to the male's wings; widespread and common along hedgerows and in woods.

Orchid (Monocotyledones: Orchidaceae). A large family of flowering plants, with simple leaves (usually narrow, but sometimes broad) and spikes of often spurred and always two-lipped flowers; over fifty species. ⇨ HELLEBORINE, LADY'S SLIPPER, LADY'S TRESSES, TWAYBLADE.

Order. One of the groups in the classification of living organisms, and consisting of closely related Families. Closely related orders are grouped into a Class.

Ordovician Period. ⇨ GEOLOGICAL PERIODS.

Oregon Grape (Berberidaceae: *Mahonia aquifolium*). Evergreen undershrub, with fragrant yellow flowers, purple-black fruits like tiny grapes, and holly-like leaflets on the pinnate leaves; widely planted and sometimes naturalized.

Orfe, Golden (Cyprinidae: *Leuciscus idus*). A popular ornamental fish related to the dace, breeding freely in ponds but nowhere really feral. Like the goldfish, it is a domesticated form of a wild greenish-brown fish; the Silver Orfe is another variant.

Organism. Any living animal or plant.

Orientation. (1) Birds, and other animals, are able to orientate themselves on their home territories, and also apparently on their regular wintering grounds, either on their annual migratory journeys or when displaced by strong winds, other natural factors, or even experimental scientists. For how they navigate, ⇨ BIRD NAVIGATION. (2) The turning of a stationary organism or part of an organism in a particular direction, as with twining plants, or plants which turn towards the light.

Oriole, Golden (Passeriformes: *Oriolus oriolus*). A medium-sized song bird, whose cock is strikingly coloured black and yellow. Uncommon annual spring visitor to Britain; has occasionally bred.

Ormer (Prosobranchia: *Haliotis tuberculata*). A large (to 4 ins.) and handsome relative of the limpets, noted for the mother-of-pearl interior of its shell, common in the Channel Isles, where it is harvested at extreme low spring tides.

Ornithology. The scientific study of birds.

Orpine or Livelong (Crassulaceae: *Sedum telephium*). One of our two pink-flowered stonecrops, a succulent that is widespread but local in woods and hedgebanks and on rocks and cliffs.

Orthoptera (Arthropoda: Insecta). The order of insects containing the grasshoppers and locusts (Acrididae), ground-hoppers (Tetrigidae), crickets (Gryllidae), bush-crickets (Tettigoniidae) and mole-crickets (Gryllotalpidae); thirty-eight native British species and eight introduced species. They have an active larval stage, resembling the adults, and are specially adapted to jumping by a large development of the femur of the hind leg. Their stridulating 'song' is made either by rubbing a comb-like structure on the femur against a projecting vein on the fore-wing (grasshoppers, crickets) or by rubbing the two fore-wings together (bush-crickets, mole-cricket).

Ortolan (Passeriformes: *Emberiza hortulana*). A bunting, much relished for the pot on the Continent; a scarce passage migrant to Britain.

Osier or Withy (Salicaceae: *Salix viminalis*). A shrubby willow, noted for its long narrow leaves and very flexible twigs, which are much valued for basket-making. For this purpose the osier and its hybrids with other willows have been widely planted, often in special osier-beds.

Osprey or Fish Hawk (Falconiformes: *Pandion haliaetus*). A large buzzard-like bird of prey, feeding on fish, which it captures by plunging to the surface of the water. An uncommon passage migrant from Scandinavia; one pair has nested in the Spey valley in the Scottish Highlands since about 1953, and a second pair since 1963.

Ostariophysi (Vertebrata: Pisces). The order of bony fishes containing the Carp Family (Cyprinidae) and the loaches (Cobitidae).

Ostracod (Crustacea: Ostracoda). A sub-class of primitive bean-like crustaceans of both the freshwater and marine zooplankton: *Cochoecia* is a common marine genus, and *Cypria* a common freshwater one.

Otter (Carnivora: *Lutra lutra*). Our only native freshwater mammalian carnivore, living in holes in river-banks (holts). Widespread on rivers and streams though not commonly seen.

Otter Shell. ⇨ GAPER.

Ouzel. Old-fashioned name for the Blackbird, Water Ouzel, ⇨ DIPPER. ⇨⇨ RING OUZEL.

Ova. Plural of ovum: egg.

Ovary. The future seed-box or fruit of a flowering plant, at the base of a carpel, and containing the ovules or future seeds.

Ovipary. Producing young by means of eggs (ova) which do not hatch until after they have been laid. Birds, some reptiles, amphibians and most insects are oviparous.

Ovipositor. A specialized egg-laying organ in insects.

Ovule. ⇨ OVARY.

Ovum (pl. Ova). Egg, especially of an invertebrate.

Owl (Strigiformes: Strigidae). Medium to large, mainly nocturnal birds of prey, feeding largely on mice and voles, noted for their silent flight and disc-like faces, the hooked bill being almost buried in the feathers. Five British breeding species, and five irregular visitors or stragglers. Barn Owl *Tyto alba*, widespread and now less common than fifty years ago, has white underparts. Little Owl *Athene noctua*, our smallest owl, mainly diurnal in habit, was introduced from the Continent in 1889–96 and is now widespread and frequent in England and Wales and just spreading into Scotland. Long-eared Owl *Asio otus*, widespread but uncommon, has long ear-like tufts on the head. Short-eared Owl *A. flammeus*, with shorter tufts, a rather local breeder but more widespread winter visitor and passage migrant, especially on coastal marshes; locally known as the Woodcock Owl because it immigrates at the same time as the woodcock. Tawny Owl *Strix aluco*, also known as Brown Owl or Wood Owl, is the commonest owl in Great Britain, but absent from Ireland, frequently seen and heard in towns and suburbs. The most frequent of the irregular visitors is the Snowy Owl *Nyctea scandiaca*, a very large and almost pure white owl from far northern Europe, which used to be not infrequent in the northern isles of Scotland in winter, but has become scarcer.

Owlet. (1) A young owl. (2) A local name for noctuid moths.

Oxbow. A loop in a river, cut off from the main stream by silt banks and eventually becoming a pond.

Oxlip (Primulaceae: *Primula elatior*). A yellow spring flower, closely related to the primrose, locally common in woods in eastern England, near Cambridge. False Oxlip is a name given to the primrose-cowslip hybrid, which is rather like the true oxlip, which is also known as Bardfield Oxlip.

Ox-Tongue (Compositae: *Picris*). Two plants with yellow dandelion-type flowers: Bristly Ox-tongue *P. echioides*, covered with rough bristles based on goose-fleshy pimples, growing on clayey soils, and Hawkweed Ox-tongue *P. hieracioides*, with wavy-edged leaves, growing on calcareous soils and easily mistaken for a hawkweed.

Ox-Tongue Fungus. ⇨ POOR MAN'S BEEFSTEAK.

Oyster (Lamellibranchia: Ostreidae). Flat-shelled gregarious marine bivalves of shallow muddy shores, much esteemed by gourmets, especially the Flat or Native Oyster *Ostrea edulis*, found mainly in the estuaries of the Thames and Essex rivers, where the famous Colchester and Whitstable 'natives' have been farmed, probably since Roman times. Nowadays the Portuguese Oyster *Crassostrea* (*Gryphaea*) *angulata*, with deeper shells, is often laid down, but only infrequently spawns. The several species of Saddle Oyster (Anomiidae), with small roundish shells, are more closely related to the mussels; the commonest, found on rocks on the shore, is *Anomia ephippium*.

Oystercatcher (Charadriiformes: *Haematopus ostralegus*). A large noisy black and white wader with red legs and bill, widespread and common round most of the British coasts, increasingly breeding inland in the north. Feeds on many molluscs, such as mussels, cockles and limpets but not especially on oysters.

Oyster Fungus or Oyster Mushroom (Basidiomycetes: *Pleurotus ostreatus*).

An edible toadstool, with a cap supposedly resembling an oyster shell, growing on broad-leaved trees.

Oyster Plant (Boraginaceae: *Mertensia maritima*). Our only smooth-leaved member of the Borage Family, with blue flowers and glaucous, oval leaves said to taste like oysters; decreasing on coastal shingle in the north; Northern Shore-wort is an old book-name.

P

Paddle Worm (Polychaeta: Phyllodocidae). A family of brightly coloured green or blue free-swimming marine bristle worms, with paddle-shaped lobes, widely distributed on the shore.

Paedogenesis. Breeding while in a larval or other pre-adult stage, e.g. young male pied flycatchers and other birds breeding in first-year plumage before assuming normal adult breeding plumage. ⇨ NEOTENY.

Painted Lady (Papilionina: *Vanessa cardui*). A migratory vanessid butterfly, with similar habits to the red admiral, but not known to overwinter.

Palaeontology. The scientific study of fossils.

Palaeozoic Era. ⇨ GEOLOGICAL PERIODS.

Pale or Palea. One of the bracts in a grass flower. ⇨ GRASS.

Palm. Country name for sallow catkins, with which churches used to be decorated on Palm Sunday.

Palmate. Divided like the fingers of a hand. ⇨ LEAF-SHAPES.

Palmer Worm. The caterpillar of the yellow-tail moth.

Palpicornia (Coleoptera: Polyphaga). The super-family containing the scavenger beetles.

Palps. The second pair of appendages in arachnids, with varying functions, tactile in the spiders and harvestmen, claws in the false-scorpions.

Palynology. ⇨ POLLEN ANALYSIS.

Pandora. ⇨ BREAM, SEA.

Pandora Shell (Lamellibranchia: *Pandora albida*). A marine bivalve with unequal oval shells, living on the surface of the sand in quiet bays in the south.

Panduriform. (Of leaves), waisted.

Panicle. A branched inflorescence, especially one which grows upwards so that the youngest flowers are nearest the tip.

Pansy (Violaceae: *Viola*). Low-growing plants with flowers like those of garden pansies, yellow, purple or a mixture of both: three species, the common arable weed Field Pansy *V. arvensis*, the less common Hearts-ease *V. tricolor*, and the handsome Mountain Pansy *V. lutea*, confined to the north and west.

Panther or Panther Cap (Agaricaceae: *Amanita pantherina*). A highly poisonous and sometimes lethal toadstool, closely related to the death cap and liable to be confused with its edible relative the blusher.

Papilionaceae. ⇨ LEGUMINOSAE.

Papilionina. ⇨ BUTTERFLY.

Papilla. Botanical term signifying a small elongated projection from a leaf or other organs.

Pappus. A tuft of hairs, in fact the modified calyx, surmounting certain fruits, e.g. dandelions, and enabling them to be carried away on the wind.

When the pappuses form a rounded head they are called a clock because of the old country custom of blowing them off the seed-head to tell the time.

Parasite. An animal or plant living on or in another animal or plant, and feeding on its host, e.g. an ichneumon fly in a caterpillar or a broomrape on the roots of another plant. An Obligate Parasite can only live on its host, but a Facultative Parasite can if need be gain its living in other ways, e.g. as a saprophyte. A Kleptoparasite is an animal that robs another of its food, as a skua habitually does with gulls. ⇨ COMMENSAL, HEMIPARASITE, SYMBIOSIS.

Parasite Fly (Cyclorrhapha: Larvaevoridae). A large family of more than 250 species of two-winged flies, whose larvae are internal parasites of other arthropods, mainly insects and especially butterflies and moths, but also spiders, woodlice and centipedes. The shining black *Phryxe vulgaris* frequently emerges from larvae collected by hopeful lepidopterists.

Parasitica (Hymenoptera: Apocrita). The group of hymenopterous superfamilies containing ichneumons and other small and very small wasps with parasitic larvae including the Ichneumonoidea (ichneumons) Cynipoidea (gall wasps), Chalcidoidea (chalcid wasps) and Proctotrupoidea: some 5,230 species in all.

Parasitic Bee (Aculeata: Apidae). Several genera of bees have specialized in laying their eggs in the nests of other bees, instead of making their own. The Cuckoo Bees parasitize the social bumblebees, which they much resemble, but the rest, sometimes called Homeless Bees, parasitize solitary bees, sometimes specializing in one or two species, as the wasp-like *Nomada* does with the mining bees of the genus *Andrena*.

Parasol (Basidiomycetes: *Lepiota*). A conical toadstool, especially the edible Shaggy Parasol *L. rhacodes* of rich soil in woods and gardens, and the Parasol Mushroom *L. procera*.

Paratype. In taxonomy, a specimen described at the same time as the holotype, and available as a substitute if the latter is mislaid or destroyed.

Parazoa. ⇨ SPONGE.

Parent Bug (Heteroptera: *Elasmucha grisea*). A small brown shieldbug, widespread on birch, one of the few insects exhibiting parental concern for its eggs and small larvae.

Paripinnate Leaf. A pinnate leaf with no single leaflet at the tip, as opposed to an Imparipinnate leaf.

Parr. The young salmon after absorbing its yolk-sac and until it becomes a smolt, usually two years later.

Parrot Toadstool (Basidiomycetes: *Hygrophorus psittacinus*). A small yellowish toadstool, common in grassy places.

Parsley (Umbelliferae). A name for many white-flowered umbellifers besides the true Parsley *Petroselinum crispum*, the garden herb with greenish-yellow flowers, which is also established locally on rocks and banks. The best known are Cow Parsley or Wild Chervil *Anthriscus sylvestris*, abundant on hedge-banks in spring; Fool's Parsley *Aethusa cynapium*, a common garden weed; Stone Parsley *Sison amomum*, with a repellent smell suggesting a mixture of nutmeg and petrol, which closely resembles the parsley-scented Corn Parsley or Corn Caraway *Petroselinum segetum*; the rare Milk Parsley *Peucedanum palustre*, food-plant of the swallow-tail butterfly; and the various Bur Parsleys *Torilis* and *Caucalis*, named from their spiny fruits.

Parsley Fern (Adiantaceae: *Cryptogramma crispa*). A locally common fern of acid soils in the north, with bright green fronds crisped somewhat like parsley.

Parsley Piert (Rosaceae: *Aphanes arvensis*). A small relative of the lady's mantles with tiny green flowers, common in bare and disturbed ground.

Parsnip, Wild (Umbelliferae: *Pastinaca sativa*). A large yellow-flowered umbellifer of dry, mainly calcareous grassland, locally common in the south; closely related to the garden plant. Great Water Parsnip *Sium latifolium* and Lesser Water Parsnip *Berula erecta* are respectively rather local and widespread white-flowered waterside umbellifers. Cow Parsnip. ⇨ HOGWEED.

Parthenogenesis. Reproduction without fertilization, as with dandelions and hawkweeds (apomixis) among plants; or in animals with certain aphids and plant-gall insects, which have no males at certain stages of their life history.

Partridge (Galliformes: Phasianidae). Medium-sized game birds, of which two species breed: the native Common, Grey or English Partridge *Perdix perdix* widespread and common in farming districts, stocks being artificially maintained by game preservers; and the larger Red-legged or French Partridge *Alectoris rufa*, introduced from the Continent at the end of the eighteenth century and now common in eastern England. A small flock or family party of partridges is a covey. Hungarian Partridge is a name for common partridges imported from the Continent.

Pasque Flower (Ranunculaceae: *Pulsatilla vulgaris*). One of our most handsome wild flowers, closely related to the anemones, with rich violet-purple petals and golden-yellow stamens; named from flowering at about Easter-time (French *Pâques*), in mid or late April; now increasingly scarce in chalk and limestone turf in the south-eastern half of England.

Passeriformes. The order containing the song birds or perching birds, the largest order of birds, both in Britain and the world. The families of the order represented in the British avifauna are: Alaudidae (larks), Hirundinidae (swallows, martins), Oriolidae (oriole), Corvidae (crows), Paridae (tits), Sittidae (nuthatch), Certhiidae (creeper), Troglodytidae (wren), Cinclidae (dipper), Turdidae (thrushes), Sylviidae (warblers), Regulidae (goldcrest), Muscicapidae (flycatchers), Prunellidae (hedge-sparrow), Motacillidae (wagtails), Bombycillidae (waxwing), Laniidae (shrikes), Sturnidae (starling), Fringillidae (finches, buntings), and Passeridae (sparrows). Males are called cock and females hen.

Passerine. A member of the order Passeriformes.

Pastor, Rosy. ⇨ Rose-coloured STARLING.

Pasture. A grass field used for grazing cattle, sheep or horses.

Pauropoda (Arthropoda). A class of tiny myriapods, measuring only about $\frac{1}{20}$ in., with branched antennae and twelve body segments; found in damp places under decaying leaves, wood, etc.

Pawnbroker's Plant (Rubiaceae: *Galium tricornutum*). A white-flowered bedstraw, named from its three roundish fruits arranged like a pawnbroker's sign; now rather local as a cornfield weed in the south.

Pea (Leguminosae: *Lathyrus*). A group of herbaceous peaflowers, with tendrilled pinnate leaves, distinguished from the vetches (*Vicia*) by their winged stems. Much the commonest is Meadow Pea or Meadow Vetchling *L. pratensis*; Everlasting Pea *L. sylvestris*, Sea Pea *L. japonicus* and Marsh Pea *L. palustris* being more local, in woods, on coastal shingle and in marshes respectively. ⇨ VETCHLING. The Garden and Field Peas, which sometimes occur in arable and waste ground are *Pisum sativum* and *P. arvense*.

Typical Peaflower

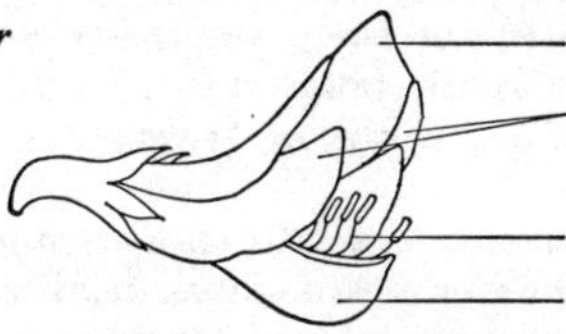

Peacock Butterfly (Papilionina: *Nymphalis io*). A handsome vanessid named from the 'eye' on each wing like those on a peacock's tail; the larvae feed on stinging nettles; widespread and common but mainly in the south. The Peacock Moth *Semiothisa notata* is a geometrid and has no such eyes.

Peacock's Tail (Phaeophyta: *Padina pavonia*). A beautiful brown seaweed, whose fan-shaped fronds are marked with concentric rings; locally common on our southern coasts.

Peacock Worm (Polychaeta: *Sabella pavonina*). A common fan worm, named from the orange and violet colouring at its tail end; attaining 4 – 10 ins.

Peal. ⇨ GRILSE.

Pear, Wild (Rosaceae: *Pyrus*). Two species: *P. communis*, usually bird-sown from an orchard, normally a tree and probably not native; *P. cordata* only near Plymouth, always a shrub and probably native. Whitty Pear ⇨ SERVICE TREE.

Pearl. A foreign body in a mollusc, especially an oyster or a freshwater mussel, surrounded and rendered harmless by an outpouring of nacre (mother-of-pearl) from the shell. The pearls of freshwater mussels may be white, pink, green or brown.

Pearl-Fish. ⇨ FIERASFER.

Pearl-Side or Sheppy Argentine (Pisces: *Maurolicus muelleri*). Small deep-sea fish, sometimes cast ashore in Scotland.

Pearlwort (Caryophyllaceae: *Sagina*). Small white- or green-flowered plants growing mostly in bare places: the commonest are the perennial Mossy Pearlwort *S. procumbens* and Annual Pearlwort *S. apetala*.

Peat. A pure organic soil, the result of plant material accumulating in waterlogged conditions. Because the soil is waterlogged and so airless, there are no earthworms, bacteria or other animals to draw the dead vegetation on top down into the soil and hasten decomposition. If the water contains mineral salts (which neutralize the acids produced by decomposition) the result is fen peat, which is neutral or slightly alkaline. Where there are no basic salts in the water (e.g. rain) to neutralize the acids the result is acid peat. Both types lack oxygen. Few plants will grow in acid peat.

Peck Order. The social hierarchy in a flock of gregarious birds, named from the habit of dominant domestic fowls pecking other fowls about the head. The top bird may peck all others; the second all but the top one, and so on down the line, although there may be a certain amount of cross-pecking in the middle reaches.

Pectinate. Toothed like a comb.

Pedate. Of a leaf with leaflets arranged palmately.

Pedicel. The stalk of a single flower.

Pedology. The scientific study of the soil.

Peduncle. The common stalk of a cluster of flowers.

Peewit. ⇨ LAPWING.

Pelage. The fur or hair of a mammal. (*Pel*, Old French for hair).

Pelagic. Inhabiting the open water of the sea, as distinct from the sea-bed or shore. For pelagic animals and plants, ⇨ NEKTON, PLANKTON. Also applied to lakes.

Pelecaniformes. An order of large aquatic birds including the pelicans (Pelecanidae), gannets (Sulidae), frigate-birds (Fregatidae), cormorants and shags (Phalacrocoracidae). No wild pelicans have been recorded in Britain in historic times, but their bones have been recovered from prehistoric peat deposits.

Pelican's Foot Shell (Prosobranchia: *Aporrhais pes-pelecani*). A marine mesogastropod operculate mollusc, common in muddy gravels off-shore, whose distinctive narrowly spiral empty shell has a 'wing' fancifully resembling a bird's foot and is not infrequent on beaches.

Pellitory of the Wall (Urticaceae: *Parietaria diffusa*). A sprawling plant with tiny greenish flowers in whorls along the leafy reddish stems, locally common on walls and rocks.

Peloria. An irregularly shaped flower which becomes regular by an unusual development of some of its parts, e.g. foxgloves sometimes produce a flat, saucer-shaped peloric flower at the tip of the spike.

Peltate. Botanical term applied to leaves or other flat organs. whose stalks are attached on the under surface.

Pen. The female swan.

Peneplain or Peneplane. A land surface weathered down until it is almost a plain, perhaps with a few hills still projecting from it.

Penicillin. ⇨ ANTIBIOTIC, MOULD.

Penny-Cress (Cruciferae: *Thlaspi*). Four white-flowered annuals, only Common Penny-cress *T. arvense*, a common arable weed, being at all widespread; named from the shape of the seed-pods.

Pennyroyal (Labiatae: *Mentha pulegium*). Our smallest native mint, now very scarce in damp heathy places.

Pennywort, Marsh (Hydrocotylaceae: *Hydrocotyle vulgaris*). A common marshland plant named from its round pennylike leaves, held aloft like tiny umbrellas, and sheltering the very inconspicuous flowers beneath.

Pennywort, Wall or Navelwort (Crassulaceae: *Umbilicus rupestris*). A succulent with round penny-like leaves with a central navel-like dimple, a spike of greenish flowers and a mainly western distribution on rocks and walls.

Pentaploid. ⇨ POLYPOIDY.

Peony, Wild (Paeoniaceae: *Paeonia mascula*). An old established garden escape on Steep Holm island in the Bristol Channel.

Peppered Moth (Notodontoidea: *Biston betularia*). A large geometrid moth famous for having first demonstrated Industrial Melanism.

Peppermint (Labiatae: *Mentha x piperita*). A mint cultivated for its essence, occasionally escaping on to waste ground; the hybrid of water mint and spearmint.

Pepperwort. Three cruciferous weeds, named from the acrid root of the first: Common Pepperwort *Lepidium campestre*, with white flowers, a frequent arable weed in the south; green-flowered Narrow-leaved Pepperwort *L. ruderale*, common on rubbish-tips and waste ground in the south-east; and the white-flowered Hoary Pepperwort or Hoary Cress *Cardaria draba*, a perennial which has spread widely on waysides and waste ground in England since it was introduced in Kent about 1810.

Peppery Furrow Shell (Lamellibranchia: *Scrobicularia plana*). A widespread and locally common marine bivalve of inter-tidal mud, resembling and related to the tellins.

Perch (Percomorphi: *Perca fluviatilis*). A strikingly barred gregarious freshwater fish, widespread and common in rivers, lakes, and ponds in Britain.

Perch, Sea (Percomorphi: Serranidae). Mediterranean fishes of which three are rare visitors to our south-western coasts: the small Comber *Serranus cabrilla*, and the two large fishes much sought after by skin-divers, who call them Groupers, the Stone Bass or Wreck-fish *Polyprion americanus* and the Dusky Perch *Epinephelus gigas*.

Percomorphi (Vertebrata: Pisces). A large and varied order of bony fishes, including the bass, perches, sea breams, wrasses, sand-eels, weevers, mackerel, tunny, gobies, dragonets, blennies and mullets.

Peregrine (Falconiformes: *Falco peregrinus*). Our largest breeding falcon, especially favoured by falconers. Widespread and formerly not infrequent on coastal and inland cliffs, but now rare as a result of secondary poisoning from toxic farm pesticides. Known as Duck Hawk in North America.

Perennial. A plant that continues to grow from year to year. A herbaceous perennial is one in which the leaves and stems (the parts above ground) die down each autumn, the new shoots coming up from the roots each spring, and the plant increasing in size each year. In woody perennials such as shrubs and trees the plant makes new growth each year from the permanent woody stems, thus growing taller or broader every year; ⇨ ANNUAL, BIENNIAL, EPHEMERAL.

Perfoliate. Of a pair of leaves completely joined round the stem which thus appears to project through a single leaf.

Perianth. The calyx (sepals) and corolla (petals) of a flower as a whole.

Pericarp. The outer covering of a fruit, either hard or soft.

Periwinkle. (1) (Apocynaceae: *Vinca*). Two blue-purple flowers, both normally garden escapes, but Lesser Periwinkle *V. minor* may sometimes be native in woods. (2) ⇨ WINKLE.

Permian Period. ⇨ GEOLOGICAL PERIODS.

Persicaria (Polygonaceae: *Polygonum*). Two common weeds with dark blotches on their leaves, allied to the bistorts: Common Persicaria or Redleg *P. persicaria*, usually with pink flowers, and Pale Persicaria *P. lapathifolium*, usually with whitish flowers.

Pesticide Effects. The regular and frequent use of chemical pesticides has resulted in the widespread contamination of the environment with residues of certain of the more persistent of them, both in the soil and in animal tissues. Birds of prey, especially the peregrine and the golden eagle which are at the end of the Food Chain, declined sharply in numbers in the early 1960's, and there is strong circumstantial evidence that this was due to pesticide residues. Many other birds have been found dead apparently poisoned by these residues. The Government therefore imposed a voluntary ban on the use of some of the persistent pesticides, notably aldrin, dieldrin and heptachlor for certain agricultural and horticultural purposes, such as seed-dressings against wheat bulb-fly (except in the autumn) and pest control by amateur gardeners. Indiscriminate use of insecticides is liable to kill off useful insects, such as honeybees and the predators of insect and other pests. A plague of red-spider mite in orchards once resulted from an early application of a pesticide which killed off its predators.

Petal. A divided portion of a corolla, lying inside the calyx (sepals) of a flower. ⇨ PERIANTH. *Petaloid* is used of sepals and other organs which are brightly coloured and look like petals.

Petiole. A leaf-stalk.

Petrel (Procellariiformes). The shorter-winged sea-birds of the family Procellariidae, of which two breed in holes on islands off our northern and western coasts: the Storm Petrel *Hydrobates pelagicus* and the Fork-tailed or Leach's Petrel *Oceanodroma leucorrhoa*, and seven more are rare or occasional visitors.

pH Scale. The scale (from 0 to 14) used to denote the acidity or alkalinity of soil by measuring the negative logarithm of the hydrogen ion concentration. An ion is a particle of an element carrying a charge of electricity. pH 7 indicates neutral soil. Higher numbers show alkalinity; lower ones acidity. More than 8 is rare in Britain. pH means potential of Hydrogen.

Phaeophyceae, Phaeophyta. ⇨ BROWN SEAWEEDS.

Phage or Bacteriophage. A virus that destroys bacteria.

Phalarope (Charadriiformes: *Phalaropus*). Small aquatic waders, spending most of their lives at sea. Three British species: Red-necked Phalarope *P. lobatus*, breeding sparsely in the Hebrides and the northern isles and in one place in Ireland, a rare visitor elsewhere; Grey Phalarope *P. fulicarius*, a straggler, sometimes storm-blown from the Atlantic in some numbers; Wilson's Phalarope *P. tricolor*, a rare straggler from North America.

Phanerogam. Outmoded Linnean term for the Spermatophytes or Seed-plants.

Phanerophyte. Woody plants with winter buds at least a foot above the level of the soil, divided into nanophanerophytes from 1 to 6 ft, microphanerophytes from 6 to 24 ft, and mesophanerophytes and megaphanerophytes from 24 ft upwards, i.e. roughly undershrubs, shrubs and trees.

Phantom Crane-Fly (Nematocera: Ptychoptera). A small group of large crane-flies, the two commonest being *P. contaminata* and *P. albimana.*

Phantom Larva. The transparent larva of the phantom midge.

Phantom Midge. ⇨ GNAT.

Pheasant (Galliformes: Phasianidae). Large, long-tailed game birds, the cocks brightly coloured, of which several species have been introduced, the most successful being the familiar *Phasianus colchicus*, with two main strains, the so-called 'Old English' pheasant *P.c. colchicus*, probably established by the Normans, and the Chinese Ring-necked Pheasant *P.c. torquatus*, introduced in the eighteenth century; both are now well established in wooded districts, stocks being maintained by liberal releases by game preservers. The Bohemian Pheasant is an albinistic variant. Several other races of *P. colchicus* have also been introduced, and the misleadingly named 'melanistic mutant' probably derives from the interbreeding of one or more of these with the Japanese Pheasant *P. versicolor.* In some districts the Golden Pheasant *Chrysolophus pictus* and Lady Amherst's Pheasant *C. amherstiae* are also established.

Pheasant's Eye (Ranunculaceae: *Adonis annua*). Decreasing cornfield weed with bright red flowers.

Pheasant Shell (Prosobranchia: *Tricolia pullus*). A small, winkle-like marine operculate mollusc related to the top shells, with a brightly coloured red and yellow shell, frequent on our south and west coasts.

Phenology. The study of seasonal changes in animals and plants, e.g. the dates when birds arrive and depart on migration, or start to nest, and the dates when plants begin to bloom.

Phenotype. The characters of an individual organism as determined by the effect of the environment on the genotype.

Philomel. Literary name for the nightingale.

Phoenicopteriformes. ⇨ FLAMINGO.

Phoresy. The use by animals of other animals as a means of transport, e.g. the false-scorpion *Chernes nodosus*, which often holds on to the legs of harvestmen and two-winged flies.

Phoronida or Phoronidea. A small phylum of small worm-like marine animals, of which only one species is British: the widespread but scarce *Phoronis hippocrepia*, which lives gregariously in membranous tubes in soft rocks or shells, and is so shy that it withdraws within the tube at the slightest vibration or even shadow; its larvae inhabit the plankton.

Phosphorescence. The emission of light produced by previously absorbed radiations, and therefore not strictly applicable to the so-called phosphorescent animals, whose light is produced by Bioluminescence.

Photosynthesis. The process by which plants synthesize carbohydrates from water and carbon dioxide, using energy absorbed from sunlight with the aid of the green colouring matter chlorophyll.

Photonasty. ⇨ NASTIC MOVEMENT.

Phototaxis. ⇨ TAXIS.

Phototropism. ⇨ TROPISM.

Phreatophyte. A plant with very long roots that reach down to the water table.

Phthiraptera. ⇨ LOUSE.

Phycomycetes. A class of minute, mostly one-celled and aquatic fungi, sometimes known as Algal Fungi; many species resemble flagellates in certain respects, others are known as Moulds. Well-known phycomycetes include the Downy Mildews, e.g. *Mucor mucedo*, which produces mould on bread; *Phytophthora infestans*, the cause of potato blight; and *Saprolegnia ferox*, the fungus disease which shows as white patches on salmon and other fish.

Phyllode. A green leaf-stalk that is flattened and so looks like a leaf.

Phylogeny. The history of the evolution of a species.

Phylum (pl. Phyla). The top group in the zoological Classification of living things, subdivided into Classes. For plants the equivalent term is Division.

Physiography. The study of the physical features of the earth's surface, formerly regarded as more or less synonymous with physical geography, but now usually regarded as more strictly confined to geomorphology.

Phytogeography. The study of the geographical distribution of plants.

Phytography. Writing about plants.

Phytophaga (Coleoptera: Polyphaga). A large superfamily of vegetarian beetles, containing some 350 species in three families, the longhorn beetles, the leaf beetles and the seed-eating Bruchidae, some of whose members are miscalled weevils.

Phytophagy. The eating of plants or vegetable matter.

Phytoplankton. Plants in the plankton.

Piciformes. The order of birds containing the woodpeckers (Picidae).

Piddock (Lamellibranchia: Pholadidae). Marine bivalves which specialize in boring into the softer rocks, such as chalk, slate, shale and New Red Sandstone; five species, of which the most widespread is the Oval Piddock *Ziphaea crispata*, the rest being confined to our south and south-west coasts, where the commonest is the Common Piddock *Pholas dactylus*. The superficially similar American Piddock *Petricola pholadiformis*, probably introduced with oysters, is common in the Thames estuary. The unrelated wood-boring Wood Piddock *Xylophaga dorsalis* looks like a shipworm.

Pigeon. A name given to some larger doves, especially the Woodpigeon and the Feral Pigeon, the escaped domesticated form of the rock dove, now common on buildings in large towns and frequent on sea cliffs. All domesticated pigeons are held to be descended from the rock dove.

Pig-Nut or Earth-nut (Umbelliferae: *Conopodium majus*). A common white-flowered umbellifer of grassy places, named from its edible root-tuber. The similar but larger Great Earth-nut *Bunium bulbocastanum* is a rarity at the foot of the chalk hills from the Chilterns towards Cambridge.

Pigweed (Amaranthaceae: *Amaranthus*). Goosefoot-like weeds of which many species occur as casuals in waste and disturbed ground, but only three are at all frequent: Common Pigweed *A. retroflexus*, Green Pigweed *A. hybridus* and White Pigweed *A. albus*.

Pike (Pisces: *Esox lucius*). Ferocious freshwater fish, with an elongated snout, feeding on other fish, water birds, water voles, etc., sometimes attaining 30–40 lb. or even more. Small pike are known as jacks. Gar Pike, Sea Pike, ➪ GARFISH. Saury Pike, ➪ SAURY.

Pilchard or Sardine (Clupeidae: *Sardina pilchardus*). A small gregarious marine fish, replacing the sprat in southern seas. Formerly gave rise to an important fishery in Cornish waters.

Pill Beetle (Diversicornia: *Byrrhus pilula*). A common beetle of low vegetation.

Pill Bug. ➪ Pill WOODLOUSE.

Pillwort (Pteridophyta: *Pilularia globulifera*). An aquatic non-flowering plant with bright green grass-like leaves and pill-like spore-cases; widespread but scarce.

Pilose. Botanical adjective meaning covered with rather long soft hairs.

Pilot Fish (Percomorphi: *Naucrates ductor*). An oceanic fish, well known for its habit of accompanying ships and sharks; rare in British waters.

Pimpernel (Primulaceae). Small prostrate plants of the Primrose Family, especially Scarlet Pimpernel *Anagallis arvensis*, a common weed named Shepherd's or Poor Man's Weather-glass from its habit of closing up when the sun goes in; the much less common Blue Pimpernel *A. foemina*; Bog Pimpernel *A. tenella* growing in wet places; and Yellow Pimpernel *Lysimachia nemorum* growing in woods.

Pimplet. Glaucous and Red-speckled Pimplets, ➪ SEA-ANEMONE.

Pine (Coniferae: *Pinus*). The Scots Pine *P. sylvestris* is one of our three native conifers, but is genuinely so only in the Scottish Highlands, the numerous Scots pines on the heaths of southern England being long naturalized. The Maritime Pine *P. pinaster* is naturalized around Poole Harbour, Dorset, and the Corsican or Austrian Pine *P. nigra* more or less so in various places. Several other species are also planted.

Pineapple Weed or Rayless Mayweed (Compositae: *Matricaria matricarioides*). An abundant weed, related to the daisy-like mayweeds, but lacking any ray florets, which smells of pineapple when crushed and has spread all over the British Isles since 1871.

Pin-eyed Flowers. Individual flowers, especially in the Primrose Family, in which the stigma is borne at the end of a long style and occupies the mouth of the corolla tube, while the anthers on short stamens are half-way down, as opposed to Thrum-eyed.

Pinhole Borer. ⇨ BARK BEETLE.

Pink-foot (pl. Pink-feet). Birdwatchers' slang for pink-footed goose.

Pink (Caryophyllaceae: *Dianthus*). Three native species of a group widely grown in gardens, two scarce; Maiden Pink *D. deltoides* and Deptford Pink *D. armeria*, and one rare, the fragrant Cheddar Pink *D. gratianopolitanus*, confined to Somerset. The allied Childling Pink *Kohlrauschia nanteulii* is also very scarce. Sea Pink, ⇨ THRIFT.

Pinnate. Of a compound leaf, with the leaflets arranged in opposite pairs on either side of the midrib, like a feather. In a *Pinnatifid* leaf the leaflets are joined at the base. In a *Pinnatisect* leaf the upper leaflets are pinnatifid, but the lower ones almost but not quite pinnate. ⇨ LEAF-SHAPES.

Pinnipedia. An order of marine carnivorous mammals, comprising the seals and walruses.

Pintail (Anatidae: *Anas acuta*). A surface-feeding duck, whose drake has a long pointed tail; widespread and frequent as a winter visitor, breeding mainly in the north.

Pinworm. ⇨ ROUNDWORM.

Pioneer Community. The first plants to colonize soil that is bare of other vegetation, usually annuals. On bare rock the first colonists are usually lichens.

Pipe-Fish (Pisces: Syngnathidae). A family of curious little elongated marine fishes found in shallow water, the commonest being the Greater Pipe-fish, Needle-fish or (in Lancashire) Horn-eel *Syngnathus acus* and the Broad-nosed Pipe-fish *Siphonostoma typhle*.

Pipewort (Eriocaulaceae: *Eriocaulon septangulare*). An aquatic monocotyledon, with a spike of small white flowers above the water surface; very local in western Ireland and on Skye and Coll in the Inner Hebrides.

Pipistrelle (Chiroptera: *Pipistrellus pipistrellus*). The smallest British bat and the commonest of the smaller ones.

Pipit (Passeriformes: *Anthus*). Small song birds allied to the wagtails. Seven British species: Meadow Pipit *A. pratensis*, one of the commonest British birds, widespread on moors and mountains, formerly known as Titlark; Richard's Pipit *A. novaeseelandii*, an irregular visitor; Rock Pipit *A. spinoletta*, common on rocky coasts; Tawny Pipit *A. campestris*, an irregular visitor; Tree Pipit *A. trivialis*, widespread and frequent where there are scattered trees, from which it starts the distinctive song flight that is its best distinction from the very similar meadow pipit; Water Pipit a Continental race of the rock pipit, scarce winter visitor.

Pirri-Pirri Bur or New Zealand Burweed (Rosaceae: *Acaena anserinifolia*). A widely established but very local alien from New Zealand, with round heads of whitish flowers that turn to bur-like fruits with soft reddish spines.

Pisces (Chordata: Vertebrata). The true or bony fishes, a class of cold-blooded, gill-breathing aquatic vertebrates, characterized by their bony skeletons and scaly outer covering, represented in British waters by twenty orders containing twenty-eight exclusively freshwater species, 246 marine ones, a great many of which are mere stragglers from deep water or distant seas, and a dozen species which can spend either all or part of their lives in both fresh and salt water. All except the primitive sturgeon (Palaeopterygii) belong to the sub-class Neopterygii. The most important classes are the Isospondyli (Herring and Salmon Families), Ostariophysi (Carp Family), Apodes (eels), Anacanthini (Cod Family), Percomorphi (perches, wrasses, gobies, blennies), Scleroparei (gurnards, sticklebacks, bullheads) and Heterosomata (flatfish).

Piscivorous. Fish-eating.

Pistil or Gynoecium. The female parts of a flower, consisting of one or more carpels.

Pitcher Plant (Sarraceniaceae: *Sarracenia purpurea*). A remarkable purple-flowered North American insectivorous plant, naturalized in a few Central Irish bogs, whose tubular leaves have a flap at the top and are filled with a fluid in which insects drown. The fluid contains enzymes that digest the insects.

Plaice (Pleuronectidae: *Pleuronectes platessa*). A common right-sided shallow-water flatfish, one of the most important commercial species, characterized by the red spots on its upper surface.

Plain. A large area of land with a more or less level surface, usually low-lying but sometimes an upland plateau, broken by gentle declivities, as with Salisbury Plain. A Flood Plain or Alluvial Plain is found along river valleys, where the river in spate overflows its banks and spreads out an alluvial sediment which tends to produce an almost flat surface. A Coastal Plain is the flat area often found between the hills and the sea, e.g. between the South Downs and the English Channel in West Sussex, and may represent a former beach from a time when sea-level was higher.

Planarian (Turbellaria: Tricladida). Free-living flatworms, mainly aquatic and freshwater, but some living on the seashore, and a few on land. Our largest planarians, the freshwater *Bdellocephala punctata* and the marine *Prostheceraeus vittatus*, may be over an inch long, but most species are much smaller. Our commonest land planarian is *Microplana terrestris*, found in damp shady places.

Plane, London (Platanaceae: *Platanus hybrida*). Tall deciduous tree with palmate leaves, widely planted as a street tree because its flaking bark enables it to withstand atmospheric pollution, but not naturalized.

Planipennia (Insecta: Neuroptera). The sub-order containing the green lacewings (Chrysopidae), the brown lacewings (Hemerobiidae), the giant lacewing (Osmylidae) and the sponge flies (Sisyridae).

Plankton. Animals, often larvae, and plants which float in the sea and are drifted by tides and currents, as distinct from the Nekton which swims.

Plant. ⇨ VEGETABLE KINGDOM.

Plantain (Plantaginaceae: *Plantago*). Five rather dull-looking plants with cylindrical or ovoid spikes of small flowers with conspicuous stamens. The two commonest and most widespread are Ribwort Plantain or Rib Grass *P. lanceolata*, with short, blackish flower-spikes and lanceolate leaves, abundant in grassland; and Ratstail Plantain *P. major*, with long greenish flower-spikes and oval leaves, common in bare and waste places, especially footpaths. More eclectic are the pink-flowered Hoary Plantain *P. media*, of dry, especially calcareous grassland; Buckshorn Plantain *P. coronopus*, with divided leaves, of bare places on sand and gravel; and Sea Plantain *P. maritima*, in saltmarshes and grassy places near the sea.

Plantation. A wood in which the trees have been planted by man, as distinct from a natural wood, where the trees have regenerated.

Plant Bug (Hemiptera: Heteroptera). Most of the bugs of the sub-order Heteroptera; including the assassin bugs (Reduviidae), barkbugs (Aneuridae), bedbugs and flower bugs (Cimicidae), beetbugs (Piesmidae), capsid bugs (Saldidae), damsel bugs (Nabidae), firebug (Pyrrhocoridae), flatbugs (Aradidae), groundbugs (Lygaeidae), lacebugs (Tingidae), four families of shieldbugs, squashbugs (Coreidae), spurgebug (Stenocephalidae), stiltbugs (Bertinidae). The Tarnished Plantbug *Lygus rugilipennis*, Lucerne Plantbug *Adelphocoris lineolatus* and Meadow Plantbug *Leptopterna dolabrata* are all capsids.

Plant Community. A recognizably distinct group of plants growing together in the same habitat and with some degree of interdependence. ⇨-ETUM.

Plasmodium. The vegetative body of slime fungi.

Plastron. The lower part of the shell of a tortoise or turtle, covering the animal's belly.

Plateau. A tract of more or less level and fairly uniformly elevated land.

Platyhelminthes. ⇨ FLATWORM.

Plecoptera. ⇨ STONE-FLY.

Pleistocene Period. The most recent geological period, down to the end of the last Ice Age. ⇨ GEOLOGICAL PERIODS.

Pleuronectidae (Pisces: Heterosomata). A family of flatfish with both eyes on the right side of the body, containing the dabs, halibut, flounders, plaice and lemon sole.

Pleurotremata (Vertebrata: Selachii). The class of marine fishes containing the sharks and dogfishes.

Pliocene Period. The penultimate geological period; ⇨ GEOLOGICAL PERIODS.

Ploughman's Spikenard (Compositae: *Inula conyza*). A tall plant with foxglove-like leaves and clusters of yellow brush-like flowers, locally frequent in calcareous grass and scrub in England and Wales; named from its fragrant root.

Plover (Charadriiformes: *Charadrius*). Long-legged, short-billed waders, of which five species breed in Britain: Lapwing or Green Plover; Golden Plover *C. apricarius*, breeding on moors, common in the north, more local in the west, widespread on the seashore in winter; Ringed Plover *C. hiaticula*, widespread and common on the coast, very local inland; Little Ringed Plover *C. dubius*, an increasing summer visitor to south, east and central England since 1938, breeding especially at gravel-pits; and the now extremely rare Kentish Plover *C. alexandrinus*, perhaps still breeding on shingle banks in one or two places in the south. The Grey

Plover *C. squatarola* is a widespread and common shore bird in winter, and four other plovers are rare stragglers. Norfolk Plover, ⇨ STONE-CURLEW.

Plum; Cherry Plum. ⇨ PRUNUS.

Plumage. ⇨ FEATHER.

Plume Moth (Pyraloidea: Pterophoridae). A family of micros, whose forewings have plume-like lobes; one of the commonest and most widespread is the pure white *Alucita pentadactyla*. The Many-plume Moth *Orneodes hexadactyla*, also widespread and common, is our sole representative of the Orneodidae.

Plutonic Rocks. Igneous rocks which have solidified deep down in the earth.

Pochard (Anatidae: *Aythya ferina*). A diving duck, common as a winter visitor, but rather local in the breeding season. The Red-crested Pochard *Netta rufina* is an uncommon but increasing winter visitor from the Continent. The White-eyed Pochard or Ferruginous Duck *A. nyroca* is a straggler from southern Europe.

Pod. A dry fruit containing one or more fruits and splitting down two sides. ⇨ CAPSULE, FOLLICLE.

Podicipediformes or Podicipitiformes. ⇨ GREBE.

Podsol. A Russian word meaning 'ash', used to describe a soil in which the minerals have been leached out of the surface layer by water and deposited lower down, because the leached soil is often ashy in colour and consistency. On the surface of a podsol is a layer of decomposing plant material, very acid, the raw humus or Mor; below that is the leached layer of sand, very poor in salts; below that again is the layer in which the minerals are deposited, the Hard Pan, sometimes so thick and hard that even tree roots cannot penetrate it. Podsols are common on sandy heaths, because of the quick drainage, and in coniferous forest, especially in the cooler and wetter northern and western parts of the British Isles. Birch, spruce and pine, and the heath and moor plants grow on podsols.

Pod-Weed (Phaeophyta; *Halidrys siliquosa*). A bushy brown seaweed, somewhat resembling the wracks, with elongated pods, widespread but rather local, mainly on rocky shores.

Pogge or Armed Bullhead (Scleroparei: *Agonus cataphractus*). A small, inshore fish, grotesque-looking from its angular head and body, which are covered with bony plates; common and mainly northern.

Pogonophore (Pogonophora). A little known phylum of tubular marine animals, not discovered until 1914, nor recognized as a separate phylum till 1937. They live within a fibrous tube on the sea bottom, usually in very deep water, but a few species at only moderate depths near the shore, even in the North Sea.

Poikilothermic. 'Cold-blooded', with temperature varying according to the surroundings, as with reptiles and fishes; as opposed to Homoiothermic or 'warm-blooded'. The state of torpidity into which swifts and a few other birds occasionally fall is known as temporary poikilothermy.

Poisonous Animals. Animals in many groups use poison as a means of catching their food or deterring their enemies: among them reptiles (adder), amphibians (toad), fishes (weevers), insects (especially bees and wasps) and spiders.

Poisonous Fungi. There are fewer poisonous than edible fungi, but some of the few are lethal, such as the death cap, responsible for most of the thirty-nine human deaths from fungus poisoning in England and Wales between 1920 and 1950, and its relatives, the destroying angel and the fool's mushroom. Fly agaric, the panther, and the yellow-stainer are examples of toadstools which are definitely poisonous but not necessarily lethal. ⇨ EDIBLE FUNGI. Fungi poisonous to man are not necessarily poisonous to other animals, especially insects.

Poisonous Plants. Many native flowering plants are poisonous, in the sense that eating a small portion, whether stem, root, leaf or fruit, will upset the normal health of creatures susceptible to the poisons they contain. Members of thirty-two different families of flowering plants fall within this definition; buttercups, e.g., are poisonous to cattle. The principal poisons secreted are volatile oils and acrid substances, glycosides and plant alkaloids. The most dangerous poisonous plants are monkshood, white bryony*, hemlock, cowbane, water dropworts, ivy*, ragwort, privet*, thorn-apple*, henbane, belladonna*, bittersweet*, black nightshade*, foxglove, black bryony* and yew*. Those starred have poisonous berries. The nectar of rhododendrons is also poisonous. ⇨ POISONOUS FUNGI.

Polecat (Carnivora: *Mustela putorius*). Resembles a stoat, and is now local and uncommon, except in parts of central Wales. From its objectionable scent it used to be known as the foumart or foul-marten, in contrast to the pine marten, which was the sweet-mart. In lowland districts most reports of polecats can be attributed to escaped polecat-ferrets, the hybrid between the polecat and the ferret. In some districts substantial numbers of polecat-ferrets are at large.

Pole Dab, Pole Flounder, ⇨ WITCH.

Policeman's Helmet. ⇨ Himalayan BALSAM.

Pollack (Gadidae: *Gadus pollachius*). A marine fish of the Cod Family, like a dark green coal-fish but with a more inshore and southern distribution.

Pollan (Salmonidae: *Coregonus*). Three Irish species of whitefish; two fished commercially in L. Neagh, *C. pollan* and *C. altior*; and the third, *C. elegans*, rare in the Shannon lakes.

Pollard. A tree which has been beheaded at about 6–7 ft from the ground, so as to produce a crop of thin branches suitable for withies, fencing or firewood; its large root system in relation to its low height gives a pollard great stability, so that it is useful in stabilizing river banks. Willows, especially the crack and white willows, are the trees most often treated in this way; pollard willows line many hundreds of miles of the banks of English lowland streams.

Pollen. Minute spores produced in the anthers of flowers and transferred by wind or insects to the stigmas to start the process of fertilizing the ovules.

Pollen Analysis or Palynology. Pollen grains of many plants, if embedded in peat, keep their characteristic appearance and can be identified after thousands of years. Thus an examination of the pollen grains in the thick layers of peat in fens, bogs and moors can indicate what plants were growing in the surrounding countryside when the peat was laid down. By this method it has been possible, e.g., to prove that Julius Caesar was wrong in saying that the beech did not grow in Britain in the first century B.C.

Pollinia. Aggregations of pollen, found especially in orchids.

Polychaeta. ⇨ BRISTLE WORM.

Polyclad (Turbellaria: Polycladida). Aquatic (marine) flatworms, the order containing the largest Turbellaria.

Polyembrony. The division of one egg into several individual animals, as with identical human twins and certain chalcid wasps.

Polygamy. Of animals, having more than one mate; of plants, having male, female and hermaphrodite flowers either on the same or on different individual plants.

Polymorphism. The occurrence together in the same habitat of two or more sharply contrasted varieties of an animal or plant species in proportions large enough to make it clear that the rarest of them is not being maintained by recurrent mutation. Examples in the British flora and fauna are the bridled form of the guillemot, the greenish variety *valesina* of the silver-washed fritillary butterfly, and the pink variety *lancastriense* of the bloody cranesbill.

Polyp. ⇨ ZOOID.

Polyphaga (Insecta: Coleoptera). The sub-order containing the Omnivorous Beetles, the great majority of the British species, grouped in eight super-families; Palpicornia (scavenger beetles), Staphylinoidea (burying, carrion and rove beetles), Diversicornia (glow-worms, soldier and sailor beetles, click beetles), Clavicornia (ladybirds, powder post and furniture beetles), Heteromera (cardinal, oil and blister, and nocturnal ground beetles), Lamellicornia, (chafers, dung and stag beetles, Phytophaga (leaf and long-horn beetles), and Rhynchophora (weevils, bark and ambrosia beetles).

Polyphagous. Of an animal that eats many kinds of food.

Polyplacophora. ⇨ AMPHINEURA.

Polyploidy. The condition of organisms having three or more times the haploid (basic) number of chromosomes in each cell nucleus; common in plants, especially the flowering plants, but rare in animals. Triploids have their chromosomes in threes; tetraploids in fours; pentaploids in fives; hexaploids in sixes; and octoploids in eights. ⇨ ALLOPOLYPLOID, DIPLOID.

Polypody (Polypodiaceae: *Polypodium vulgare*). One of our commonest ferns, with simply pinnate leaves, in woods, where it is often an epiphyte on trees, and on walls, rocks and dunes. Two closely allied species have recently been shown also to occur in Britain, the widespread ***P. interjectum*** and the very local and lime-loving ***P. australe***. Limestone Polypody, ⇨ OAK FERN.

Polypore. ⇨ BRACKET FUNGUS.

Polyzoa. ⇨ SEA-MAT.

Pome. A fruit, such as an apple, in which the seeds are surrounded by a more or less fleshy layer.

Pond. A small sheet of inland fresh water, not clearly distinguished from a small lake.

Pond Skater or Water Strider (Heteroptera: Gerridae). Nine slender, elongated predatory surface water bugs; *Gerris lacustris* is the one most often seen gliding over the surface water film on ponds, ditches and slow streams.

Pond Snail (Gastropoda: Pulmonata). A general term for the freshwater pulmonate snails, more specifically applicable to those of the genus *Limnaea*, especially the widespread and common Great Pond Snail *L. stagnalis*, which has a narrowly spiral shell up to 2 ins. long and is sometimes carnivorous, attacking even newts and small fish, and the much smaller Dwarf Pond Snail *L. truncculata*, a dweller in damp grassland, which plays an unhappy role as the secondary host of the Liver-fluke. The Eared Pond Snail *L. auricularia* is named from the ear-like shape of the opening of its shell. ⇨ MARSH SNAIL, MUD SNAIL, WANDERING SNAIL.

Pondweed. Name of waterweeds of four monocotyledonous families, especially the Potamogetonaceae, with two dozen species, the commonest being the Broad-leaved Pondweed *Potamogeton natans*, whose pointed oval leaves float on the surface of the water. The two Tassel Pondweeds (Ruppiaceae: *Ruppia*) grow only in brackish water; Horned Pondweed (Zannichelliaceae: *Zannichellia palustris*) and Fennel Pondweed *P. pectinatus* grow in both fresh and brackish water. The widespread alien Canadian Pondweed (Hydrocharitaceae: *Elodea canadensis*) was a serious pest of waterways soon after it was first introduced, but has now settled down as a regular member of our aquatic flora.

Pondweed Bug (Heteroptera: *Mesovelia furcata*). A small surface water bug frequenting the floating leaves of water plants, mainly in the south-east.

Pony. A horse not more than 13 or 14 hands high (a hand equals 4 ins.). Horses *Equus caballus* belong to the Perissodactyla order of the ungulates. The ponies of Exmoor, Dartmoor and the New Forest are not truly wild, but domestic animals that are allowed to run wild. There are also a few feral ponies in the Lake District.

Poor Man's Beefsteak, or Beefsteak Fungus (Basidiomycetes: *Fistulina hepatica*). An edible bracket fungus, up to a foot across, whose flesh looks not unlike meat; also known as Ox-tongue Fungus.

Poor Man's Weatherglass. ⇨ Scarlet PIMPERNEL.

Pope or Ruffe (Percomorphi: *Acerina cernua*). A smallish gregarious freshwater fish, allied to the perch, rather local in still or slow-moving waters in the south.

Poplar (Salicaceae: *Populus*). Medium to tall deciduous catkin-bearing trees, preferring damp soil. Three native species: Black Poplar *P. nigra*, Grey Poplar *P. canescens* and Aspen; and three frequently planted species: the more or less naturalized White Poplar *P. alba*, the Black Italian Poplar *P. x canadensis*, of which only male trees are known, and the aromatic

Balsam Poplar *P. gileadensis*. The Lombardy Poplar is an often planted variety of *P. nigra*.

Poppy (Papaveraceae). Our seven native poppies have large, handsome red or yellow flowers, with two sepals and four petals, and white juice in the stem, except for *P. lecoqii*, which has yellow juice. Corn Poppy *Papaver rhoeas* and Long-headed Poppy *P. dubium* are the commonest of the four red ones, all growing in cultivated or disturbed ground. The related Opium Poppy *P. somniferum*, whose unripe seed-capsules provide the drug, with pale purple flowers, is rare in cornfields and also escapes from gardens. The two yellow poppies are the Welsh Poppy *Meconopsis cambrica*, a local native in Wales and south-west England but widespread as a garden escape; and Yellow Horned Poppy *Glaucium flavum*, frequent on coastal shingle, with the longest seed-pods (6–12 ins.) of any British plant.

Porbeagle (Pleurotremata: *Lamna cornubica*). One of the less common large sharks in British waters.

Porifera. ⇨ SPONGE.

Porpoise (Cetacea: *Phocaena phocaena*). The smallest (4–6 ft) and also much the commonest whale in British seas, often seen offshore in small schools, characteristically surfacing two or three times and then disappearing. Allied to the dolphins, but lacking their beak.

Porrect. Botanical adjective meaning pointing both outwards and forwards.

Portuguese Man-o'War (Siphonophora: *Physalia physalia*). An oceanic siphonophore, floating on the surface of the sea, with numerous polyps, some several feet long and others with ferocious stinging cells (never touch one with bare hands) hanging down from its gas-filled 6-in. bladder. An erratic visitor from the open ocean, especially to our western coasts after long periods of south-westerly winds.

Potato Blight (Phycomycetes: *Phytophthora infestans*). A serious fungal disease of potatoes, the cause of the great Irish famine of the 1840's.

Pot Hole or Sink Hole. An aperture in the ground in limestone districts, such as the White Peak of Derbyshire or the Craven Pennines of West Yorkshire, through which streams flow deep underground, often for considerable distances before emerging again. On the way, the solvent chemical reaction of the carbon dioxide in rain water on the lime in the rock enables them to carve out substantial underground caverns, the exploration of which forms the scientific sport of Pot-holing. ⇨ SPELAEOLOGY.

Potter Trap. A simple trip-release trap used by bird ringers for catching birds.

Potter Wasp (Aculeata: *Eumenes coarctata*). A small black and yellow solitary wasp named from its habit of making a nest shaped like a flask or pot from dabs of clay cemented with saliva and attached to the shoots of heather or other plants.

Pot Worm. ⇨ WHITE WORM.

Pout. The name of several members of the Cod Family (Gadidae), the most frequent in British seas being the mainly southern and western Pout, Bib or Whiting Pout *Gadus luscus* and the small Norway Pout *G. esmarkii* of the North Sea. ⇨ EEL-POUT.

Poutassou. ➪ Couch's WHITING.

Powan (Salmonidae: *Coregonus clupeoides*). A species of whitefish found in Lochs Esk and Lomond in Scotland, also in North Wales and the Lake District, where the local races are known as Gwyniad and Schelly respectively.

Powder-Post Beetle (Clavicornia: Lyctidae). Six species of wood-boring beetles, named from the fine powdery 'sawdust' from their borings; *Lyctus fuscus* is the commonest species.

Power. ➪ Poor COD.

Pratincole (Charadriiformes: *Glareola pratincola*). A tern-like wader, a straggler from the Mediterranean region.

Prawn (Malacostraca: Decapoda Natantia). Large shrimps, frequenting rock and other pools on the shore, the common species including the Aesop Prawn *Pandalus montagui* and the Common Prawn *Leander serratus*.

Pre-Adapted. Of an animal or plant already adapted in such a way that it is also adapted to some new habitat into which it comes by natural spread, or more often by human intervention, e.g. the introduced little owl in England. ➪ ECOLOGICAL NICHE.

Pre-Cambrian Era. ➪ GEOLOGICAL PERIODS.

Predator. A living organism that preys on another, and thus keeps the environment healthy (by preventing excess population) and species fit (by removing the least well adapted). The line between predators and ectoparasites is very difficult to draw. The term is often restricted to vertebrate predators on other vertebrates.

Preening. The process of cleaning a bird's feathers, and restoring them to a smooth condition after ruffling. False-preening is preening done at an apparently inappropriate moment as a Displacement Activity.

Premorse. Botanical adjective used of rootstocks and other organs which end abruptly and look as if they had been bitten off, e.g. devil's bit scabious, common catsear.

Pricket. A two-year-old stag or fallow buck.

Prickle. A short, sharp type of thorn, irregularly scattered over the stems and leaves of especially roses and brambles. *Pricklets* are small prickles.

Primrose (Primulaceae: *Primula vulgaris*). A widespread and common yellow spring flower of woods and hedge-banks. Birdseye Primrose *P. farinosa* and Scots Primrose *P. scotica* are smaller pink-flowered plants of north England and north Scotland respectively. Evening Primrose is unrelated. ➪ PIN-EYED and THRUM-EYED.

Prisere. The primary sere or complete succession of plant communities in an area starting with bare ground and progressing to the climatic Climax.

Privet (Oleaceae: *Ligustrum vulgare*). Widespread and common half-evergreen shrub, with white flowers and black berries. The privet of garden hedges is the Chinese *L. ovalifolium*.

Procellariiformes. A fairly primitive order of sea birds, including the petrels, shearwaters, fulmars (Procellariidae) and albatrosses (Diomedeidae). They come to land only to breed. On account of their tubular nostrils they are sometimes called tube-noses.

Procumbent. Botanical adjective used of plants that lie rather loosely along the ground, as distinct from prostrate plants which lie more closely.

Profile. The series of different layers to be seen in a vertical cross-section of soil down to the subsoil or parent rock, as in the vertical sides of a pit. The main layers are called Horizons which are subdivided into Zones. ⇨ SOIL.

Prominent Moth (Notodontoidea: Notodontidae). A family of fairly large moths, characterized by an erect tuft of scales on each forewing, prominent when they are at rest; thirteen of the twenty-five members of the family are so named, the commonest being the Coxcomb Prominent *Lophopteryx capucina*; ⇨⇨ BUFF-TIP MOTH, KITTEN MOTH, PUSS MOTH, LOBSTER MOTH.

Propolis. A resinous substance exuded by certain plants, used by honey-bees to seal up tiny crevices in the hive.

Prosobranchia. ⇨ OPERCULATE Molluscs.

Prostigmata (Arachnida: Acari). The largest and most important order or sub-order of mites, including most of the mites which feed on higher plants, as well as many predatory species and parasites on both vertebrates and invertebrates. Includes the earth mites (Trombiidae), eriophyid mites (Eriophyinae) and snout mites (Bdellidae). ⇨⇨ BIG BUD MITE, BULB MITE, GALL MITE, HARVEST MITE, ISLE OF WIGHT DISEASE, ITCH MITE, MANGE, RED SPIDER, STONE MITE, VELVET MITE, and WATER MITE.

Protophyta. An outmoded grouping in which the Algae were included with the Flagellata.

Protoplasm. The physical basis of life found in all living cells, a protein jelly.

Protozoan (Protozoa). The most primitive phylum of animals, mostly microscopic, consisting of a single cell or unit of protoplasm, which performs all the vital functions; usually classed as a separate sub-kingdom. Most of our protozoans live in fresh water, but some live on the shore and others in the soil, where they prey on bacteria. They are divided into four classes: the flagellates (Mastigophora), the Rhizopoda, the parasitic Sporozoa and the Ciliata (Infusoria).

Protura (Arthropoda: Insecta). An order of tiny wingless insects, usually less than 0.05 in. long, found under bark and stones, and in turf and light soils; seventeen British species.

Pruinose. Botanical adjective meaning covered with a whitish bloom.

Prunus (Rosaceae: *Prunus*). A group of shrubs or small trees, bird-sown or planted in hedgerows, including the Cherry Plum *P. cerasifera*, which flowers before the blackthorn; the Bullace *P. insititia*, now more often cultivated as the damson; the Greengage *P. italica* and the Plum *P. domestica*. ⇨⇨ BLACKTHORN, CHERRY.

Pseudoscorpiones or Chelonethi. ⇨ FALSE-SCORPION.

Psittacosis. A virus disease of birds, especially parrots, which also occurs in some British wild birds, including sparrows, fulmars and feral pigeons.

Psocid (Insecta: Psocoptera). An order of tiny insects whose sixty-eight British species fall into two main groups: a large one of mainly winged species living on the bark or leaves of trees or in undergrowth, feeding on fungi, lichens and algae and called bark-lice; and a smaller one of wingless species inhabiting buildings and feeding on moulds, stored products and starchy materials, known as book-lice.

Psocoptera. ⇨ PSOCID.

Psychoidea (Lepidoptera: Heteroneura). A superfamily of moths containing three families of which the best known is the Zygaenidae (burnets and foresters). The two others are the Limacodidae with only two British species, and the Psychidae, a family of 'micros' with ten species.

Psyllid. ⇨ SUCKER (2).

Ptarmigan (Galliformes: *Lagopus mutus*). The grouse of high mountain tops in the Scottish Highlands, whose plumage turns almost completely white in winter.

Pteridophyta. A phylum or major grouping of vascular plants containing the ferns (Pteropsida) and their allies, the horsetails (Sphenopsida), club-mosses and quillworts (Lycopsida), differing from the other vascular plants in reproducing from spores instead of seeds.

Pteropsida. ⇨ FERN.

Pubescent. Botanical adjective meaning covered with short, soft, downy hairs. *Puberulent* means slightly or minutely pubescent.

Pudding-Stone. A conglomerate rock, probably the remains of an ancient shingle beach, consisting of pebbles firmly cemented together.

Puff-Ball (Basidiomycetes: *Lycoperdon*). Roundish or pear-shaped fungi, growing in woods and grassland, whose ripe brown spores can be puffed out in a cloud by pressing the fruit-body. Giant Puff-ball *L. giganteum*, the largest species, averages about a foot in diameter, and has been known to grow to 5 ft across. Common smaller species are Common Puff-ball *L. perlatum* and Mosaic Puff-ball *L. caelatum*, easily told by the mosaic of hexagonal cracks on its surface.

Puffin (Charadriiformes: *Fratercula arctica*). Our smallest breeding auk, with grotesque bill and head adornments in the breeding season. Nests in burrows on cliffed coasts and marine islands, mainly in the north and west; winters at sea.

Pug Moth (Notodontoidea: Hydriomenidae). Many of the small geometrid moths of this family are so named, often from their food-plant, e.g. the Toadflax Pug *Eupithecia linariata*, whose larvae feed on the flowers of yellow toadflax *Linaria vulgaris*.

Pullus. ⇨ YOUNG BIRD.

Pulmonata (Mollusca: Gastropoda). The sub-class of gastropods containing the land snails and slugs and most of the freshwater snails, characterized by having lungs instead of gills, and, unlike the superficially similar operculates, no operculum. The shells of the snails are mostly recognizably akin, either to the familiar garden snail or to the marine winkles.

Punctate. Botanical adjective meaning dotted with tiny excrescences or pitted with tiny depressions.

Pungent. Botanical adjective meaning equipped with a point sharp enough to prick the skin.

Pupa. The third, non-mobile but actively changing stage of the metamorphosis of many insects; verb, *to pupate.* ⇨ CHRYSALIS.

Purple Emperor (Papilionina: *Apatura iris*). A handsome purple-barred vanessid butterfly, frequenting the tops of trees; scarce in oakwoods in the Midlands and south.

Purple Worm. ⇨ CHESTNUT WORM.

Purse-Web Spider. ⇨ TRAP-DOOR SPIDER.

Purslane. A name given to various small and mostly unrelated plants, especially Water Purslane (Lythraceae: *Peplis portula*), frequent on wet mud on acid soils, and Sea Purslane, which is either sea sandwort, or *Halimione portulacoides*, a relative of the oraches which is often abundant in saltmarshes.

Puss Moth (Notodontoidea: *Cerura vinula*). A large Prominent, noted for its grotesquely shaped caterpillars, with forked 'tails', which can adopt a remarkable attitude presumably in order to frighten potential enemies; widespread and common on poplars and willows.

Pwdre Ser. ⇨ STAR-SLIME.

Pycnogonida. ⇨ SEA SPIDER.

Pyralid Moth (Pyraloidea: Pyralidae). A family of large 'micros', including the small magpie moth, the grass moths and several important pests of stored products. ⇨ HONEYCOMB MOTH, HONEY MOTH.

Pyraloidea (Lepidoptera: Heteroneura). A superfamily of small moths consisting entirely of Microlepidoptera, some of which are actually larger than the so-called Macrolepidoptera, and roughly divided into the Pyralid and Plume Moths.

Pyramidal Orchid (Orchidaceae: *Anacamptis pyramidalis*). A locally frequent bright pink orchid of calcareous turf and dunes, whose flower-spike is arranged in a flattish pyramid.

Pyrrophyta. A phylum or major botanical division into which botanists put certain flagellates which they regard as algae but which zoologists class as protozoans of the sub-class Phytomastiginia. ⇨⇨ EUGLENOPHYTA.

Q

Quail (Galliformes: *Coturnix coturnix*). The smallest British game bird, a local summer visitor breeding mainly in the south. A small flock or family party is known as a bevy.

Quaking Grass, Quaker Grass or Totter Grass (Gramineae: *Briza media*). A common grass, especially of calcareous grassland, with rounded individual flower spikelets on very thin stalks, which shake in every air current.

Quartzite. A hard white metamorphic rock, originating from sandstone and entirely composed of grains of quartz cemented together with other forms of silica.

Queen. The fertile female of ants, bees and wasps.

Quekett Microscopical Club. A national society for the furtherance of microscopy, founded in 1865; publishes a *Journal.*

Quick; Quickthorn. The hawthorn when planted in a hedge.

Quillwort (Lycopsida: *Isoetes*). Three species of flowerless plants with tufted leaves, two growing on lake or pool bottoms in acid water, in the north and west, and the third very locally in turf near the sea in the Lizard peninsula; *I. lacustris* is the most frequent.

R

Rabbit (Lagomorpha: *Oryctolagus cuniculus*). (Male, buck; female, doe). This burrowing native of the Mediterranean countries was introduced to England for its flesh and fur at the end of the twelfth century, but escaped from its warrens and spread over the whole countryside, where it has formed one of the principal biotic factors affecting the vegetation, especially on chalk downs and southern heathlands. During the hundred years prior to 1953 it became a severe pest in many parts of the country, but the myxomatosis epidemic which began in that year severely reduced its numbers. Though it has since recovered somewhat, myxomatosis has become more or less endemic in many districts, and breaks out again as soon as the rabbit population begins to increase. ⇨ MYXOMATOSIS.

Rabbit-Fish (Bradyodonti: *Chimaera monstrosa*). The commonest of the three British Chimaeridae, found mainly off the northern isles; named from its large cutting teeth and also known as Rat-fish from its long tail and King of the Herrings from its supposed appearance at the same time as the herring shoals.

Race. A geographical subspecies.

Raceme. ⇨ INFLORESCENCE.

Rack. A path worn by deer.

Radish (Cruciferae: *Raphanus*). Wild Radish *R. raphanistrum* is a common arable weed with white or yellow flowers. Sea Radish *R. maritimus*, a yellow-flowered perennial, grows mainly on the south and west coasts. Garden Radish *R. sativus* sometimes escapes from gardens. ⇨⇨ HORSE-RADISH.

Ragged Robin (Caryophyllaceae: *Lychnis flos-cuculi*). A widespread pink-flowered ally of the campions, with petals appearing tattered, common in fens and damper grassland.

Ragstone. A sandstone hard enough to be used for building; the calcareous form found in Kent is known as Kentish Rag.

Ragworm (Polychaeta: Nereidae). A family of active marine bristle worms, often brightly coloured, much the largest being the green *Nereis virens*, usually over a foot long and sometimes attaining 3 ft and the thickness of a man's finger, which is locally common on our north and west coasts. Tube-dwelling bristle worms of the genus *Nerine* are also called ragworms. The White Ragworm is *Nephthys caeca*, allied to the catworm, much used as bait by fishermen under the names of Lurg, Lurcher and Glutton.

Ragwort (Compositae: *Senecio*). Composites with yellow daisy-like flowers closely related to the groundsels, the commonest being *S. jacobaea*, widespread in dry grassland, which is poisonous to farm stock, and a common food plant of the cinnabar moth; also widespread and common are Marsh Ragwort *S. aquaticus* in wet grassy places, Hoary Ragwort

S. erucifolius, in grassland on heavy soils, and Oxford Ragwort *S. squalidus*, a Mediterranean species, which escaped from the Oxford Botanic Garden, and has spread, largely along the railway system, to waste ground in all parts of the British Isles, though still rather local in Scotland and Ireland. Silver Ragwort *S. cineraria*, with whitely woolly stems and leaves, is naturalized on sea cliffs in the south. ⇨ GROUNDSEL.

Rail. An Irish name for caddis-flies.

Rail, Water (Ralliformes: *Rallus aquaticus*). A small shy moorhen-like marsh bird, widespread but local in marshes, swamps and fens. Land Rail, ⇨ CORNCRAKE. Sora Rail, ⇨ Carolina CRAKE.

Rain. Water, condensing on a minute particle of dust in a cloud, and coalescing with other nearby particles into a drop heavy enough to fall to the ground. The average rainfall over Great Britain is 34.2 ins. over England and Wales and 52.4 ins. over Scotland, the spring months being relatively dry and the autumn months relatively wet. In the west the winter tends to be wet, but in the east there is a tendency to a wetter summer.

Rainbow. An arc of prismatic colours formed in the sky when the sun, at an altitude (in British latitudes) of 42° or less, shines on a rainstorm, its rays being reflected and refracted on the raindrops. The colours are red, orange, yellow, green, blue, indigo and violet. The fainter double rainbow sometimes seen results from the double reflection of the rays within the raindrop. ⇨⇨ FOG-BOW.

Rainworm (Nematoda: Mermithidae). A family of mainly terrestrial roundworms up to 8 ins. long, named from their habit of emerging from the soil and crawling on vegetation after rain in summer.

Raised Beach. A beach that has been raised above the level of the sea by a rise in the level of the land; often the old sea-cliffs, now inland, can be seen behind it. Clearly seen, for instance, round the coasts of Devon and Cornwall, about ten feet above the present high-water level.

Ralliformes. The order of birds which includes the cranes (Balearicidae), rails, crakes, moorhen and coot (Rallidae) and bustards (Otididae).

Rampion (Campanulaceae: *Phyteuma*). Two local southern plants, with compound heads like composites: Round-headed Rampion *P. tenerum*, with globular blue flowerheads, growing in chalk turf is much the commoner.

Ramshorn Snail or Trumpet Snail (Pulmonata: *Planorbis*). A dozen freshwater snails with characteristically flattened spiral shells, much favoured by aquarists: *P. planorbis* is the commonest and most widespread species, and the more local Great Ramshorn Snail *P. corneus*, which may be 1 in. across, the largest. The Button Ramshorn Snail *P. spirorbis* is like the marsh bladder snail in being able to survive the drying up of its pond or ditch.

Ramsons (Amaryllidaceae: *Allium ursinum*). Our only native woodland garlic, with heads of white star-like flowers, widespread and locally common in damp woods and copses.

Rape, Cole or Coleseed (Cruciferae: *Brassica napus*). A relative of the cabbage, widely naturalized in disturbed ground; the swede is a cultivated form.

Raptores or Raptorial Birds of Prey. An old term for the diurnal birds of prey, the hawks, falcons, eagles, buzzards and their allies (Falconiformes).

Raspberry (Rosaceae: *Rubus idaeus*). The wild raspberry is very similar to the cultivated forms; widespread and frequent in woods and bushy places.

Raspberry Beetle (Clavicornia: *Byturus urbanus*). A well-known pest of raspberries.

Rat (Rodentia: *Rattus*). The two species found in Britain are both aliens. The Black or Old English Rat *R. rattus* arrived in the Middle Ages and the Brown, Norway or Hanoverian Rat *R. norvegicus* in the early eighteenth century, both as stowaways on board ship. At present the brown rat is much the commoner pest, in both towns and countryside, the black rat being largely confined to ports and the West End of London. The water rat is a vole.

Rat-Fish. ⇨ RABBIT-FISH.

Rat Mite (Acari: Mesostigmata). The Spiny Rat Mite *Laelaps echidninus* occurs mainly on brown rats, and the Tropical Rat Mite *Ornithonyssus bacoti* on black rats.

Raunkiaer's Life Forms. A classification of plants by the Danish botanist Raunkiaer according to the position of the resting buds in winter in relation to the level of the soil; ⇨ CHAMAEPHYTE, GEOPHYTE, HELOPHYTE, HEMICRYPTOPHYTE, HYDROPHYTE, PHANEROPHYTE and THEROPHYTE.

Raven (Passeriformes: *Corvus corax*). The largest British crow, all black in plumage, and noted for its hoarse croaking voice. Widespread and frequent on the west side of the British Isles; rare in the east.

Ravine. A narrow, steep-sided small valley, larger than a Gully.

Ray. (1) (Selachii: Hypotremata). Large, primitive cartilaginous fishes, allied to the sharks, and characterized by their heart-shaped or lozenge-shaped bodies, which are flattened vertically, not laterally like the flatfish, and their spiny tails. Some twenty species recorded from British waters, but half are only stragglers. The five long-nosed species are known as Skates. Cuckoo Ray *Raja naevus* is common in the North Sea. Electric Ray *Torpedo nobiliana*, not infrequent in British waters, can give an electric shock of up to 70 volts. Manta Ray ⇨ DEVIL-FISH. Sting Ray *Trygon pastinaca*, uncommon off Britain, can inflict serious wounds with its spined tail. Thornback Ray *Raja clavata*, named from its spiny 'wings', is our commonest species. (2) A botanical term applied both to the stalk of a partial umbel and to the strap-shaped florets of a composite flower.

Ray Society. A national society for the promotion of British natural history by the printing each year of an original work or translation in zoology or botany; founded in 1844, and named after the great seventeenth-century English naturalist John Ray.

Razorbill (Charadriiformes: *Alca torda*). An auk with a flattened bill somewhat resembling the old-fashioned cut-throat razor. Breeds widely on cliffed coasts in the north and west, wintering at sea.

Razor Shell (Lamellibranchia: Solenidae). A family of marine bivalves with markedly elongated shells shaped like the old-fashioned cut-throat razor, living mainly on sandy shores; the Pod Razor Shell *Ensis siliqua*, up to 8 ins. long, is our largest species; *E. ensis* and *E. arcuata* are also widespread and common. The similarly shaped *Pharus legumen*, related to the tellins, is also called a razor shell.

Razor-Strop Fungus or Birch Polypore (Ascomycetes: *Piptoporus betulinus*). A common bracket fungus on birch.

Receptacle. The cup-shaped, conical or flat surface from which the parts of a flower arise.

Red Admiral (Papilionina: *Vanessa atalanta*). A handsome, red-barred, migratory vanessid butterfly, reaching Britain annually from the Mediterranean in some numbers and breeding here during the summer; only rarely overwintering, in south England in mild winters; a return migration is often observed from our shores, but it is not known whether any insects actually reach the Mediterranean again.

Redbreast. Old name for the robin.

Redd. The gravelly spawning bed of salmon at the headwaters of rivers.

Redleg. (1) A frequent country name for common persicaria. (2) Slang term for red-legged partridge.

Red Nose. Fisherman's name for the spiny cockle.

Redpoll (Passeriformes: *Carduelis flammea*). A small finch named from its red crown; commoner in the north in the breeding season, but frequent in the south in winter, often feeding on birches and alders. The British breeding race is the Lesser Redpoll; two other races are irregular winter visitors or passage migrants: the Continental Mealy Redpoll and the Greater or Greenland Redpoll. The Arctic Redpoll *C. hornemanni* is a rare straggler from the high north.

Red Rattle (Scrophulariaceae: *Pedicularis palustris*). Hemiparasite with pink flowers, closely related to lousewort and named as for yellow rattle; widespread and frequent in marshes and damp meadows.

Red Seaweeds (Rhodophyta). The largest major grouping or phylum of marine algae, all contained in the single class Rhodophyceae and coloured red or occasionally blue. They are widespread and often abundant on rockier shores, and include dulse, Carragheen or Irish moss and the nullipores. ⇨ LAVER.

Redshank. (1) (Charadriiformes: *Tringa*). Medium-sized waders with red legs. The Common Redshank *T. totanus* is widespread and common on the shore in winter, also breeding inland, but more local in the south. The Spotted or Dusky Redshank *T. erythropus* is a regular but uncommon passage migrant. (2) Country name for common persicaria.

Red Spider (Acari: Prostigmata). Misnamed mites which are serious pests of fruit. The Fruit Tree Red Spider *Panonychus ulmi*, which attacks apples, pears, plums and damsons, has become the most important orchard pest in Britain since its predators were killed off by the misguided use of insecticides which were not also acaricides. A complex of sibling species under the names of the Hop or Glasshouse Red Spider *Tetranychus urticae* are important pests of tomatoes, cucumbers, hops and various soft fruits. Not to be confused with the Flat Scarlet Mite *Cenopalpus pulcher*, found on unsprayed apple trees.

Redstart (Passeriformes: *Phoenicurus*). Small song birds allied to the thrushes, named from their fiery red tails (Old English *steort*, tail). The Common Redstart *Ph. phoenicurus* is a summer visitor, widespread and common in the north and west, more local in the south, in woodland, parkland and by rivers lined with old trees. The Black Redstart *Ph. ochrurus*, named from the sooty black plumage of the cock, is mainly a passage migrant, but some also spend the winter, and a few pairs breed in London and some south-eastern coastal towns, frequenting built-up areas, waste ground and cliffed coasts.

Red-Thread (Polychaeta: Cirratulidae). Marine bristle worms, living in mucus-lined tubes in muddy gravel under stones, and characterized by red gill-threads.

Red Underwing (Caradrinoidea: *Catocala*). A handsome large moth with two red bars on the underwing; four species recorded, of which *C. nupta* is fairly frequent in southern England, its caterpillars feeding on willows and poplars.

Redwing (Passeriformes: *Turdus musicus*). A rather small thrush, named from its reddish-chestnut flanks and underwings, often associating with fieldfares as a widespread and frequent winter visitor from northern Europe. Occasionally breeds in northern Scotland.

Red Worm. ⇨ MARSH WORM.

Reed (Gramineae: *Phragmites communis*). A tall grass with purplish plume-like flower-heads, forming extensive swamps known as reed-beds in and by fresh and brackish water. In East Anglia dried reed stems are used for thatching. The three species of Small-reed *Calamagrostis* are closely related to Bush Grass; the least uncommon is Purple Small-reed *C. canescens*, in fens mainly in eastern England. Reed Grass *Phalaris arundinacea* is another common waterside grass.

Reedmace or False Bulrush (Typhaceae: *Typha*). Tall, stout, swamp-forming monocotyledons, sometimes known (from their decorative sausage-shaped flowerhead) by their old English and modern American name of Catstail, and more often (because the Victorian artist Alma-Tadema in a famous painting put the cradle of the infant Moses in a reed-mace swamp) as Bulrushes, a name which properly belongs to a common plant of the Sedge family; two species, the common *T. latifolia* and the less common *T. angustifolia*.

Reeve. The female ruff.

Reflex. In animal behaviour, a relatively simple and stereotyped response involving the central nervous system and occurring very shortly after the stimulus which evokes it. (W. H. Thorpe).

Refugium. Any area which has escaped drastic physical or climatic change, e.g. by glaciation, and so provides a refuge for animals and plants, which may or may not subsequently spread out from it; often a Nunatak.

Regeneration. The process whereby a woodland or any other habitat perpetuates itself, or recovers after felling, fire or some natural disaster, by the natural growth of seedlings from the trees, shrubs or other plants already present.

Reindeer (Artiodactyla: *Rangifer tarandus*). The only deer in which females are antlered. A circumpolar arctic species, extinct in Britain since the twelfth century, but experimentally reintroduced in the Cairngorms in 1952.

Reindeer Moss (Lichenes: *Cladonia rangerinifera*). A large branched lichen, up to 3 ins. high, growing abundantly on the higher Scottish moors, also on lower moors, and sand dunes, and forming a staple diet of reindeer in Lapland.

Releaser. In animal behaviour, any specific feature or complex of features in a situation eliciting an instinctive activity or mood (W. H. Thorpe). For instance, the appearance of a hawk is a releaser for instinctive avoiding behaviour in young birds.

Relict (Adj.). Applied to fauna, flora or their distribution wherever a local population represents the relics of a formerly much wider distribution, e.g. mountain avens in Teesdale.

Remora or Sucking-Fish (Percomorphi: *Echineis remora*). A curious fish, with an adhesive disc on the back of its head and neck, by which it attaches itself to sharks, ships and other moving objects; a tropical species not infrequently occurring in British waters.

Rendzina. A shallow soil formed from limestone, e.g. chalk and the Jurassic limestones, and therefore alkaline. It is typically found on fairly steep slopes and supports grassland. The characteristic tree in the south is beech, in the north and west ash.

Reniform. Botanical adjective meaning kidney-shaped.

Reptile (Vertebrata: Reptilia). A class of 'cold-blooded' horny-skinned vertebrates, poorly represented in the native British fauna by three snakes (Serpentes) and three lizards (Sauria) in the order Squamata, and only by stranded stragglers in the order Testudines (turtles). Like amphibians their temperature varies with that of their surroundings.

Rest-Harrow (Leguminosae: *Ononis*). Three pink-flowered peaflowers named from the tough resistance to the medieval harrow of the two commoner species, *O. repens* and *O. spinosa*.

Retuse. Botanical adjective meaning blunt with a slight indentation.

Revolute. Botanical adjective meaning rolled downwards, of a petal or leaf-edge.

Rhachis. The main stem of a feather or a pinnate leaf or a panicled inflorescence.

Rhaetic Period. ⇨ GEOLOGICAL PERIODS.

Rhine. A fairly broad, water-filled ditch in flat, drained marsh country in the south of England.

Rhizome or Rootstock. An underground storage stem, usually stout and sometimes, as in irises, showing on the surface.

Rhizomorph. White mycelial strands composed of numerous hyphae, which in parasitic fungi act as an inimical organ. ⇨ MYCELIUM.

Rhizopoda (Protozoa). A large class of minute freshwater, marine, soil and parasitic animals, including the well known Amoeba, the shelled Foraminifera, and the organisms which cause blackhead disease *Histomonas* of turkeys and amoebic dysentery *Entamoeba* in man.

Rhododendron (Ericaceae: *Rhododendron ponticum*). A large alien evergreen shrub, with handsome mauve flowers, so thoroughly established in many places on acid soils as to constitute a serious weed.

Rhododendron Bug (Heteroptera: *Stephanitis rhododendri*). A lacebug, first discovered in Britain in 1901, which has since spread to become a minor pest of rhododendrons in many areas.

Rhodology. The scientific study of wild roses, a very variable group of plants.

Rhodophyceae, Rhodophyta. ⇨ RED SEAWEEDS.

Rhynchophora (Coleoptera: Polyphaga). A large superfamily of beetles, with 590 species, mainly weevils, but also including the bark and ambrosia beetles.

Ribbon-Fish (Pisces: *Regalecus glesne*). A very elongated northern deep-sea fish, attaining twenty feet or more; rare in British waters.

Ribbon Worm (Nemertina). A rather small phylum of unsegmented worms, with the power of substantially contracting the length of their often very long bodies; and progressing by means of rhythmic waves along their bodies. The great majority are marine, and occur especially on rocky and stony shores, including the remarkably long Bootlace Worm and the Pink Ribbon Worm *Amphiporus lactifloreus. Prostoma graecense* is our only freshwater nemertine, and the Australian alien *Geonemertes dendyi* our only terrestrial one.

Rib Grass. Farmers' name for the ribwort plantain.

Rice Grass or Cut Grass (Gramineae: *Leersia oryzoides*). An uncommon grass of wet places in the south, whose flower spikes only expand in hot summers. ⇨⇨ CORD GRASS.

Rictal Bristle. Bristles at the gape or mouth of a bird.

Ride. An open unmade track through a wood.

Riffle. A shallow stretch in a stream where the water flows swiftly (H. J. Egglishaw).

Rift Valley. ⇨ FAULT.

Rigg. A common northern term for a ridge or narrow elongated hill, especially in Yorkshire.

Rime. ⇨ FROST.

Ringing. ⇨ BIRD RINGING.

Ringlet (Papilionina: *Aphantopus hyperanthus*). One of the commoner Brown Butterflies, frequenting woodland in summer, named from being spotted with open circles. The allied Mountain Ringlet *Erebia epiphron* is found on mountains in the Lake District and Scotland.

Ring Ouzel (Passeriformes: *Turdus torquatus*). A black thrush with a conspicuous white breast, hens being browner; a summer visitor frequent in the hill country of the north and west.

River Snail. ⇨ Freshwater WINKLE.

River Worm or Bloodworm (Oligochaeta: *Tubifex*). Freshwater oligochaete worms, living upside down in muddy tubes, often making red patches in river mud with their red tails waving in unison as they seek oxygen; their blood is coloured red by the haemoglobin that tints human blood. They can live in quite foul water, even in the Thames mud in central London.

Roach (Cyprinidae: *Rutilus rutilus*). A widespread and common fish of ponds and streams much favoured by anglers; rare and introduced in Ireland, where the closely related rudd is confusingly known as roach.

Roast-Beef Plant. ⇨ GLADDON.

Robber Fly or Assassin Fly (Brachycera: Asilidae and Empididae). Two families of predatory two-winged flies, containing respectively 27 and over 300 species. The largest asilid is *Asilis crabroniformis* and the largest empid *Empis tessellata*.

Robin (Passeriformes: *Erithacus rubecula*). Familiar small song bird, allied to the thrushes, noted for its red breast (it used to be known as Redbreast), its aggressive habits in defence of its territory, and its song period lasting almost throughout the year, being interrupted only by the moult period in late June and July. In 1961 it was selected by the British Section of the International Council for Bird Preservation as the British national bird. Widespread and abundant, penetrating right into the middle of towns. The American Robin *Turdus migratorius*, a much larger bird, but also with a red breast, is a rare straggler from North America.

Robin's Pincushion. A red and green gall of wild roses, covered with moss-like excrescences and made by the gall-wasp *Rhodites rosae*; also known as Bedeguar or Moss Gall.

Rock. As a technical geological term rock includes sands, clays and other soft geological deposits not included in the common use of the word, which covers only hard rocks, such as granites and limestones.

Rock Cook (Labridae: *Centrolabrus exoletus*). One of the rarer wrasses.

Rock-Cress (Cruciferae: *Arabis*). Small white-flowered crucifers; four species. the commonest being Hairy Rock-cress *A. hirsuta*, frequent on rocks and bare ground on limy soils. Northern Rock-cress *Cardaminopsis petraea* grows on mountains in Snowdonia and the Highlands.

Rocket (Cruciferae). Old name for salad plants, still applied to several members of the Cabbage Family; London Rocket *Sisymbrium irio* with yellow flowers, abundant after the Great Fire of London in 1666 but now rare; Eastern Rocket *S. orientale*, a spreading alien; Sea Rocket *Cakile maritima*, with lilac flowers, common on sandy shores; and Wall Rocket *Diplotaxis muralis*, with yellow flowers, on old walls.

Rocket-Netting. A method of trapping flocks of birds for research purposes by firing rockets which pull a net over the resting birds before they can take off; especially used by the Wildfowl Trust in its studies on wild goose populations in Britain.

Rockling (Gadidae: *Onos*). Three smallish marine fishes of the Cod Family, the commonest being the coastal Five-bearded Rockling *O. mustelus*, named from its five barbels; the two others are deep-water species.

Rock-Rose (Cistaceae: *Helianthemum*). Low-growing undershrubs, two yellow-flowered, Common Rock-rose *H. chamaecistus*, especially on calcareous grassland in the south, and the rare Hoary Rock-rose *H. canum*;

one white-flowered, the rare White Rock-rose *H. apenninum*. Annual Rock-rose *Tuberaria guttata* is also rare.

Rock Tripe (Lichenes: Gyrophoraceae). A family of grey or brown lichens growing on rocks and walls in hill districts, supposedly resembling tripe, especially *Umbilicaria pustulata*.

Rodent (Mammalia: Rodentia). A large group of mostly small vegetarian mammals with teeth specially adapted for gnawing. Represented in our native fauna by the squirrels (Sciuridae), the extinct beaver (Castoridae), the dormouse (Muscardinidae) and the mice and voles (Muridae); the unusually high proportion of naturalized rodents includes the house mouse, the black and brown rats, the fat dormouse, the South American coypu, and the fortunately extinct muskrat.

Roding. (1) ⇨ WOODCOCK. (2) Clearing a Fenland dyke of vegetation.

Roe Deer (Artiodactyla: *Capreolus capreolus*). The smaller of our two native deer, widely distributed in woodlands. The stock in southern England was reintroduced over 150 years ago. Male, buck; female, doe; young, kid.

Rogue. A single abnormal animal that does persistent damage to crops, stock or game, e.g. a badger that takes to raiding chicken runs.

Roller (Coraciiformes: *Coracias garrulus*). Brightly coloured, crow-sized bird, uncommon straggler to Britain from southern Europe. Named from its tumbling courtship display flight.

Roman Snail (Pulmonata: *Helix pomatia*). Much the largest British land snail, 1½–2 ins. in both height and breadth, locally frequent on chalk and limestone in southern England, and requiring at least 5 per cent of calcium carbonate in the soil. This is the edible *escargot* of France, and was also eaten by the Romans who are popularly believed to have introduced it; if they did, they were bringing coals to Newcastle, for it has been found in pre-Roman deposits.

Rond. A small strip of marsh separating some of the Norfolk Broads from the adjacent rivers.

Rook (Passeriformes: *Corvus frugilegus*). The most familiar of the British crows, all black in plumage, with a hoarse, cawing voice, highly gregarious and nesting in trees in colonies known as rookeries; widespread and common.

Rookworm. Farmers' name for the larvae of cockchafers, which are much relished by rooks.

Root. The part of a vascular plant that absorbs water and nutrient salts from the soil, normally extending downwards into the soil and serving also to anchor the plant; a tap-root is a stout main root extending vertically downwards. The absorbing region is that of root-hairs. Rootstock, ⇨ RHIZOME.

Rooting-Shank (Basidiomycetes: *Oudemansiella radicata*). A rather small yellowish-brown toadstool, growing in broad-leaved woods, especially of beech, and named from the hard downward extension of its stem into the soil, which resembles a tap-root.

Root Rot. A disease of trees and shrubs caused by honey fungus.

Rorqual (Cetacea: *Balaenoptera*). Species of large whalebone whale, of which the Common Rorqual or Fin Whale *B. physalus*, the second largest known whale (up to 78 ft) and the Lesser Rorqual or Piked Whale *B. acutorostrata* are the most frequent. They migrate well out to sea off our shores, but sometimes get stranded. Also known as Finners.

Rosaceae (Angiospermae: Dicotyledones). An important family including the brambles, hawthorn, sorbs and other berry-bearing trees and shrubs besides the roses; also includes many common herbs, such as wild strawberry, agrimony and lady's mantle.

Rose, Wild (Rosaceae: *Rosa*). Thorny shrubs or clamberers, with flowers recognizably like single garden roses; fruits called hips, usually red. Burnet Rose *R. pimpinellifolia*, with white flowers and purple-black hips, low growing on heaths and dunes, especially near the sea. Dog Rose *R. canina*, with pink or white flowers; widespread and common, mainly in the south; Downy Rose *R. villosa* with deeper pink flowers, widespread and frequent, mainly in the north; Field Rose *R. arvensis*, flowers white, widespread and common. Scotch Rose, the Burnet Rose.

Rosebay or Rosebay Willow-herb (Onagraceae: *Chamaenerion angustifolium*). A tall showy gregarious plant, with bright pinkish-purple flowers, of woods and waste places; known as Fireweed from its habit of growing in burned clearings in woods and other recently burned ground.

Rose Beetle or Rose Chafer (Lamellicornia: *Cetonia aurata*). A handsome golden-green beetle, allied to the dung beetles, sometimes seen in roses and other large blooms.

Rose of Sharon (Hypericaceae: *Hypericum calycinum*). A low shrubby St John's wort, with large solitary yellow flowers, often planted and sometimes naturalized.

Rose-Root (Crassulaceae: *Sedum rosea*). Succulent plant, named from the smell of its stout rootstock; flowers small, yellow; frequent on hills and cliffs in north-west Scotland; more local elsewhere and absent from south and east England.

Rostrum. A beak-like prolongation of a beetle's head.

Rotate. Botanical adjective used of a corolla, whose petals or corolla-lobes spread out to make a flat, wheel-like surface.

Rotifer or Wheel Animalcule (Aschelminthes: Rotifera). Tiny aquatic animals, mostly living in fresh water but with a few marine species, allied to the roundworms. Most are free-swimming, but some attach themselves to other aquatic animals or plants or are even parasitic. There are 530 British species, none with vernacular names, but some are common and rotifers are present in almost all fresh water out of doors.

Roundworm (Aschelminthes: Nematoda). The great class of nematode worms consists largely of endoparasites, but also contains many free living freshwater and terrestrial forms, the latter living in the soil, e.g. rainworms and eelworms. The parasitic forms infest both plants, e.g. eelworms, and animals, mainly vertebrates. Many of the most troublesome animal diseases are caused by roundworms, e.g. gapes by gapeworms, some of which are known as threadworms or pinworms. Roundworms are mostly very small – our largest freshwater one *Dorylamius* *stagnalis* being only $\frac{1}{5}$ in. long, with unsegmented cylindrical bodies usually pointed at both ends.

Rove Beetle (Polyphaga: Staphylinoidea). All members of the superfamily other than the carrion beetles, comprising some 1,210 species in 11 families, much the largest of which (954 species) is the Staphylinidae, characterized by their very short elytra, which includes the devil's coach horse.

Rowan or Mountain Ash (Rosaceae: *Sorbus aucuparia*). A small deciduous tree growing in acid soils on moors and heaths; its orange berries are much relished by thrushes and other birds.

Royal. A stag with twelve points (six on each antler), with the top three on each forming a cup or crown, hence the name.

Royal Entomological Society of London. The national society for the study of entomology, founded in 1833; publishes *Transactions* and *Proceedings.*

Royal Fern (Pteropsida: *Osmunda regalis*). A handsome large fern that bears its golden-brown spores on central spikes resembling flower-spikes; now rather local in damp and wet habitats, mainly in the west.

Royal Microscopical Society. The national society for the study and promotion of microscopy, founded in 1839; publishes a *Journal.*

Royal Society for the Protection of Birds. Founded in 1889 and made 'Royal' in 1904, its early successes included the saving of the great crested grebe, the great skua and the goldfinch as breeding birds in Britain, and since 1945 the avocet, black-tailed godwit and osprey. The Society has taken active measures against egg collectors, including prosecution, and helped to secure the passing of the Wild Birds Protection Act in 1954; has twenty-one reserves, including the famous Minsmere and Havergate Island in Suffolk; publishes *Bird Notes.*

Ruby-tail Wasp (Apocrita: Chrysididae). A family of parasitic solitary wasps, some of which have the hinder end of the abdomen an attractive bright red, which lay their eggs in the nests of solitary bees and other solitary wasps.

Rudd (Cyprinidae: *Scardinius erythrophthalamus*). A common freshwater fish, similar in most respects to the roach, but native in Ireland, preferring slower waters, and with redder fins, from which it gets its name.

Rudder-Fish. ⇨ BARREL-FISH.

Ruff (Charadriiformes: *Philomachus pugnax*). A medium-sized wader, whose males are noted for the fantastic neck-ruffs they assume in the breeding season, when they gather on display grounds for mock battles. Females are known as reeves. Frequent passage migrant; formerly bred in eastern England.

Ruffe. ⇨ POPE.

Rugose. Botanical adjective meaning wrinkled.

Runcinate. Botanical adjective used of pinnately lobed leaves whose lobes point backwards.

Runner or Stolon. A creeping stem above ground that roots at the tip to form a new plant that may eventually become independent of its parent.

Rupture-Wort (Caryophyllaceae: *Herniaria*). Two low-growing rather weedy, green-flowered plants, both rare.

Rush (Juncaceae: *Juncus*). A large genus of monocotyledons, related to the lilies, but with heads of small greenish or brownish flowers, stems often filled with pith and sometimes no leaves, generally growing in damp situations. The commoner British species include Soft Rush *J. effusus*, Compact Rush *J. conglomeratus*, Hard Rush *J. inflexus*, Heath Rush *J. squarrosus* and Toad Rush *J. bufonius*. Wood-rushes are related, but bog-rush, bulrushes, club-rushes and spike-rushes are Cyperaceae.

Rush, Flowering (Butomaceae: *Butomus umbellatus*). Tall pink-flowered monocotyledon, unrelated to the true rushes, frequent by fresh water in England.

Rush Veneer Moth (Pyraloidea: *Nomophila noctuella*). A smallish pyralid, well known as an immigrant from the Continent, its numbers varying greatly from year to year.

Rust (Basidiomycetes: Uredinales). An order of small parasitic, often rust-coloured fungi, many of which cause plant diseases, one of the best known being the Wheat Rust *Puccinia graminis*, whose intermediate host is the Barberry. Other species of *Puccinia* produce rusts on snapdragons, hollyhocks, mints and gooseberries, while the broad bean rust is *Uromyces fabae*.

Rusty-Back Fern (Aspleniaceae: *Ceterach officinarum*). A distinctive small fern, whose pinnately lobed fronds are heavily encrusted with rust-coloured scales, locally frequent on rocks and walls, mainly in the west.

Rut. The mating period of deer and other ungulates, when the male deer bellow or roar. Rutting in the red deer lasts from mid-September to late October, in the fallow deer from early October to mid-November, and in the roe deer from late July to mid-August.

Rye (Gramineae: *Secale cereale*). A long-awned barley-like grain crop, now rather infrequently cultivated on light soils, so only rarely occurring as a relic of cultivation.

Rye-Grass (Gramineae: *Lolium perenne*). A nutritious grass, common in natural grassland and frequently sown in leys. Italian Rye-grass *L. multiflorum*, also sown in leys, is now widely established on waste ground.

S

Saccate. Botanical adjective meaning pouched.

Saffron. ⇨ MEADOW SAFFRON.

Sage (Labiatae: *Salvia*). Two relatives of the garden Sage *S. officinalis*: the brilliant blue Meadow Sage *S. pratensis*, scarce in calcareous grassland in southern England; and Clary.

Sagittate. Botanical adjective meaning arrow-shaped; ⇨ LEAF-SHAPES.

Sailfish. ⇨ Basking SHARK.

Sail-Fluke. ⇨ MEGRIM.

Sainfoin (Leguminosae: *Onobrychis viciifolia*). A peaflower with elongated heads of bright pink flowers, grown as a fodder crop and widely established in calcareous districts, also a very local native.

St Elmo's Fire. A small electrical brush discharge sometimes seen on the masts and rigging of ships during storms; the amount of electricity involved is too small to produce any risk of actual fire.

St John's Wort (Hypericaceae: *Hypericum*). A distinctive group of mainly herbaceous yellow-flowered perennials, with several common and widespread species, including *H. perforatum*, *H. hirsutum* and *H. pulchrum* on dry soils, and *H. tetrapterum* on damp ones. ⇨⇨ ROSE OF SHARON, TUTSAN.

St Mark's Fly (Nematocera: *Bibio marci*). A small black hairy two-winged fly related to the gnats, named from the time of its emergence, around St Mark's Day, 25 April; the name is also loosely applied to other members of the genus.

St Olaf's Candlestick. ⇨ WINTERGREEN.

St Patrick's Cabbage (Saxifragaceae: *Saxifraga spathularis*). A saxifrage, one of the parents of London pride; as a wild plant confined to the west of Ireland.

St Swithin's Day. An ancient but quite inaccurate legend prognosticates that if it rains on 15 July, it will rain for forty days thereafter, but though July and August are often wet they are rarely as wet as this, irrespective of what happens on 15 July. St Swithin was a Saxon Bishop of Winchester, who is said posthumously to have delayed his own reburial following canonization by just such a forty-day period of rain.

Saithe. ⇨ COAL-FISH.

Salicaceae (Angiospermae: Dicotyledones). The family containing the willows, sallows and poplars.

Salientia or Anura (Chordata: Amphibia). The order of tailless amphibians containing the frogs (Ranidae) and toads (Bufonidae).

Sallow (Salicaceae: *Salix*). A name for some of the broader-leaved willows, especially those called Pussy Willow from their golden male catkins in early spring, also known as Palm. Three common and widespread species:

Great Sallow or Goat Willow *Salix caprea*, Common Sallow *S. cinerea* and Eared Sallow *S. aurita*.

Salmon (Salmonidae: *Salmo salar*). A large anadromous fish, the premier game fish, known by different names at successive stages of its life history. When hatched the young are alevins; once the yolk-sac is absorbed they become parr; after two years most parr turn silvery and become smolts, when they migrate to the sea. After a year or more they return as grilse, and ascend the rivers to spawn in the redds, or gravelly spawning beds far upstream. After spawning they return downstream as spent fish or kelts, some dying but others returning to spawn again.

Salmonidae (Pisces: Isospondyli). The important family of bony fishes containing the game fish: salmon, trout, char, grayling and the various species of whitefish.

Salsify (Compositae: *Tragopogon porrifolius*). A purple-flowered plant like goatsbeard, formerly cultivated as a vegetable, now established in waste and grassy places in various parts of England.

Salting. ⇨ SALTMARSH.

Saltmarsh. An area of sand and mud, covered by the sea at high tide and exposed at low tide. It has a distinctive flora because all the plants must tolerate salt; marsh samphire is a typical plant.

Saltwort (Chenopodiaceae: *Salsola kali*). A seaside plant with minute green flowers and spiny leaves.

Samara. A winged achene or key.

Samphire. The name of several unrelated coastal plants eaten for their fleshy leaves, among them the rather local Golden Samphire (Compositae: *Inula crithmoides*) with yellow daisy-like flowers; the true Samphire (Umbelliferae: *Crithmum maritimum*), with yellow-green carrot-like flowers, growing on cliffs and rocks (the plant mentioned in *King Lear*) and Marsh Samphire, ⇨ GLASSWORT.

Sand. Minute mineral particles, most commonly quartz. Sandy soils do not hold water and tend to be dry, the coarser ones especially being deficient in plant foods partly because the good drainage leads to leaching. Bagshot Sand is a coarse and infertile soil found mainly to the south of London (Bagshot area of Surrey), giving rise to a heathy vegetation. ⇨⇨ DUNE.

Sand Eel or Launce (Percomorphi: *Ammodytes*). The Greater *A. lanceolatus* and Lesser *A. tobianus* Sand Eels are small eel-shaped fish of sandy shallows, much preyed on by terns and other sea-birds; not related to the true eels.

Sanderling (Charadriiformes: *Crocethia alba*). A small wader, very white in winter, when it is frequent on sandy shores; an Arctic breeder.

Sand Fly (Trichoptera: *Rhyacophila dorsalis*). A caddis fly of fast streams.

Sand-Grouse (Columbiformes: Pteroclidae). Long-tailed ground-living birds not native to the British Isles. Pallas's Sand-grouse *Syrrhaptes paradoxus* occasionally irrupts from the Central Asian steppes to Western Europe, though not to Britain since 1909; after the great invasion of 1888, pairs bred in Yorkshire and Morayshire.

Sandhopper (Amphipoda: *Orchestia* and *Talitrus*). The abundant small jumping crustaceans of seaweed and other tide-wrack on the shore, related to the sand shrimps; *O. gammarella* and *T. saltator* are the commonest species.

Sandpiper (Charadriiformes: Scolopacidae). Small waders of which only one species, the Common Sandpiper *Tringa hypoleucos*, breeds regularly in Britain, mainly in the north and west, as a summer visitor. The Green Sandpiper *T. ochropus* and the Wood Sandpiper *T. glareola* are regular passage migrants which have occasionally bred in the north. The Curlew-Sandpiper *Calidris testacea* is a less common passage migrant and the Purple Sandpiper *C. maritima* is a local winter visitor; both are related to the dunlin. Nine species or subspecies of sandpiper, of which the most frequent is the Pectoral Sandpiper *C. melanotos*, are stragglers on migration from across the Atlantic, and four more are stragglers from Europe or Asia.

Sand Shrimp (Amphipoda: *Gammarus*). A group of small, slightly shrimp-like marine crustaceans, closely related not to the true shrimps but to the freshwater shrimps; *G. locusta* is the commonest species on our shores.

Sand Star (Ophiuroidea: *Ophiura*). A genus of brittle-stars.

Sandstone. A rock composed of compressed sand, usually bound together by other substances. Old Red Sandstone is a Devonian rock. New Red Sandstone is a Permian formation. ⇨ GRIT, QUARTZITE.

Sand Wasp (Aculeata: Sphegidae). A large family of solitary wasps, including many digger wasps, noted for the female's habit of capturing and paralysing caterpillars much larger than herself and dragging them to her nest hole as provisions for her unborn young; nest holes are usually in sandy soils, but occasionally in rotten wood; two of the commonest species are *Ammophila sabulosa* and *A. campestris*. ⇨ BLACK WASP, WOOD-BORING WASPS.

Sandwort (Caryophyllaceae). Name given to about a dozen small weedy white-flowered plants with undivided petals, formerly all in *Arenaria*, which is now split into five genera: most are local or rare, but Thyme-leaved Sandwort *A. serpyllifolia* and Wood Sandwort *Moehringia trinervia* are widespread and common inland, and Sea Sandwort or Sea Purslane *Honkenya peploides* is so by the sea. ⇨ LEADWORT.

Sanicle (Umbelliferae: *Sanicula europaea*). A widespread and common white-flowered wood-carpeting plant, with ivy-like leaves, liking calcareous soils.

Saprophagous. Of an animal that feeds on decaying plant matters.

Saprophyte. A plant that derives its nourishment from decayed organic matter, e.g. the bird's nest orchid *Neottia nidus-avis* and most fungi.

Sap-Wood. The layer of wood surrounding the heart-wood in a tree.

Sardine. ⇨ PILCHARD.

Sarsen Stone. Large boulder of grey siliceous sandstone from Tertiary deposits which have otherwise been wholly denuded away from the surface of the southern chalk, especially in Wiltshire, where they are most frequent on the Marlborough Downs; also known as Grey Wethers from their likeness, at a distance, to a sheep.

Saucer Bug (Heteroptera: *Ilyocoris cimicoides*). A squat, sluggish water bug of muddy stagnant waters, mainly southern in distribution.

Sauria (Reptilia: Squamata). The sub-order of reptiles containing the lizards.

Saury, Saury Pike or Skipper (Pisces: *Scomberesox saurus*). An elongated fish, related to the garfish but with a shorter beak and less common off our western coasts.

Saw-Bill. ⇨ MERGANSER.

Saw-Fly (Symphyta: Tenthredinoidea). A superfamily of Hymenoptera mostly belonging to the Tenthredinidae, named from the saw-like ovipositors which they use to lay eggs in twigs or leaves, some species, such as the Willow Saw-fly *Pontonia vesicator* producing galls. Their larvae are remarkably like those of moths, with which they can be confused, though many species can readily be told by their habit of holding the hind part of their body in the air, or curling it up under themselves. The Palisade Saw-fly *Pteronidea compressicornis* is named from its curious habit of erecting a 'palisade' of small posts made of hardened saliva around the spot where it feeds, presumably as a protective measure. Saw-flies feed on most cultivated plants of economic importance, and several species are noted pests of the gardener, farmer or forester, among them the Gooseberry Saw-fly *Nomatus ribesii*, Turnip Saw-fly *Athalia spinarum*, Apple Saw-fly *Hoplocampa testudinea*, Pear Saw-fly *Eriocampa limacina*, Rose Slug *E. rosae* (the larvae of *Eriocampa* spp. are known as Slugworms), Rose Saw-fly *Argeodiropus*, Pine Saw-fly *Diprion pini*, and Larch Saw-fly *Pristiphora erichsonii*. Stem Saw-flies, ⇨ BORER.

Sawwort (Compositae: *Serratula tinctoria*). Plant with purple thistle-like flowers and saw-toothed pinnate leaves, local in grassy places in England and Wales.

Saxifrage (Saxifragaceae: *Saxifraga*). A genus of fifteen, mostly mountain plants, some very rare; twelve with white flowers, including the widespread Meadow Saxifrage *S. granulata*, Rue-leaved or Fingered Saxifrage *S. tridactylites*, growing in grassy and bare places respectively, away from mountains, and the rock-loving Mossy Saxifrage *S. hypnoides*; also two with yellow flowers, *S. aizoides* being much the commoner; and one, *S. oppositifolia*, with purple flowers. Golden Saxifrages *Chrysosplenium*, in damp places on acid soils, are closely related. ⇨⇨ London Pride and St Patrick's Cabbage, which are included in the group known as Robertsonian saxifrages. All saxifrages are named from their supposed capacity, in the medieval pharmacopoeia, to 'break' or dissolve kidney stones. Hence three umbellifers, fairly common in grassland, are also called saxifrage: the white-flowered Burnet Saxifrage *Pimpinella saxifraga* and Greater Burnet Saxifrage *P. major*, and the yellow-flowered Pepper Saxifrage *Silaum silaus*.

Scabious (Dipsacaceae). Several blue-flowered plants allied to the Teasel, with compound flower-heads like composites: Field Scabious *Knautia arvensis*, common in grassy places; Small Scabious *Scabiosa columbaria* in calcareous grassland; and Devilsbit Scabious *Succisa pratensis*, of damp grassy places, named from its truncated root formerly believed to have been bitten off by the Devil. The superficially similar Sheepsbit Scabious (Campanulaceae: *Jasione montana*) of grassy places, preferring acid soils, is so named because sheep like it.

Scabrid. Botanical adjective meaning rough to the touch.

Scad. ⇨ Horse MACKEREL.

Scald Fish (Bothidae: *Arnoglossus*). Three small left-sided flatfish, named from the ease with which the scales come off when caught, making the flesh look as if scalded; two are not uncommon off our coasts: *A. laterna* and *A. thori*.

Scale Insect (Homoptera: Coccidae). Small plant-bugs with singular life-histories, the tiny adult males being mouthless, with a single pair of wings, but the wingless females, covered with a scale and unable to move, remain fixed to their food plant for life, often appearing to be a mere excrescence on the bark of a tree. Some species, such as the Apple Scale *Mytilaspis pomorum*, are regarded as pests. Common scale insects include the Oyster-shell Scale *Aspidiotus ostreaformis* and Mussel Scale *Leoidosaphes ulmi* on fruit trees, and the Felted Beech Scale *Cryptococcus fagi* on beeches. Mealy-bugs *Dactylopius* are coccids named from their mealy or waxy exudations.

Scale Worm (Polychaeta: Aphroditidae). A large family of free-moving marine bristle worms, with scales folded over their backs, including the aberrant sea-mouse and *Sthenelais boa*, which may be 8 ins. long.

Scallop (Lamellibranchia: Pectinidae). A family of marine bivalves, with five common but mainly offshore species, whose fan-shaped shells are often washed ashore: the Great or Edible Scallop or Clam *Pecten maximus*, the small Tiger Scallop *Chlamys tigerina* and Hunchback Scallop *C. distorta*, and the large Queen Scallop *C. opercularis* and Variegated Scallop *C. varia*.

Scape (adj.: Scapigerous). A leafless flowering stem.

Scaphopoda. The smallest of the five molluscan classes found in British seas, represented by the tusk shells.

Scar. A common term in the North of England, especially Yorkshire, for cliffs, crags or other rocky outcrops inland.

Scarious. Botanical adjective used of leaves, bracts and other organs which are thin, dry and devoid of green colouring matter.

Scarp. The short steep slope at one end of a line of hills where the underlying rock strata are tipped at an angle; can also be formed by a geological fault. ⇨ DIP-SLOPE.

Scaup (Anatidae: *Aythya marila*). A diving duck, frequent as a winter visitor off our northern coasts; has occasionally bred in Scotland.

Scavenger Beetle (Polyphaga: Palpicornia). A family (Hydrophilidae) of 118 species feeding on decomposing vegetable matter on land or under the water, though they are not truly aquatic. *Hydrophilus piceus* is larger than any of the carnivorous water beetles.

Scent. The physical basis of the sense of smell consists in minute droplets of volatile oil given off by certain animals and plants, which carry better in a moist than in a dry atmosphere.

Schelly (Salmonidae: *Coregonus clupeoides stigmaticus*). The race of the powan found in three lakes draining into the River Eden in Cumberland, Haweswater, Ullswater and Red Tarn.

Schist. An easily flaking metamorphosed rock consisting mainly of mica and quartz.

Schizomycetes. The sole class in the Bacteriophyta.

Scleroparei (Vertebrata: Pisces). The mail-cheeked fishes, an order of bony fishes including the gurnards, bullheads, sea snails and sticklebacks.

Sclerophyll. A plant with tough or leathery leaves, such as holly.

Scorpion-Fly (Insecta: Mecoptera). A small order of weakly flying predatory winged insects, named from the male's habit of carrying its body curled up over its back, like a scorpion's tail; four species, the common ones being *Panorpa communis* and *P. germanica.*

Scotch Bonnets. ⇨ FAIRY RING Champignon.

Scoter (Anatidae: *Melanitta*). Black or blackish diving ducks. The Common Scoter *M. nigra* is common offshore in winter and can be found throughout the year; breeds locally inland in north Scotland and Ireland. The Velvet Scoter *M. fusca* is a less common winter visitor. The Surf Scoter *M. perspicillata* is a rare straggler from the Arctic.

Scottish Field Studies Association. The Scottish counterpart of the Field Studies Council, with a field centre at Kindrogan, Perthshire.

Scottish Ornithologists' Club. The national bird-watching organization of Scotland, founded in 1936; publishes *Scottish Birds* quarterly.

Scrawbog. A lough covered with a floating mat of vegetation, sometimes thick enough for a man to walk on.

Scree or Talus. A mass of stones, boulders and loose fragments of rock detached from rocks or cliffs above as a result of weathering and rolled downhill. Screes are commonest at high altitudes where they have a typical mountain flora, especially on gentle slopes where there is little movement of the stones.

Screech Beetle or Squeak Beetle (Adephaga: *Hygrobia hermanni*). A common water beetle, which makes a squeaking sound by stridulation.

Scrophulariaceae (Angiospermae: Dicotyledones). The Figwort Family, containing two quite different types of flower: flat like the speedwells and mulleins, and two-lipped like the toadflaxes and eyebrights. Some of its members are hemiparasites, e.g. lousewort, red rattle, eyebright.

Scrub. Land dominated by low shrubs and bushes with no trees more than twenty-six feet high. Usually the bushes are colonizing open ground and will be followed by trees, but where the area is too windswept or cold for trees to grow the scrub is the natural climax.

Scurvy-Grass (Cruciferae: *Cochlearia*). A group of white- and lilac-flowered plants owing their anti-scorbutic name to the high vitamin-C content of their leaves and stems; four or five species, the commonest being Common *C. officinalis* and Early Scurvy-grasses *C. danica.*

Scyphozoa, or Scyphomedusae. ⇨ JELLYFISH.

Sea-Anemone (Anthozoa: Actiniaria). An order of carnivorous coelenterate marine animals, looking like small truncated cones of jelly, but resembling a flower when their tentacles are expanded, whence their name. There are about fifteen British species, some comparatively large, to about 5 ins. tall, mostly adhering to rocks and stones. The commoner ones include the widespread Beadlet Anemone *Actinia equina*, whose 200 tentacles retreat quickly when touched; the western Snakelocks or Opelet Anemone *Anemonia sulcata*, whose 2-in. tentacles do not contract; the Dahlia and Wartlet or Gem Anemones *Tealia felina* and *Bunodactis verrucosa*, and the Glaucous and Red-specked Pimplets *Anthopleura thallia* and *A. ballii*, whose vertical columns are covered with small warts; the feathery looking Plumose Anemone *Metridium senile*; and the small 300-tentacled Daisy Anemone *Cereus pedunculatus*. Some of the soft corals (⇨ CORAL) look like small sea-anemones and are sometimes referred to as such.

Sea Beet (Chenopodiaceae: *Beta maritima*). The wild parent of cultivated beetroot, widespread and locally common on bare ground by the sea.

Seablite (Chenopodiaceae: *Suaeda*). Two seaside plants, the widespread Common Seablite *S. maritima* with fleshy leaves in saltmarshes and on mud, and the local Shrubby Seablite *S. fruticosa* with woody stems on shingle in the south-east.

Sea-Cat. ⇨ CAT-FISH.

Sea-Cucumber (Echinodermata: Holothuroidea). A class of marine animals shaped like small cucumbers or gherkins, with no 'arms' but a ring of tentacles around the mouth, including the sea-gherkin and the nigger.

Sea-Fir or Hydroid (Cnidaria: Hydrozoa). Colonial marine coelenterate animals, belonging to the orders Athecata (or Gymnoblastea or Anthomedusae) and Thecata (or Calyptoblastea or Leptomedusae), which spend most of their lives in branched sedentary colonies, looking remarkably like seaweeds or miniature flowering plants or fir-trees. Among the commonest species on our shores are *Obelia geniculata* and *Dynamena pumila*, which grow on sea wracks, *Hydractinea echinata*, which grow on shells inhabited by hermit-crabs, and White Weed.

Sea-Gherkin (Holothuroidea: *Cucumaria saxicola*). A sea-cucumber, common among rocks and stones off the south-west coasts of England and Ireland.

Sea-Gooseberry. (1) (Ascidiacea: *Dendrodoa grossularia*). A red or yellow sea squirt, common off our south and west coasts. (2) (Ctenophora: *Pleurobrachia pileus*). A bright yellow egg-shaped comb-jelly, common offshore, especially in late summer.

Sea-Hare (Tectibranchiata: *Aplysia punctata*). A widespread and common large (up to 6 ins.) tectibranch sea slug, with an internal shell, changing from red to brown and finally olive-green in colour as it grows older, and able to eject a slimy purple substance when disturbed.

Sea-Heath (Frankeniaceae: *Frankenia laevis*). A local plant of saltmarshes from the Wash to the Solent, with small pink flowers and leaves like the heaths *Erica*, to which it is quite unrelated.

Sea-Hen. ⇨ LUMP-SUCKER.

Sea-Holly (Umbelliferae: *Eryngium maritimum*). A seaside plant, mainly in the south, with rounded flower-heads of a beautiful greyish powder blue, and prickly leaves, whence its name.

Sea-Horse (Pisces: *Hippocampus hippocampus*). A singular little fish, shaped like a knight in chess with a curly tail, related to the pipe-fishes and very rare in British waters.

Sea Kale (Cruciferae: *Crambe maritima*). Cabbage-leaved, white-flowered plant of coastal shingle, mainly in the south-east and north-west; the origin of the vegetable.

Seal (Pinnipedia: Phocidae). Marine mammals, feeding on fish and shell-fish. Two species breed round our coasts: the Common Seal *Phoca vitulina*, generally distributed, and the Grey Atlantic Seal *Halichoerus grypus*, virtually confined to the north and west coasts. There is a famous colony of grey seals on the Farne Islands, Northumberland. Three Arctic species are rare visitors, mainly to our northern coasts.

Sea-Lace (Phaeophyta: *Chorda filum*). A brown seaweed, with very narrow fronds somewhat resembling a bootlace and anything up to 8 – 9 yards long, common on rocky, stony and sandy shores.

Sea-Lavender (Plumbaginaceae: *Limonium*). Four species of maritime plant with lavender-purple flowers, of which the most frequent are Common Sea-Lavender *L. vulgare*, which empurples saltmarshes in August, and Rock Sea-Lavender *L. binervosum*, on cliffs, rocks and shingle, mainly in the west.

Sea-Lemon (Nudibranchiata: *Archidoris pseudoargus*). Our second largest nudibranch sea slug, 2–3 ins. long, widespread and common on the shore, feeding on the breadcrumb sponge.

Sea-Lettuce or Green Laver (Chlorophyceae: *Ulva lactuca*). A common green seaweed with broad, supposedly lettuce-like leaves.

Sea Level. The level which the sea would assume if there were no tides or similar factors at work, calculated for any given place as the average between high and low tide levels. In Great Britain all land heights are calculated from the mean sea level at Newlyn, near Penzance in Cornwall. This is known as the Ordnance Survey Datum, and is used in all altitude calculations on O.S. maps, e.g. the summit of Ben Nevis, Britain's highest mountain is 4,406 ft O.D.

Sea-Mat. A colonial aquatic animal of the phylum Bryozoa or Polyzoa with two divisions, Polyzoa proper or Ectoprocta, and the much smaller group of Entoprocta or Endoprocta, now commonly regarded as a separate phylum. Most species are marine, but some are also found in fresh water. They either form crusts on rocks or seaweed, or themselves resemble seaweeds, each animal occupying a separate cell in the colony, and feeding by means of moss-like tentacles. Hornwrack is one common species, another is *Flustrella hispida*, the greyish encrustation often seen on sea-weeds. The small sea-mats of the Pedicellinidae are our principal representatives of the Entoprocta. Sea-mats were formerly classed with the quite different but also plant-like Sea-firs as zoophytes.

Sea-Milkwort (Primulaceae: *Glaux maritima*). Prostrate plant with small pink flowers, widespread on the coast; unrelated to the true milkworts.

Sea-Mouse (Polychaeta: *Aphrodite aculeata*). A quite unwormlike, 3–4 ins. long and nearly 2 ins. broad worm, covered with green and gold hairs, frequent offshore and found on sandy shores at extreme low water of spring tides.

Sea-Potato. ⇨ SEA-URCHIN.

Sea-Scorpion. Two marine species of bullhead.

Sea Slater (Isopoda: *Ligia oceanica*). Our largest woodlouse, common on the upper shore, hiding by day under stones and in seaweed, or in cracks in walls.

Sea Slug (Gastropoda: Opisthobranchia). A sub-class of marine molluscs, with the shell very much reduced or absent, so somewhat resembling land slugs; often attractively coloured, divided into the tectibranchs, which may have a shell, and the nudibranchs, which never have one.

Sea-Snail (Scleroparei: *Liparis*). Two blunt-nosed, inshore fishes, slug-like rather than snail-like: *L. liparis* and *L. montagui*, both fairly common but rather northern in distribution.

Sea Spider (Arachnida: Pycnogonida). Small, superficially spider-like marine arachnids, also known as Nobody Crabs, mostly feeding on coelenterates, their larvae parasitic on hydroids; their habit of the males caring for the young is unique among arthropods. Some two dozen species live between tide marks or in shallow water, and a good many more in deeper water; the commonest species is *Pycnogonum littorale*.

Sea Squirt (Urochordata: Ascidiacea). A class of tunicates, also known as ascidians, common on our shores, often colonial, the adults always fixed to rocks or seaweeds; including the Star Sea-squirt *Botryllus schlosseri* in many colour varieties, and the Sea-Gooseberry.

Sea-Swallow. A general name for terns.

Sea-Urchin (Echinodermata: Echinoidea). A class of spiny marine animals, roundish, oval or heart-shaped with a brittle limy external skeleton. The commonest species on the shore are the globular Common Sea-Urchin *Echinus esculentus*, the small, oval cake-urchin called the Green Sea-Urchin *Echinocyamus pusillus*, the local Purple Heart-Urchin *Spatangus purpureus* and the straw yellow Sea-Potato *Echinocardium cordatum*, the two last-named both heart-shaped.

Seaweed. Marine Alga. ⇨ BROWN SEAWEEDS, GREEN SEAWEEDS, RED SEAWEEDS.

Sea Wrack. ⇨ WRACK.

Secund. Botanical adjective meaning nodding or drooping to one side.

Sedge (Cyperaceae: *Carex*). A large genus of grass-like plants, growing in grassy places and by fresh water, but distinguishable from grasses by their solid stems; among the commoner of our seventy-five species are the Spring Sedge *C. caryophyllea*, Fox Sedge *C. otrubae*, Hammer Sedge *C. hirta*, Pond Sedge *C. riparia*, Wood Sedge *C. sylvatica* and Pendulous Sedge *C. pendula*. The few others with vernacular rather than book names include Hop Sedge *C. pseudocyperus*, Bottle Sedge *C. rostrata*, Tussock Sedge *C. paniculata*, Sand Sedge *C. arenaria* and Flea Sedge *C. pulicaris*. Some other members of the Cyperaceae are also loosely called sedges, most notably Saw Sedge *Cladium mariscus*, the tall saw-leaved sedge used for thatching, especially on the Norfolk Broads. Carnation Sedge, ⇨ CARNATION GRASS. ⇨ BEAK-SEDGE.

Sedge Fly. Caddis-flies, including the Grey Sedge or Grey Flag, the Great Red Sedge or Murragh, and the Cinnamon Sedge *Limnephilus lunatus*.

Sedimentary Rock. Rock that was originally laid down in layers or beds, usually under water in the sea or a lake or river, such as sandstones, shales and limestones. ⇨ IGNEOUS ROCK.

Seed-Leaf. ⇨ COTYLEDON.

Seed-Plant (Spermatophyta, formerly Phanerogamia). The major division of the Plant Kingdom containing the flowering plants (angiosperms) and conifers (gymnosperms), and comprising all our trees and shrubs and all our herbaceous plants except the ferns and their allies; they are all vascular plants, differentiated from the ferns and their allies, the vascular cryptogams, in reproducing by seeds which develop from ferilized ovules, not from spores.

Seiche. A phenomenon of lakes, after a strong wind has piled the warm surface layer of water (epilimnion) up on to one side of the lake. When the wind drops the warm water 'slops' back to the other side, as if it were in a tilted bath that had been righted, and continues to oscillate over the colder water (hypolimnion) for some time.

Selachii (Chordata: Vertebrata). The class of primitive cartilaginous marine fishes whose gill-slits open directly on to the water instead of being covered by a gill-cover, including the sharks and dogfishes (Pleurotremata) and skates and rays (Hypotremata).

Selection. ⇨ ADAPTATION; NATURAL SELECTION.

Self-Heal (Labiatae: *Prunella vulgaris*). Abundant purple flower of grassy places, named from its former medicinal use. Cut-leaved Self-heal *P. laciniata* is scarce, with creamy-white flowers.

Semi-natural Vegetation. Vegetation which has been subject to human interference or management in the past, but has now taken on an apparently natural aspect, e.g. a chalk down, a beech plantation, or a mown fen. Almost all the apparently natural vegetation of the British Isles is in fact semi-natural.

Sepal. A divided portion of a calyx, lying outside the petals of a flower. ⇨ PERIANTH.

Sere (adj. Seral). A natural succession of plant communities leading to a Climax. ⇨ SUBSERE.

Serin (Passeriformes: *Serinus canarius*). A small yellowish-green finch, the origin of the canary; straggler from the Continent. The domesticated canary originated in a yellower race from the Canary Islands.

Serpentes (Reptilia: Squamata). The sub-order of reptiles containing the snakes.

Serpentine. A beautiful marbled red or green rock, found especially in the Lizard peninsula of Cornwall and in the north of the Shetlands, in both of which it supports a distinctive flora.

Serrate. Botanical adjective meaning saw-toothed.

Service Tree (Rosaceae: *Sorbus domestica*). A relative of the rowan and whitebeam, known in the wild in Britain only from a single tree in the Wyre Forest, Worcestershire, where it is called the Whitty Pear and was probably originally planted. ⇨ WILD SERVICE TREE

Sessile. Botanical adjective meaning unstalked.

Set. The burrow or burrows of a badger family group.

Setaceous. Bristly.

Sewage Fungus. An unpleasant looking deposit, which is not a fungus but produced by bacteria close to any source of water pollution by sewage.

Sewall Wright Effect. The tendency for small populations to have useless or even harmful mutations accidentally fixed in their constitutions, and so to die out, as originally suggested by the American geneticist, Sewall Wright.

Sewin or Sewen. The sea trout of Ireland and the west coast of Great Britain.

Sexton Beetle. ⇨ BURYING BEETLE.

Shad (Clupeidae: *Alosa*). Two smallish species of anadromous fish: the Thwaite Shad *A. finta*, which spawns in estuaries, and the scarcer Allis Shad *A. alosa*, which ascends rivers to spawn.

Shag (Pelecaniformes: *Phalacrocorax aristotelis*). A large greenish-black sea-bird, sometimes called the green cormorant, nesting on cliffs all round the British Isles except between Northumberland and the Isle of Wight. It has a crest in the breeding season.

Shaggy Cap. One of the ink cap fungi.

Shaggy Parasol. ⇨ PARASOL.

Shale. A rock made up of thin layers of consolidated and compressed clay, often found in the Coal Measures.

Shamrock. The national plant of Ireland, of indeterminate identity, but on St Patrick's Day (17 March) Irishmen wear sprigs of small clover-like plants such as yellow trefoil and black medick.

'Shank. Bird-watchers' slang for those waders of the genus *Tringa* named after their leg colour, notably the redshank and greenshank.

Shanny. ⇨ Common BLENNY.

Shark (Selachii: Pleurotremata). Elongated cartilaginous marine fishes with very sharp teeth; mostly with their mouths crescent-shaped and set well back below the tip of the snout, unevenly forked tails and sandpapery skins; often attaining a great size and sometimes ferocious, though man-eating is not known in British waters. The smaller sharks are known as dogfishes or hounds. Some two dozen species of seven families occur in British waters, and nine are rare or uncommon stragglers. The Basking Shark or Sailfish *Cetorhinus maximus*, the largest European shark and the second largest fish in the world, is a harmless plankton feeder named from its habit of basking on the surface of the water; common off our western coasts. Blue Shark *Carcharinus glaucus*, a man-eater in other parts of the world, is not uncommon off our south-western coasts in autumn. Fox Shark or Thresher *Alopias vulpes*, also not uncommon, is noted for its extremely long pointed tail, occupying half its length of 10–20 ft. Hammerhead Shark *Sphyrna zygaena*, with its singular hammer-shaped head, is one of the rare stragglers. ⇨ DARKIE CHARLIE, DOGFISH, PORBEAGLE, TOPE.

Shearwater (Procellariiformes: *Puffinus*). The longer-winged sea-birds of the family Procellariidae, of which one, the Manx Shearwater *P. puffinus*, breeds in burrows on islands off our northern and western coasts; three, the Great Shearwater *P. gravis*, which breeds on Tristan da Cunha, the Sooty Shearwater *P. grisea*, which breeds in the South Pacific, and the North Atlantic Cory's Shearwater *P. diomedea* are more or less regular migrants through British seas, and two more are rare visitors. The name derives from their habit of shearing the waves on stiffly outspread wings, tilting alternately from side to side.

Sheep (Artiodactyla: *Ovis aries*). There are no wild sheep in the British Isles, though sheep of the primitive dark brown Soay breed, resembling the mouflon of the Mediterranean islands, are allowed to run wild on St Kilda and several islands off the Welsh coast.

Shelduck or Sheld-duck (Anatidae). Large goose-like surface-feeding ducks. The native species is *Tadorna tadorna*, which breeds in burrows all round our coasts and migrates to the Heligoland Bight to complete its moult. A smaller moult-migration rendezvous exists in Bridgwater Bay, Somerset, but the origin of the birds there is at present unknown. The Ruddy Shelduck *Casarca ferruginea* is a straggler from the Mediterranean, and sometimes also escapes from collections of waterfowl.

Shell. (1) The outer covering or exo-skeleton of tortoises and turtles, composed of bony plates covered with horny shields. The upper part, covering the back, is the carapace, and the lower part, covering the belly, the plastron. (2) The hard outer covering of most molluscs, lacking in some groups such as the sea slugs and squids or concealed as in the land slugs. It consists of conchiolin, a horny chitinous substance, mixed with various mineral salts, with an inner lining of nacre or mother-of-pearl, which adds greatly to the aesthetic attraction of some shells. Snails and most other gastropods have a unitary shell, but the mussels and other bivalves have two, and the chitons have eight valves. (3) The outer covering of a bird's egg.

Shellfish. Name given by fishmongers and other traders to both shelled molluscs, such as oysters and mussels, and edible crustaceans, such as lobsters and prawns.

Shepherd's Needle or Venus's Comb (Umbelliferae: *Scandix pecten-veneris*). A somewhat untypical white-flowered umbellifer, having only one or two spokes to the umbel, named from its 1–2 in. fruits; a frequent arable weed in the south.

Shepherd's Purse (Cruciferae: *Capsella bursa-pastoris*). Abundant annual weed of gardens and arable; flowers white; named from the shape of the seed-pods.

Shepherd's Rod (Dipsacaceae: *Dipsacus pilosus*). A close relative of the teasel, with round white flower-heads, rather local in bushy places in the south.

Shepherd's Weatherglass. ⇨ Scarlet PIMPERNEL.

Shieldbug (Hemiptera: Heteroptera). A group of flattened, shield-shaped plant-bugs of the families Acanthosomidae, Cydnidae, Scutelleridae and Pentatomidae: about forty species, including the handsome widespread Green Shieldbug *Palomena prasina*, the Hawthorn Shieldbug *Acanthosoma haemorrhoidale*, common on hawthorn, and the Birch Shieldbug *Elasmostethus interstinctus* common on birch. ⇨⇨ BISHOP'S MITRE, BLUE BUG, FOREST BUG, NEGRO BUG, SLOEBUG, TORTOISE BUG and TURTLEBUG.

Shield Fern (Aspidiaceae: *Polystichum*). Two of our larger tufted ferns allied to holly fern, somewhat like male fern but with more leathery, spine-toothed fronds: the widespread Hard Shield Fern *P. aculeatum* and the more local Soft Shield Fern *P. setiferum.*

Shiner, The. Folk name for the German cockroach, from its shining tawny colour.

Shingle. Found by rivers and on the sea shore, and consisting of small smooth water-worn pebbles, the result of hard rocks or flints being eroded by water. On the coast shingle is carried along by the currents and deposited in banks or on flat shores, or builds up into spits, e.g. Calshot Spit in Southampton Water, or long bars such as Chesil Bank in Dorset. Shingle beaches are most frequent on the south and east coasts. Below the high tide line they have no plant life because of the constant movement of the pebbles. Shrubby seablite is a typical plant of shingle.

Shipworm (Lamellibranchia: *Teredo*). Highly specialized marine bivalves, whose long worm-like bodies end in a much reduced shell, with which they bore into the wood of ships, piers, piles, etc; they can be serious pests. The largest and commonest species is *T. norvegica*, interesting in being one of the few animals possessing the enzyme cellulase which can hydrolyse cellulose to glucose.

Shoddy Alien. ⇨ WOOL ALIEN.

Shorebug (Heteroptera: Saldidae). Large-eyed predatory bugs found by both fresh and salt water, and in damp hollows; a score of species include the Common Shorebug *Saldula saltatoria*, the Marine Bug, and one dry-land species, *S. orthochila.*

Shoreweed (Plantaginaceae: *Littorella uniflora*). An aquatic plant forming thick swards under water in lakes and tarns, but flowering only on the dried-out mud nearby.

Shore Zonation. Marine biologists divide the sea shore into five main zones, each with its characteristic flora and fauna: First the *splash zone*, above the shore proper, but liable to be drenched with spray at high tide in a gale, and so requiring its denizens to be able to withstand some salt in their environment. The *upper shore* lies between the average high tide level and the extreme high water level of spring tides, and is therefore only occasionally covered by the sea and then only for a short time. The *middle shore* lies between average high tide level and average low tide level, and comprises the main part of the shore, mostly covered by the sea twice a day. The *lower shore* is the strip between average low tide level and the extreme low water level of spring tides, rarely uncovered by the sea and then only for comparatively short periods. Finally, the *sublittoral fringe* lies below the shore proper, but its inhabitants must be able to stand living in shallow water for short periods at regular intervals, while the large seaweeds that grow in it, such as oarweed, may actually project above the surface.

Shothole Borer. ⇨ BARK BEETLE.

Shoveler (Anatidae: *Spatula clypeata*). A surface-feeding duck, distinctive for its large spatulate bill; widespread and frequent in winter, but breeding mainly in the north.

Shrew (Insectivora: Soricidae): Small short-lived mouse-like insectivorous mammals. Four species in Britain, of which three are widespread: the Common Shrew *Sorex araneus*, the Pigmy Shrew *S. minutus*, which is less than 4 ins. long, and the Water Shrew *Neomys fodiens*; the fourth, *Crocidura suaveolens*, not discovered till 1924, is confined to the Scilly Isles.

Shrike (Passeriformes: *Lanius*). Small to medium-sized song birds, noted for their predatory habits, including the impaling of bees and other small prey animals in 'larders' on thorn bushes, Four British species: Great Grey Shrike *L. excubitor*, scarce winter visitor from northern Europe; Lesser Grey Shrike *L. minor*, straggler from southern Europe; Red-backed Shrike or Butcher Bird *L. cristatus*, decreasing summer visitor, now almost confined to south-east England. Woodchat Shrike *L. senator*, increasing straggler from southern Europe.

Shrimp (Decapoda Natantia: Crangonidae). Familiar crustaceans of sandy shores, the Common Shrimp being *Crangon vulgaris*, with five species of *Philocheras*. Several other crustaceans named shrimp are not closely related; ⇨ BURROWING SHRIMP, FAIRY SHRIMP, FRESHWATER SHRIMP, OPOSSUM SHRIMP, SAND SHRIMP, SKELETON SHRIMP.

Shrub. A woody plant with several fairly stout stems from a single rootstock, usually less tall than a small tree.

Sibling Species or Species Pair. Species which are similar or even identical in outward appearance, but yet are quite distinct in genetical make-up and never interbreed, e.g. common and heath spotted orchids, the chiffchaff and the willow warbler, or the marsh and willow tits. They have probably evolved apart in the comparatively recent past, but are now kept apart by geographical, ecological or behavioural factors.

Sickener, The (Basidiomycetes: *Russula emetica*). A strikingly red-capped toadstool, growing in coniferous woods, and liable to cause vomiting if eaten raw, though apparently harmless if cooked.

Sika (Artiodactyla: *Cervus nippon*). An Asiatic deer, resembling a small red deer, originally introduced for hunting, and now widely distributed in scattered colonies from Dorset to the Highlands. Male, stag; female, hind; young, calf.

Sike. ⇨ SYKE.

Silicula, Siliqua. Fruiting capsules of crucifers; respectively less and more than three times as long as broad.

Sill. A sheet of igneous rock formed during the Alpine earth movements of the Tertiary era, when molten lava forced its way up and then horizontally between the layers of older bedded rocks. The vertical sheet is a dyke; the horizontal a sill, e.g. the Great Whin Sill. Sills may be of any thickness. ⇨ WHINSTONE.

Silt. An alluvial deposit laid down in water with particles finer than sand but coarser than clay, i.e. between 0.02 and 0.002 in. in diameter.

Silurian Period. ⇨ GEOLOGICAL PERIODS.

Silver-Fish (Thysanura: *Lepisma saccharina*). A bristle-tail, named from its bright silvery scales, common indoors, especially in kitchens, bakehouses, libraries, etc., and capable of seriously damaging books by eating the glue in their bindings.

Silverhorn (Trichoptera: *Mystacides* and *Leptocerus*). Caddis-flies of lakes and rivers, whose larvae are relished by trout, including the Black Silverhorn *M. azurea* and the Brown Silverhorn or Grouse Wing *M. longicornis*.

Silver-Leaf. A common disease of plum trees, caused by the basidiomycete fungus *Stereum purpureum*.

Silver Tommy (Prosobranchia: *Gibbula cineraria*). The grey top shell, widespread and common on the shore.

Silverweed (Rosaceae: *Potentilla anserina*). A common yellow-flowered cinquefoil of dampish grassy and waste places with silvery pinnate leaves.

Silver Y (Caradrinoidea: *Plusia gamma*). A common migratory moth, often flying by day, named from the y-shaped mark on the forewings.

Simple. ⇨ HERB (2).

Sinistral. Spiralling left-handed, especially of snail shells.

Sink Hole. ⇨ POT HOLE, SWALLOW HOLE.

Sinter. A siliceous or calcareous rock formed as a deposit in springs.

Sinuate. Botanical adjective meaning wavy-edged.

Sinus. The angle between two lobes or teeth of a leaf or other plant organ.

Siphon Algae (Chlorophyceae: Siphonales). A small group of freshwater green algae, with tubular strands, whose most familiar representative is *Vaucheria*, making dark green felted masses on the mud in and by ponds.

Siphonaptera. ⇨ FLEA.

Siphonophore or Chondrophore (Hydrozoa: Siphonophora). Colonial marine coelenterate animals, consisting of a large number of polyps with specialized functions: for locomotion, attacking prey, feeding and digestion, reproduction, and so on. Two oceanic species, the Portuguese man-o'war and the by-the-wind-sailor occur in British seas.

Sipunculoidea. A small phylum of thick-skinned worm-like marine animals, formerly included in the abandoned phylum Gephyrea. *Golfingia elongata* is uncommon on sandy shores; *Phascolion strombi* lives in old mollusc shells.

Siskin (Passeriformes: *Carduelis spinus*). One of our smallest finches, formerly a favourite cage bird under the name of aberdevine; breeds in coniferous woods in Scotland and Ireland, more widespread in winter, when many birds arrive from the Continent.

Site of Special Scientific Interest, abbreviated S.S.S.I. An area which has been notified to the local planning authority by the Nature Conservancy as of special interest for its fauna, flora, or geological or physiographical features, in order that the authority shall know to consult the Conservancy before giving permission for any development on the site. There is at present no legal obligation on owners or occupiers of such sites, nor any right of entry or public access.

Skate (Selachii: Hypotremata). The five long-nosed rays of the genus *Raja*, of which the commonest and shortest-nosed is the Common Skate *R. batis*. The longest-nosed, the Long-nosed Skate *R. oxyrhynchus* is mainly a deep-water fish.

Skeleton Shrimp (Amphipoda: Caprellidae). Very slender small marine crustaceans, which attach themselves to hydroids, bryozoa and seaweeds by their hind legs and adopt a curious praying-mantis-like attitude with the fore part of the body.

Skep. Old-fashioned straw or wicker beehive.

Skerry. (1) A low rocky islet. (2) Local name for a hard band of sandstone in the Keuper marls of Nottinghamshire, quarried for building, especially around Tuxford.

Skipjack. ⇨ CLICK BEETLE.

Skipper. ⇨ SAURY.

Skipper Butterfly (Papilionina: Hesperiidae). Our most primitive family of butterflies, with eight species, all small, brown or grey-brown, somewhat moth-like, and mainly frequenting grassy places. Large Skipper *Ochlodes venata* and Small Skipper *Thymelicus sylvestris* are the commonest; Dingy Skipper *Erynnis tages* and Grizzled Skipper *Pyrgus malvae* are locally common in the south.

Skua (Charadriiformes: Stercorariidae). Medium to large gull-like sea-birds, noted for their aerial attacks on gulls and terns to deprive them of fish they have caught; known as Jaegers in North America. The two British breeding species, both confined to the far north of Scotland and its islands, are the Great Skua or Bonxie *Catharacta skua* and the Arctic or Richardson's Skua *Stercorarius parasiticus*, the latter being frequent further south on passage. Two other species are uncommon passage migrants: Pomatorhine or Pomarine Skua *S. pomarinus* and Long-tailed Skua *S. longicaudus*. The three last all have two colour phases, a light and a dark.

Skullcap (Labiatae: *Scutellaria*). Two flowering plants, the common blue *S. galericulata* of stream-sides and marshes, and the less frequent pinkish Lesser Skullcap *S. minor* of damp heathland.

Skylark. ⇨ LARK.

Slack or Dune Slack. A damp hollow between the ridges of the sand dunes, with its own specialized flora.

Slate. A layered metamorphosed rock, originally a shale but compressed into a much harder substance, used for roofing, especially frequent in Cornwall and North Wales.

Slater. ⇨ WOODLOUSE. Hog and Water Slaters, ⇨ WATER-LOUSE. ⇨⇨ SEA SLATER.

Slime Fungus, or Slime Mould (Myxomycophyta). A curious phylum or division of plants with many of the characteristics of animals, consisting of a slow-moving mass of amoeboid protoplasm, which engulfs particles of solid food which it encounters, while reproducing itself by spores produced by a vegetative body called the plasmodium. The majority of slime-fungi are saprophytes. There are three classes; the Myxomycetes, Acrasiae and Plasmodiophoreae, the last being often included among the true Fungi. ⇨ FLOWERS OF TAN.

Sloe. The acrid fruit of the blackthorn.

Sloebug (Heteroptera: *Dolycoris baccarum*). A brown shieldbug, widespread but apparently not often found on blackthorn or sloes.

Slot. The tracks or footprints of deer.

Slow-Worm or Blind-worm (Sauria: *Anguis fragilis*). A legless viviparous harmless lizard, often mistaken for a snake and neither blind nor slow; widespread and locally common. The blue-spotted slow-worm is a variant found only in England and Wales in which only males are known.

Slug (Gastropoda: Pulmonata). A group of terrestrial pulmonate molluscs, closely allied to the land snails, in which the shell is oval in shape and concealed under the slimy mantle; some are destructive garden pests. The seventeen British species are contained in three unrelated families; Sea Slugs are also unrelated. The commoner species include the Black Slug *Arion ater* and the Garden Slug *A. hortensis*, both all too common in gardens; the Field Slug *Agriolimax agrestis*, another pest of gardens and farm crops; the Marsh Slug *A. laevis* and the two Keeled Slugs *Milax*, whose names are self-explanatory; the Great Slug *Limax maximus*, growing to 8 ins. long, and the even larger woodland Ash-black Slug *L. cinereoniger*, which has been known to attain over 1 ft; the Tree Slug *L. marginatus*, named from its habit of ascending tree trunks in wet weather to feed on lichens and fungi; and the three Shelled Slugs *Testacella*, which have a small external ear-shaped shell at the rear end, live in the soil and unlike all other British slugs are carnivorous. The Spotted or Kerry Slug *Geomaculus maculosus* is a Lusitanian animal of south-west Ireland. Rose Slug ⇨ SAW-FLY, SLUGWORM.

Slugworm. Larvae of saw-flies of the genus *Eriocampa*, including the pear saw-fly and the rose slug; they are both slimy and slug-like in appearance.

Smelt or Sparling (Pisces: *Osmerus eperlanus*). Anadromous herring-like fish, allied to the Salmonidae, spawning in estuaries, where it is fished commercially, and mainly northern in distribution. The name derives, not from the cucumber-like smell of its flesh, but from the same Anglo-Saxon word, meaning smooth and shining, that gives rise to *smolt*. The Sand Smelt (Percomorphi: *Atherina presbyter*) is a small fish of inshore waters in the south, related to the grey mullet.

Smew (Anatidae: *Mergus albellus*). A small diving duck, the smallest British saw-bill, not breeding, but a widespread winter visitor. The handsome black and white drakes were formerly known as White Nuns.

Smolt. The young salmon, after usually two years as parr, become smolts when they turn silvery and migrate to the sea. Sea trout also pass through a smolt stage.

Smut or Smut Fungus (Basidiomycetes: Ustiginales). An order of small parasitic fungi, named from their black spores, many of which cause plant diseases, such as anemone, onion and violet smuts caused by various species of *Urocystes*. Stinking Smut or Bunt is a serious disease of wheat caused by *Tilletia caries*.

Snail (Mollusca: Gastropoda). Univalve mollusc of the land and freshwater; most are Pulmonata, though a few freshwater ones are operculates. Our largest species is the Roman Snail. Its close relative the variable Common or Garden Snail *Helix aspersa* is common in gardens, and was formerly eaten in Britain, as it still is on the Continent. According to E. W. Swanton, it was once sold in the Bristol markets as 'wall-fish'.

Snake (Squamata: Serpentes). Scaly, legless reptiles; three native species: the poisonous Adder or viper (Viperidae) and the harmless grass and smooth snakes (Colubridae). The oviparous Grass or Ringed Snake *Natrix natrix* is locally common in England and Wales. The viviparous Smooth Snake *Coronella austriaca* is confined to heaths from Surrey to Dorset. There are no snakes in Ireland.

Snake Fly (Megaloptera: *Raphidia*). Predatory neuroptera, with long snake-like 'necks', frequenting woodlands and laying their eggs under the bark of trees; three British species.

Snakeroot. ⇨ BISTORT.

Snakeshead. ⇨ FRITILLARY (plant).

Snakestone. ⇨ AMMONITE.

Snakeweed. ⇨ BISTORT.

Snapdragon (Scrophulariaceae: *Antirrhinum majus*). The familiar garden plant, with flowers usually pink, purple or white, established on many old walls and railway cuttings. The related Lesser Snapdragon, Weasel's Snout or Calf's Snout *Misopates orontium* is a pink-flowered weed of light soils in the south.

Sneezewort (Compositae: *Achillea ptarmica*). A non-aromatic relative of yarrow, with undivided leaves and larger flowers, preferring damper grassland on acid soils and commoner in the north.

Snipe (Charadriiformes: Scolopacidae). Medium to small waders. The Common Snipe *Gallinago gallinago*, noted for its extremely long bill and zigzag flight when flushed, is widespread and common, but breeds more locally in the south. Sabine's Snipe is a melanistic variant. The smaller Jack Snipe *Lymnocryptes minimus* is a locally frequent winter visitor. The Great Snipe *G. media* is an uncommon passage migrant. Red-breasted Snipe, ⇨ DOWITCHER.

Snipe Fly (Brachycera: Rhagionidae). A family of eighteen two-winged flies, including the downlooker fly.

Snout Moth (Caradrinioidea: Hypeninae). A sub-family of moths named from their conspicuous proboscis.

Snow. Water vapour frozen at a high altitude into delicate crystals of ice and precipitated, like rain, by gravity. There are no permanent snowfields on British mountains.

Snowberry (Caprifoliaceae: *Symphoricarpos rivularis*). Low shrub with large white berries, suckering freely to form thickets; widely planted and often escaping.

Snowdon Lily. ⇨ SPIDERWORT.

Snowdrop (Amaryllidaceae: *Galanthus nivalis*). A white-flowered bulbous plant, blooming from January to March, frequently naturalized in ornamental grounds and perhaps native in damp woods in the west.

Snowflake (Amaryllidaceae: *Leucojum*). Two white-flowered bulbous plants: Spring Snowflake *L. vernum* flowering in February and March and confined to two sites in the south-west; and Summer Snowflake or Loddon Lily *L. aestivum*, flowering in April and May in damp meadows in the south, especially by the Thames and the Berkshire Loddon.

Snow-in-Summer (Caryophyllaceae: *Cerastium tomentosum*). A common garden plant related to the mouse-ears, frequently escaping and now established in many places.

Snow Mould. One of the commonest turf diseases, caused by the fungus *Fusarium nivale*.

Soapwort or Bouncing Bett (Caryophyllaceae: *Saponaria officinalis*). A pink-flowered ally of the campions, formerly widely used as a soap substitute – its leaves and stems make a lather when rubbed together – and so much planted and still surviving by many waysides; possibly also native by streams in the south-west.

Sobole. An underground sucker or creeping stem.

Social Insects. Insects which live in a society consisting of the two parents, or at least the fertilized female, and their offspring; the two generations living together in a common abode or shelter and exhibiting at least some degree of mutual cooperation (A. D. Imms); i.e. all true insect societies are basically families. The social insects found in Britain are the ants, social bees and social wasps.

Society for the Promotion of Nature Reserves. Founded in 1912. Has twelve reserves and sponsors the county naturalists' trust movement; the Society also made a large grant to help start the Council for Nature. Publishes an annual *Handbook*.

Soft Grass (Gramineae: *Holcus mollis*). A common softly downy grass of shady places on acid soils, closely related to Yorkshire fog.

Softwood. The timber of a coniferous tree. ⇨ TREE.

Soil. The surface covering of the earth in which vegetation grows and a large fauna of mammals, insects, worms, beetles, bacteria, etc., lives and gets its food. Soil is created by the erosion and disintegration of rocks, the decay of vegetable and animal matter, and the action of climate, i.e. water and heat. Water erodes rock and soil, washes the particles down and deposits them, making the basis for new soils. Water also percolates through soil and permeable rock, carrying down both soluble and the finer insoluble substances, and thus leaching the top soil and enriching the lower layers. The type of rock from which the soil is derived determines its nature, e.g. sand, clay, limestone. All natural mineral soils consist of a mixture of different-sized particles; the preponderance of one size decides its classification, e.g. coarse sand, fine sand, clay. An apparently solid clod of earth is usually only about half solid matter; the rest is air and water. A loam is a good mixture of different-sized particles. Soils are divided into acid, neutral and alkaline according to the concentration of hydrogen and hydroxyl ions they contain. Acid soil has an excess of hydrogen, alkaline an excess of hydroxyl ions, and in neutral soil they are equal. Most British soils, other than limestone ones, are more or less acid, as a result of the fairly heavy rainfall which leaches out the salts, especially calcium. Typical acid soils are the heather moors of the north and west; typical neutral soils the loams and clays of flat lowland country in the Midlands and south; typical alkaline soils those of the chalk and limestone hills, e.g. the South Downs, Cotswolds and Craven Pennines. The acidity or alkalinity of the soil is measured by the pH scale. The soil dictates to a large extent the vegetation and therefore the fauna of a region.

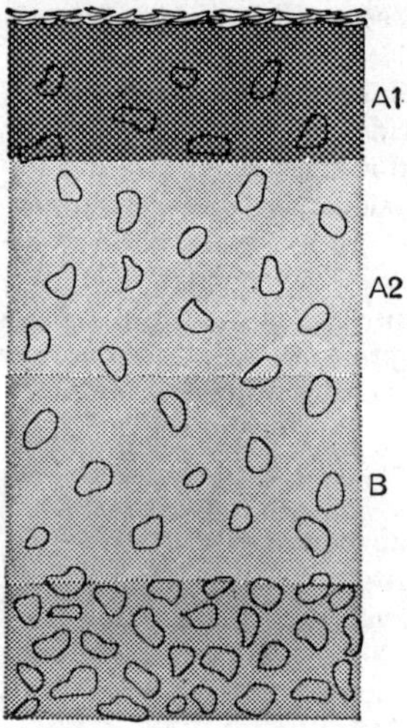

Soil Profile

A, Eluvial horizon
A1, Surface layers stained with humus, A2, lighter
B, Illuvial layer not sharply marked but showing increasing fineness of texture downwards

Soil Creep or Hill Creep. A gradual downhill movement of the surface soil, caused by gravity and induced both by the cracking of the soil surface by summer heat and the loosening of the finer soil particles by rain. ⇨ SOLIFLUCTION.

Soil Mite. Mites, especially the Cryptostigmata, are the most abundant animals in the soil, in numbers of both species and individuals, which may run into hundreds of millions per acre. They feed on algae, fungi, and decaying leaves and wood, and many species are predatory, feeding on various soil animals, including other mites.

Solanaceae (Angiospermae: Dicotyledones). A plant family containing both poisonous plants, such as the various nightshades, henbane and thorn-apple, and closely related edible ones, such as the potato and tomato.

Soldier and Sailor Beetle (Diversicornia: Cantharidae). A family of forty-one carnivorous beetles, commonly seen on flowers. By far the commonest species is *Rhagonycha fulva*, typical of umbellifer flowers. Other common species include the large *Malthinus flaveolus*, *Cantharis rustica* and *C. pallida*.

Soldier Fly (Brachycera: Stratiomyidae). A family of over fifty, mostly brightly coloured two-winged flies.

Soldier Orchid. ⇨ MILITARY ORCHID.

Sole (Heterosomata: Soleidae). Somewhat elongated right-sided flatfish, the Common or Dover Sole *Solea solea*, with highly esteemed flesh, being much the commonest, preferring muddy bottoms. The French or Sand Sole *Pegusa lascaris* and the reddish-brown Variegated or Thickback Sole *Microchirus variegatus* are scarce and mainly southern fish. The Lemon Sole, Lemon Dab or Smear Dab *Microstomus kitt* is a yellowish relative of the plaice, mainly northern in distribution.

Soleidae (Pisces: Heterosomata). A family of right-sided flatfish containing the soles (except the lemon sole) and the solenette.

Solenette (Soleidae: *Microchirus boscanion*). The smallest of our soles, common on sandy bottoms.

Solifluction. Soil creep accentuated by the existence of a subterranean layer of frozen soil, so that large masses of soil and rock fragments moistened by the spring rains slide downhill over the hard frozen surface. Solifluction during the Ice Ages caused large masses of melting snow, soil and rock to slide into valley bottoms leaving deposits of miscellaneous rock and rubble; such as head and coombe rock.

Solomon's Seal (Liliaceae: *Polygonatum*). Three species of woodland plant, with graceful arching stems and drooping greenish-white flowers; commonest and most widespread is *P. multiflorum; P. odoratum* is local in limestone woods; and *P. verticillatum* is now rare in Perthshire.

Solstice. The time of year at which the longest and shortest days occur. The summer solstice falls on or about 21 June, when in the South of England there are approximately sixteen hours of daylight and only eight of darkness, and the sun rises in the north-east and sets in the north-west. The winter solstice falls on or about 22 December, when the periods of daylight and darkness are reversed, and the sun rises in the south-east and sets in the south-west.

Song Bird. In general parlance a bird with a highly developed song, such as a blackbird, nightingale or canary, but technically a member of the large order of the Passeriformes or passerines, to which all the British birds which could be called song birds in the general sense do in fact belong. Many passerines, however, have songs of no great quality, while many non-passerines have calls or cries that perform the same biological function as the songs of song birds. ⇨ BIRD SONG.

Sorb. A tree of the genus *Sorbus* (Rosaceae), which includes the rowan, white beam and wild service tree.

Sorel or Sorrel. A third-year fallow buck.

Sori. Plural of sorus.

Sorrel (Polygonaceae: *Rumex*). Small acid-tasting relatives of the docks, with arrow-shaped leaves and red flower-heads, especially Common

Sorrel *R. acetosa*, widespread and abundant in grassland, and Sheep's Sorrel *R. acetosella*, widespread and common in dry grassy and heathy places, especially on acid soils. The related Mountain Sorrel *Oxyria digyna*, with rounded leaves, is frequent on mountains in the north. Wood Sorrel (Oxalidaceae: *Oxalis acetosella*) is a common woodland plant with trefoil leaves and solitary lilac-veined white flowers, well known because its leaflets have a drooping 'sleep' position assumed at night.

Sorus. A group of spore-heaps on the underside of a fern frond.

Sowbane (Chenopodiaceae: *Chenopodium hybridum*). One of the less common goosefoots.

Sowbread (Primulaceae: *Cyclamen hederifolium*). An old-fashioned garden plant, recognizably akin to the greenhouse cyclamens, naturalized in widely scattered woods and hedge-banks, and perhaps native in south-east England.

Sow-Thistle (Compositae: *Sonchus*). Four plants with yellow dandelion-type flowers, weakly spiny leaves, and milky juice, three being common weeds: Smooth Sow-thistle *S. oleraceus*, Prickly Sow-thistle *S. asper* and Corn Sow-thistle *S. arvensis*. Alpine Sow-thistle *Cicerbita alpina* is a rare blue-flowered mountain plant of the eastern Highlands. Blue Sow-thistle *C. macrophylla* is an increasing garden escape on waysides and waste ground.

Spanish Fly. ⇨ BLISTER BEETLE.

Sparling. ⇨ SMELT.

Sparrow (Passeriformes: *Passer*). Small finch-like seed-eating birds, related to the African weaver birds. House Sparrow *P. domesticus*, the commonest urban British bird, is the only one to have made itself a complete commensal of man. Tree Sparrow *P. montanus*, the house sparrow's chestnut-headed country cousin, is widespread but rather local. Many other quite unrelated small birds are loosely called 'sparrow', e.g. the Hedgesparrow, and the Reed Sparrow or Pit Sparrow, country names of the Reed Bunting.

Sparrowhawk (Falconiformes: *Accipiter nisus*). A small broad-winged bird of prey, feeding mainly on sparrows and other small birds. Widespread and formerly common resident, but now as a direct result of persecution by game preservers and a secondary result of the use of pesticides, rare or extinct in many districts.

Spate. Flood water in an upland river after heavy rains or melting snows.

Spathulate or Spatulate. Botanical adjective meaning shaped like a wooden spoon, broadest at the tip; ⇨ LEAF-SHAPES.

Spawn. (1) Aggregation of eggs, e.g. of frogs, toads, fishes, bivalve molluscs. (2) The mycelium of mushrooms and some other fungi.

Spearmint (Labiatae: *Mentha spicata*). A cultivated mint, the usual accompaniment of roast lamb, often escaping on to wayside and waste ground.

Spearwort (Ranunculaceae: *Ranunculus*). Three buttercups with undivided leaves, growing in marshy places, the commonest being Lesser Spearwort *R. flammula*.

Speciation. The evolutionary process by which species are created.

Species. The smallest and the only natural group in the biological Classification of living things. The members of a species are all able to interbreed, whereas members of different species normally do not. Common names often denote species, e.g. daisy, rabbit. In the scientific name, e.g. *Bellis perennis,* the second word denotes the species. ⇨ NOMENCLATURE, HYBRID, SIBLING SPECIES.

Speckled Wood (Papilionina: *Pararge aegeria*). One of the commoner Brown Butterflies, frequenting woodlands from April onwards; named from its pale spots on a dark background.

Speculum. A distinct and differently coloured patch of feathers on a bird's wing, especially in the dabbling ducks.

Speedwell (Scrophulariaceae: *Veronica*). Low-growing plants with blue or lilac flowers; twenty-one native or established alien species, one of the best known being Birdseye or Germander Speedwell *V. chamaedrys*, widespread and common in grassy and bushy places. Other common species are the common weeds Field Speedwell *V. persica*, which flowers all the year round, Thyme-leaved Speedwell *V. serpyllifolia*, Wall Speedwell *V. arvensis* and Ivy-leaved Speedwell *V. hederifolia*, together with *V. officinalis* of heaths and woods, *V. montana* of broad-leaved woods, and Brooklime and Water Speedwell *V. anagallis-aquatica* of wet places.

Spent Gnat. Angler's name for the spinners of the mayflies *Ephemera danica* and *E. vulgata*. The duns are Greendrakes. ⇨ GREY DRAKE.

Spelaeology or Speleology. Caving, the study of the natural history and physiography of caves.

Spermaceti. A valuable white waxy substance found in the huge head of the sperm whale, and from which the animal derives its name.

Spermatophyta. ⇨ SEED-PLANT.

Sphagnales (Bryophyta: Musci). The order to which the bog mosses belong.

Sphagnicolous. Growing on or inhabiting sphagnum moss.

Sphagnum Bug (Heteroptera: Hebridae). Two minute predatory bugs inhabiting sphagnum moss: *Hebrus ruficeps* is commoner than *H. pusillus.*

Sphagnum Moss. ⇨ BOG MOSS.

Sphenopsida (Pteridophyta). A class of fern allies containing the horsetails (Equisetales).

Spider (Arachnida: Araneae). Predatory arachnids with two parts to their bodies, joined by a short link (pedicel), and equipped with silk glands, which produce silk for various purposes, including the spinning of webs (⇨ SPIDER'S WEB) to catch their prey. They do not have a larval stage like insects, but after hatching from the egg pass through several moults before becoming completely adult. At least 584 species, in twenty-four families, inhabit the British Isles, occupying a wide variety of habitats. The main groups are the cribellates or lace-web spiders (Dictynidae, Eresidae, Uloboridae), the six-eyed spiders (Dysderidae, Oonopidae), the Clubionoidea (Anyphaeinidae, Clubionidae, Gnaphosidae, Sparassidae, Thomisidae), the jumping spiders, the wolf spiders, the house spiders and their allies (Agelinidae), the comb-footed spiders (Nesticidae, Theridiidae), orb-web builders (Argiopidae, Tetragnathidae) and money spiders. ⇨ CARDINAL, CRAB, DADDY-LONG-LEGS, GARDEN AND TRAP-DOOR SPIDERS.

Harvest-spiders or Harvestmen, Red Spiders and Sea Spiders are not true spiders.

Spider Beetle (Clavicornia: Ptinidae). A family of twenty somewhat spider-like beetles, some of which are pests of stored products, notably *Ptinus tectus* and the White-marked Spider Beetle *P. fur*. Other genera are *Tipnus* and *Niptus*, the three names providing a classical example of foolish scientific names derived mnemonically.

Spider-hunting Wasp (Aculeata: Pompilidae). A family of about seventeen solitary wasps which specialize in hunting spiders, which they paralyse and carry off to store in their nests against the hatching of the larvae. The largest is *Anoplius fuscus*, and *Pompilus viaticus* is another large and widely distributed species.

Spider Orchid (Orchidaceae: *Ophrys*). Two relatives of the Bee Orchid named from the supposed resemblance of their flowers to certain spiders; both are very local in calcareous turf, the Early Spider Orchid *O. sphegodes* from Kent to Dorset and the Late Spider Orchid *O. fuciflora* only in East Kent.

Spider's Web. Spiders use their silk to construct webs of various shapes, which are the only deliberately constructed traps in the Animal Kingdom apart from those made by the caddis-flies *Hydropsyche*. The most advanced type of web is the orb or geometric web, made by spiders of three families in the British fauna, some three dozen species of the Argiopidae, of which the best known is the garden spider, the Tetragnathidae and the Uloboridae. Orb-webs are woven by laying spirals of silk around a series of radial threads which are attached to an outer framework of threads. The prey caught in orb-webs mainly consists of winged insects. Other kinds of web include cobwebs, which are large sheets of silk ending in a silken tube, hammock webs, which are just sheets of silk, such as those spun by linyphiid spiders on lawns that catch the morning dew in autumn, and the webs spun by theridiid spiders, which are no more than a tangle of threads.

Spiderwort or Snowdon Lily (Liliaceae: *Lloydia serotina*). An attractive white-flowered lily, rare and confined to Snowdonia.

Spignel-Meu (Umbelliferae: *Meum athamanticum*). A white-flowered aromatic umbellifer with whorls of thread-like leaflets, local on mountains in the north.

Spikelet. The basic floral unit of the grasses.

Spike-Rush (Cyperaceae: *Eleocharis*). A group of rush-like sedges, mostly without leaves and growing in damp places, much the commonest being *E. palustris*.

Spindle-Shank (Basidiomycetes: *Collybia fusipes*). A rather small dark reddish-brown toadstool, related to wood woolly foot, growing in tufts on the roots of living broad-leaved trees, especially oak and beech, and named from its spindle-shaped stems.

Spindle Shell (Prosobranchia: *Neptunea antiqua*). An offshore marine operculate mollusc allied to the whelks, whose empty shells are often washed up on our northern coasts and appropriated by hermit crabs.

Spindle Tree (Celastraceae: *Euonymus europaeus*). Deciduous shrub or occasionally a small tree with handsome pink berries, frequent in the south, especially on chalk and limestone, its wood was formerly used to make spindles.

Spinner. The brightly coloured, shining adult mayfly, which lives only for a day or two.

Spinney. A small wood.

Spiraea, Willow (Rosaceae: *Spiraea salicifolia*). One of a group of several suckering garden shrubs, that sometimes spread to heaths and hedges, especially in Wales and the north; flowers pink-purple, in a spike.

Spire Shell (Prosobranchia: Hydrobiidae). Small or very small fresh- and brackish-water mesogastropod operculate molluscs, with spiral shells. Commonest in estuaries is the Laver Spire Shell *Peringia ulvae*, which eats laver; and in fresh water, which it has only recently invaded from the upper parts of estuaries, is *Potamopyrgus jenkinsi*.

Spleenwort (Aspleniaceae: *Asplenium*). Seven small ferns that grow on rocks and walls, mainly in the damper west; the commonest are Common or Maidenhair Spleenwort *A. trichomanes*, Black Spleenwort *A. adiantum-nigrum*, Wall-rue and, on sea cliffs, Sea Spleenwort *A. marinum*.

Spit. ⇨ SHINGLE.

Sponge Fly (Megaloptera: *Sisyra*). Small and inconspicuous relatives of the brown lacewings, whose larvae live by sucking the juices of freshwater sponges; three British species.

Sponge (Porifera or Parazoa). A phylum of plant-like aquatic animals, so primitive that though multicellular they are not usually included in the Metazoa, and divided into two classes, Nuda and Gelatinosa, which are sometimes regarded as phyla, in which case Porifera (Parazoa) becomes a sub-kingdom. They feed by passing through their gut cavity a stream of water drawn in through their many pores, and reproduce both sexually, with a tiny free-swimming larval form, and asexually by means of over-wintering buds. The great majority of our 260 or so species are marine, but a handful live in fresh water. Sponges may be stalked and seaweed-like, encrusting or cushion-like, or button-like protruding from holes bored in soft rocks or shells; none of our species look like the familiar bathroom sponge from warmer waters. The main kinds of sponge found on our sea shores are the stalked, seaweed-like Purse Sponges (Nuda: Calcarea), the commonest being *Grantia compressa*, *Sycon coronatum* and various species of *Leucosolenia*; and in the Gelatinosa the encrusting sponges, often brightly coloured, red, yellow or orange, of which the best known are the Breadcrumb Sponge *Halichondria panicea* and the equally common *Hymeniacidon sanguinea*; and the boring sponges, of which the commonest and most widespread is *Cliona celata*, frequent in old oyster shells. Our two commonest freshwater sponges are the misnamed Pond Sponge *Spongilla lacustris*, most often found in slow-moving rivers, and River Sponge *S. fluviatilis*, which prefers lakes and ponds. Both are related to the marine encrusting sponges.

Spoonbill (Ardeiformes: *Platalea leucorodia*). A large, long-legged all-white marsh bird, which formerly bred and is now a not infrequent visitor, especially on the east coast of England. Named from its spatulate bill.

Sporangium. Spore case, a receptacle containing spores.

Spore. The basic reproductive cell of plants.

Sporozoa (Protozoa). A class of minute parasitic animals, including *Eimeria* (*Coccidia*), the origin of coccidiosis, and *Plasmodium*, the cause of malaria in man, which passes part of its life cycle in mosquitoes and part in the blood of various vertebrates.

Spotted Orchid (Orchidaceae: *Dactylorhiza*). Two common orchids with spotted leaves and pale purple or white flowers; Common Spotted Orchid *D. fuchsii* avoiding, and Heath Spotted Orchid *D. maculata* preferring acid soils.

Spraints. The droppings of otters.

Sprat (Clupeidae: *Clupea sprattus*). A small gregarious marine fish abundant off the north and east coasts of Britain, and the basis of important fisheries. Norwegian sprats are sold canned under the name of brisling.

Spring Beauty (Portulacaceae: *Montia perfoliata*). North American plant with pair of leaves like one large leaf encircling the stem below the white flowers; widely established in sandy soils.

Spring-Tail (Insecta: Collembola). An order of minute wingless insects, rarely as much as $\frac{1}{4}$ in. long, found in immense abundance in the soil, of which an acre may hold as many as 230 million; 260 British species, of which one, the blue-black *Lipura maritima*, is common on the shore.

Spring Usher (Notodontoidea: *Erannis leucophaearia*). A geometrid moth, related to the mottled umber, and one of the earliest species to emerge in the year, in February.

Spruce (Coniferae: *Picea*). The Norway Spruce *P. abies*, well known as the Christmas tree, and the Sitka Spruce *P. sitchensis* are both widely planted, and the former regenerates and is more or less naturalized in various places.

Spurge (Euphorbiaceae: *Euphorbia*). A large group of plants with poisonous milky juice in their stems, undivided leaves and loose heads of small greenish flowers. The commonest species are Wood Spurge *E. amygdaloides*, frequent in woods in the south; Sun Spurge *E. helioscopia*, Petty Spurge *E. peplus* and Dwarf Spurge *E. exigua*, widespread and common arable weeds; and Sea Spurge *E. paralias* and Portland Spurge *E. portlandica*, frequent on sandy shores and dunes on our south and west coasts. The scarce Caper Spurge *E. lathyrus*, our tallest spurge, has fruits resembling capers, but not recommended as substitutes.

Spurgebug (Heteroptera: *Dicranocephalus*). Two elongated plant-bugs feeding on spurges.

Spurge Laurel (Thymelaeaceae: *Daphne laureola*). Neither a spurge nor a laurel, but an evergreen undershrub with green flowers, black berries and laurel-like leaves, closely related to mezereon; frequent in calcareous woods in the south.

Spurrey (Caryophyllaceae). Weedy plants with whorls of narrow leaves up the stem: Corn Spurrey *Spergula arvensis* with white flowers, a common arable weed; five species of pink-flowered *Spergularia*, *S. rubra* on sandy soils inland, the rest in saltmarshes or on rocks by the sea.

Spring Net. A type of net used by bird ringers that operates on the break-back mousetrap principle, but does not harm the birds, though smaller sizes are rather dangerous.

Squamata (Reptilia). The order of reptiles including the snakes (Serpentes) and lizards (Sauria).

Squashbug (Heteroptera: Coreidae). Elongated plant-bugs named after a North American species occurring on squashes *Cucurbita*; ten British species, *Coreus marginatus* on docks and sorrels being one of the commoner.

Squeak Beetle. ➪ SCREECH BEETLE.

Squid (Cephalopoda: *Loligo forbesi*). A marine mollusc allied to the cuttlefishes, widespread but uncommon offshore.

Squill (Liliaceae: *Scilla*). Two attractive blue-flowered relatives of the bluebell growing in grassland by the sea: Spring Squill *S. verna* on cliffs from south Devon west and northabout to Northumberland; Autumn Squill *S. autumnalis* almost confined to Devon and Cornwall.

Squinancywort (Rubiaceae: *Asperula cynanchica*). A low plant with pale pink flowers; local in chalk and limestone turf, mainly in the south; named from its former use as a cure for quinsy.

Squirrel (Rodentia: *Sciurus*). The native Red Squirrel *S. vulgaris* is now much less common in England and Wales than in Scotland, where it died out in the eighteenth century and was reintroduced. Its place has been largely taken by the North American Grey Squirrel *S. carolinensis*, introduced in the 1890's, which has since spread widely; the grey did not drive out the red, which had already decreased as a result of disease, but did prevent it from reoccupying its lost ground. The red squirrel prefers coniferous woods, and the grey broad-leaved ones. Both can be pests to the forester by stripping and barking young trees and eating their leading shoots. Squirrels are less active in winter but do not hibernate.

Squirrel Tail. Anglers' name for the common earthworm *Lumbricus terrestris*.

S.S.S.I. Site of Special Scientific Interest.

Stable Fly or Biting Horse Fly (Muscidae: *Stomoxys calcitrans*). If the 'harmless' house fly suddenly appears to go berserk and bite a human, it will be the stable fly or a close relative, which sucks the blood of mammals. ➪ HORN FLY.

Stack. An isolated pillar of rock, usually created by the erosion of the sea.

Stag. The adult male red or sika deer. Young males are called staggies.

Stag Beetle (Lamellicornia: Lucanidae). Three species of beetle; the males of *Lucanus cervus*, the largest British beetle, especially common in the southern suburbs of London, have large mandibles somewhat resembling a stag's antlers. The Lesser Stag Beetle *Dorcus parallipipedus* has much less conspicuously developed 'antlers'. The larvae of stag beetles feed on rotten wood.

Stagshorn Weed. ➪ NAIAD.

Stalactite. An icicle-like column, usually of calcium carbonate, hanging from the roof of a (usually limestone) cave, created as drops of rainwater, percolating through the cave roof, drip down leaving a minute mineral deposit behind which gradually, over thousands of years, accumulates into a column. Where the dripping reaches the cave floor a similar upward column (a stalagmite) is formed.

Stalagmite. ⇨ STALACTITE.

Stalk. A leaf-stalk is a *petiole* and may have a *stipule* at its base. The stalk of a single flower is a *pedicel*, of a cluster of flowers a *peduncle*; with *bracts* or *bracteoles* at their base.

Stamen. The male organ of a flower, consisting of a stalk, the filament, bearing at its tip an anther, which produces the pollen.

Staminode. A barren stamen.

Staph. Coleopterists' slang for the large rove beetle family of Staphylinidae.

Staphylinoidea (Coleoptera: Polyphaga). A large superfamily containing some 1,270 species of burying, carrion and rove beetles.

Stare. An old name for the starling.

Starfish (Echinodermata: Asteroidea). A class of free-living, flattened, star-shaped, carnivorous marine animals. The common species on our shores include the abundant reddish or yellowish Common Starfish *Asterias rubens*, the larger Spiny Starfish *Marthasterias glacialis*, both usually with five 'arms'; two species of Sunstar (*Solaster*) with more than five arms, and several species of the very short-armed Cushion Stars (Poraniidae). ⇨⇨ BRITTLE STAR, FEATHER STAR.

Star-Fruit or Thrumwort. (Alismataceae: *Damasonium alisma*) white flowered aquatic monocotyledon with star-shaped fruits; local in the south.

Starling (Passeriformes: *Sturnus vulgaris*). One of the commonest British birds, breeding everywhere from remote northern sea-cliffs to city centres; noted for its habit of noisy communal roosting in small woods in the country and on buildings in towns, including London. The Rose-coloured Starling or Rosy Pastor *S. roseus* is an irregular visitor from eastern Europe, the adults being strikingly black and pink in plumage.

Star of Bethlehem (Liliaceae: *Ornithogalum*). Name of three bulbous plants, the white-flowered *O. umbellatum*, widely naturalized and perhaps native in East Anglia; the tall green-flowered Spiked Star of Bethlehem or Bath Asparagus *O. pyrenaicum*, very local in woods in southern England, and the always introduced Nodding Star of Bethlehem *O. nutans*.

Starr Grass. Local name for marram grass in Lancashire.

Star-Slime or Pwdre Ser. A substance found lying on the ground and formerly associated with shooting stars, but now known to be the gelatinous remains of the oviducts of frogs and toads.

Stell. A grassy channel in the saltmarshes at the mouth of the River Tees.

Stellate. Botanical adjective meaning star-shaped.

Stenohaline. Tolerating only a narrow range of salinity, as opposed to Euryhaline.

Stenotopic. Of an organism having a restricted range or distribution, as opposed to Eurytopic.

Stickleback (Scleroparei: Gasterosteidae). Three small spiny fish: the widespread and common Three-spined or Tittlebat *Gasterosteus aculeatus* of fresh or brackish water, the Ten-spined or Tinker *Pygosteus pungitius* of fresh water in the south, and the widespread Fifteen-spined *Spinachia spinachia*, the largest, common in shallow coastal waters. Sticklebacks are noted for the male aggressively guarding the eggs until they hatch.

Stickworm. Country name for caddis larvae.

Stigma. The part of a flower's carpel which receives the pollen, so starting the process of fertilization; usually borne at the tip of a style.

Stigmelloidea (Lepidoptera: Heteroneura). A small superfamily of Microlepidoptera, consisting of the family Stigmellidae.

Stiletto Fly (Brachycera: Therevidae). A small family of two-winged flies with ten species, named from their long pointed bodies; *Therevia nobilitata* is the commonest.

Stilt, Black-winged (Charadriiformes: *Himantopus himantopus*). A large black and white wader with immensely long legs; a straggler from southern Europe, which has bred once, in Nottinghamshire in 1945.

Stiltbug (Heteroptera: Berytinidae). Elongated brownish plant-bugs, mostly long-legged, *Neides tipularius* being sufficiently like a crane-fly to be named after it.

Stilt-legged Fly (Cyclorrhapha: Tylidae). A small family of eight long-legged predatory two-winged flies.

Sting-Fish. ⇨ Lesser WEEVER.

Stinkhorn (Basidiomycetes: *Phallus impudicus*). A fungus which starts with a roundish ball like a puff-ball; this bursts to produce a close-fitting black cap on a stout stem, when it bears a remarkable resemblance to a human phallus; as it decays it emits a stench which attracts carrion-eating flies which bring about spore-dispersal; frequent in woods and gardens in late summer and autumn. Dog Stinkhorn *Mutinus caninus* is a smaller species with a yellowish stem.

Stinkweed (Cruciferae: *Diplotaxis muralis*). Yellow-flowered crucifer with an unpleasant smell when bruised; locally frequent on bare ground.

Stint (Charadriiformes: *Calidris*). The smallest British waders; Little Stint *C. minuta*, an uncommon passage migrant; Temminck's Stint *C. temminckii*, a scarce passage migrant which has occasionally attempted to breed; and the very rare American Stint *C. minutilla*. Also formerly used for the Dunlin.

Stipe. The stalk of a seaweed's frond.

Stipitate. Botanical adjective used of organs that have a very short stalk.

Stipule. A small, often leaf-like, appendage at the base of a leaf-stalk.

Stitchwort (Carophyllaceae: *Stellaria*). Four plants with narrow leaves and deeply cleft white petals, especially Greater Stitchwort *S. holostea*, common in hedgerows in spring, and Lesser Stitchwort *S. graminea*, common in grassy and heathy places in summer.

Stoat (Carnivora: *Mustela erminea*). A small mustelid carnivore, common and widespread in the British Isles, known for its hypnotic methods of attacking its prey, especially rabbits. In the north stoats tend to turn white in winter when they are known as ermines. Called 'weasel' in Ireland where there are no true weasels.

Stock (Cruciferae: *Matthiola*). Two scarce sea-cliff plants, Sea Stock *M. incana* with purple or white and Great Sea Stock *M. sinuata* with purple flowers only. Sea stock is the origin of the garden stocks.

Stock-Fish. Dried unsalted cod.

Stolon (adj. Stoloniferous). ⇨ RUNNER.

Stonechat (Passeriformes: *Saxicola torquata*). Small song bird allied to the thrushes, named from its call-note which sounds like two stones being struck together. Formerly widespread and common, but now rather local and mainly near the coast in the west, often associated with gorse.

Stonecrop (Crassulaceae: *Sedum*). A genus of succulent plants, with fleshy leaves; six native species, one possibly native and three widely established aliens; four with yellow flowers, and three each with pink and white flowers. The commonest natives are the yellow Wall-pepper and the white English Stonecrop. The related Mossy Stonecrop *Crassula tillaea* is tiny, prostrate, white-flowered and found mainly in East Anglia. ⇨⇨ ORPINE, ROSE-ROOT.

Stone-Curlew or Norfolk Plover (Charadriiformes: *Burhinus oedicnemus*). A large, mainly terrestrial wader, somewhat resembling a curlew with a short straight bill and having a curlew-like call-note. A summer visitor, breeding sparsely on downs and sandy heathlands in south and east England.

Stone-Fly. A fairly primitive order of winged insects, passing through an aquatic nymph stage, and usually found near fresh water, preferring rivers and streams with stony or gravelly bottoms; eaten by fish and imitated by anglers. Thirty-one species are recorded for Britain, four of which are known to anglers as the Large Stone-fly; ⇨⇨ EARLY BROWN, FEBRUARY RED, NEEDLE-FLY, WILLOW FLY, YELLOW SALLY.

Stone Mite (Prostigmata: *Petrobia latens*). A plant-feeding mite named from the appearance of its egg, said to resemble a saucepan with a lid like a pork-pie crust.

Stone-Sucker. ⇨ Sea LAMPREY.

Stonewort (Chlorophyta: Charophyceae). A specialized group of branched green algae, with whorls of thread-like leaves, sometimes regarded as a separate phylum (Charophyta), and superficially resembling higher plants. There are more than thirty British species, growing in still or slow moving fresh or brackish water, the best known belonging to the genera *Chara*, *Nitella* and *Tolypella*. Their name derives from the fact that they become encrusted with lime; some species also smell of rotten eggs when growing in mass.

Stork (Ardeiformes: *Ciconia*). Both the White Stork *C. ciconia* and the Black Stork *C. nigra*, which breed in Europe, are rare visitors to Britain. The white stork, which has once bred, is the more often seen.

Storksbill (Geraniaceae: *Erodium*). Small pink-flowered plants whose fruits have a long twisted 'bill'; three species, the commonest being Common Storksbill *E. cicutarium* in dry and rather bare grassy places.

Stout. A local name for blood-sucking tabanid flies, used especially in the New Forest for the larger species of *Tabanus.*

Strath. A valley, usually fairly broad, in the Scottish Highlands.

Stratocumulus, Stratus. ⇨ CLOUD.

Stratum (pl. Strata). Recognizable layers in soil or rock.

Strapwort (Caryophyllaceae: *Corrigiola littoralis*). A small, now very rare, white-flowered plant.

Strawberry, Wild (Rosaceae: *Fragaria vesca*). Miniature of garden strawberry, widespread and common in woods and grassy places. The related Barren Strawberry *Potentilla sterilis,* similarly widespread and common, has similar flowers but not fruits, and blooms very early in the spring.

Strawberry Tree (Ericaceae: *Arbutus unedo*). Small evergreen tree with strawberry-like fruits growing in western Ireland and belonging to the Lusitanian element in our flora.

Strepsiptera. ⇨ STYLOPS.

Striate. Botanical adjective meaning ridged.

Stridulation. The mechanical sound, often loud and harsh, made by male grasshoppers and crickets, either by rubbing a hind limb against the elytra or by rubbing the elytra together; it may perform a biological function similar to bird song.

Strigiformes. The order of birds including the owls (Strigidae).

Strigose. Botanical adjective used of plants with stiff, closely appressed hairs.

Struggle for Existence. ⇨ NATURAL SELECTION.

Sturgeon (Pisces: *Acipenser sturio*). A large anadromous fish, the sole representative in Britain of the primitive sub-class Palaeopterygii; attaining 10–14 ft; not infrequent in British rivers and inshore seas. The roes yield caviare and the air-bladder isinglass. It is a 'royal' fish in Britain, and the prerogative of the Crown when caught.

Style. The stalk arising from the ovary of a flower which bears the stigma.

Stylops (Insecta: Strepsiptera). Minute endoparasitic insects closely related to and sometimes considered to be degenerate beetles; the females look like larvae and never leave the host insect; eighteen species in one family (Stylopidae).

Subclimax. A stage in the vegetational succession where progress to the true climatic climax is more or less permanently impeded by some natural factor, such as wet soil.

Subfamily. A category in the Classification of animals and plants intermediate between a Family and a Genus, and characterized by names ending in -inae, instead of -idae as in families.

Sub-genus. A subdivision within the Genus.

Subfossil. An incompletely fossilized fossil.

Sub-Order. A category in the Classification of animals and plants intermediate between an Order and a Family.

Subsere. A secondary sere, starting from some point at which the progress of the primary sere towards its climax has been arrested, e.g. by fire, grazing or timber felling. The majority of British lowland heaths, for instance, are subseres, as is also chalk grassland.

Subsoil. An outmoded term for the parent material underlying the soil, used by farmers, gardeners and estate agents, but not by soil scientists. ⇨ ROCK.

Subsong. A quiet form of bird song, whose significance is not fully understood; the pattern of notes differs in various ways from the full song of the same species.

Subspecies. A group with similar characteristics that differs from the species in some minor respect, often as a result of being geographically isolated from the main species, particularly on islands; e.g. the St Kilda wren *Troglodytes troglodytes hirtensis*, which has slight but constant plumage differences from the mainland wren but is otherwise identical; or the British form of the swallow-tail butterfly *Papilio machaon britannicus*. Breeding between members of different subspecies of one species may occur, with the result that subspecies tend to merge into one another. ⇨ NOMENCLATURE.

Subulate. Botanical adjective meaning awl-shaped, narrow, pointed and more or less flattened. ⇨ LEAF-SHAPES.

Succession. The sequence in which plants colonize a piece of ground, each community being superseded by the next, usually taller one, and leading to the Climax.

Succory or Chicory (Compositae: *Cichorium intybus*). A plant with spikes of blue dandelion-like flowers, locally frequent in grassy and waste places; its dried and ground roots produce the chicory used with coffee. Swine's or Lamb's Succory *Arnoseris minima* is a rare plant of sandy places with small yellow dandelion-type flowers.

Sucker. (1) (Pisces: *Lepadogaster*). A group of small marine fish, 2–4 ins. long, with adhesive suckers; widespread off our coasts; *L. gouani*, the Cornish Sucker, is the largest. The name may also be applied to other fish with suckers, including the lumpsucker, remora and sea-snails. (2) Jumping Plant Lice (Homoptera: Chermidae or Psyllidae). A family of small plant-sucking bugs, able to jump with their large hind legs, exuding honey-dew and often making galls. The best known are the Apple Sucker *Psylla* (*Chermes*) *mali*, an orchard pest, the Hawthorn Sucker *P. crataegi*, common on hawthorn hedges, and the Alder Sucker *P. alni*, whose white woolly larvae may be conspicuous on alders. (3) A shoot arising from tree or shrub roots, especially of elms, poplars, and cherries, away from and often at some distance from the main stem; sometimes making substantial thickets.

Sucking-Fish. ⇨ REMORA.

Suffruticose. Botanical adjective applied to small shrubs or undershrubs.

Sugar Mite (Astigmata: *Glycyphagus domesticus*). A pest of stored products, which is believed to feed largely, if not wholly on the micro-organisms in the products rather than on the products themselves, though it often swarms in old damp houses and furniture, and is common in birds' nests.

Sulphur Tuft (Basidiomycetes: *Hypholoma fasciculare*). A bright yellow toadstool, growing in tufts on the stumps of mainly broad-leaved trees.

Sulphurwort or Narrow-leaved Water Dropwort (Umbelliferae: *Oenanthe silaifolia*). A very local white-flowered umbellifer of damp meadows in the south of England.

Sundew (Droseraceae: *Drosera*). Curious plants which catch and digest small insects on their basal rosettes of sticky red leaves. They have leafless spikes of small white flowers, and grow on wet moors and heaths and in bogs. Much the commonest of our three species is Common or Round-leaved Sundew *D. rotundifolia*.

Sun Fish (Pisces: *Mola mola*). A large oceanic fish up to 7–8 ft and 7–8 cwt, fairly often drifted into British waters on the Gulf Stream. Its almost circular shape, with two long fins projecting on either side, give it a most singular appearance. The Truncated Sun Fish *Ranzania truncata* is much rarer. Sun Fish is also a name of the opah.

Sunshine. The mean duration of bright sunshine over the year in the British Isles ranges from just over 1,600 hours per annum along the promontories of the south coast and in south Devon and south Cornwall to less than 1,200 hours per annum in north-west Ireland, the Outer Hebrides, Orkneys and Shetlands, as well as in two inland areas, the Pennines and south-western Highlands of Scotland. This represents a range from about 40 per cent of the possible maximum to less than 25 per cent.

Sunstar. ⇨ STARFISH.

Superfamily. A grouping of several closely related families within an order the name usually ending in -oidea.

Surmullet. ⇨ Red MULLET.

Survival of the Fittest. ⇨ NATURAL SELECTION.

Swaling. The Exmoor term for muirburn.

Swallow (Passeriformes: *Hirundo rustica*). A familiar, widespread and common summer visitor, with blue-black upper parts, red throat and long streamers to its forked tail, which nests on ledges inside buildings and feeds entirely on the wing, catching flying insects. One of the earliest summer migrants to arrive, often in late March. The Red-rumped Swallow *H. daurica* is a rare straggler.

Swallow Hole or Sink Hole. A shallow depression through which all or part of a stream flows underground through fissures in the underlying rocks for short distances below the main stream, which may dry right up in drought conditions, usually in chalk or limestone. Some of the best known swallow holes are those of the River Mole in Surrey. ⇨⇨ POT HOLE.

Swallow-Tail Butterfly (Papilionina: *Papilio machaon*). This handsome black and yellow species, named from the pointed 'tails' of its hind wings, is our only representative of the large family Papilionidae, which contains the largest butterflies in the world; now confined to the Norfolk Broads. The Swallow-tailed Moth *Ourapteryx sambucaria*, pale yellow with smaller 'tails', is widespread and common.

Swamp. A wet area differing from Fen and Marsh in that the summer water level is normally above the surface. Often colonized by reeds (*Phragmites*) or other tall grasses, sedges or rushes.

Swan (Anatidae: *Cygnus*). Very large, all white, aquatic birds; male, cob;

female, pen; young, cygnet. Three species: Mute Swan *C. olor*, with orange bill, the common breeding and park bird, now widely feral from semi-domesticated stock; Whooper Swan *C. cygnus*, with yellow bill, winter visitor, mainly in the north; and the smaller Bewick's Swan *C. columbianus*, with yellow bill, a less common winter visitor, mainly in the south. The Polish Swan is a variety of the mute swan in which the cygnets are white instead of ash-brown.

Swash. ⇨ WAVES.

Sweep's Brush. A vernacular name for field woodrush, from the shape of its flowerheads.

Sweet Cicely (Umbelliferae: *Myrrhis odorata*). An aromatic white-flowered umbellifer, with white-flecked leaves and fruits up to 1 in. long, otherwise resembling cow parsley, and commonest in the north.

Sweet Flag (Araceae: *Acorus calamus*). A large waterside plant, introduced in the Middle Ages for the use of its sweet-smelling flag-like leaves as a floor-covering and now well established by many watersides in England; flowers in a phallic, plantain-like spike, projecting at 45° from the stem.

Sweet Gale or Bog Myrtle (Myricaceae: *Myrica gale*). An aromatic catkin-bearing undershrub, frequent in wet places on acid soils.

Sweet Grass. ⇨ FLOTE GRASS.

Sweet Vernal Grass (Gramineae: *Anthoxanthum odoratum*). A common early-flowering grass, its strong scent of coumarin yielding the characteristic fragrance of new mown hay.

Swift (Apodiformes: *Apus apus*). Long-winged, short-tailed, superficially swallow-like bird, almost exclusively aerial in habit, catching its food of flying insects on the wing. Widespread and common summer visitor, breeding mainly in roofs of old houses and crevices in stone walls. Alpine Swift *A. melba* is a rare visitor from the Continent. Needle-tailed Swift *Chaetura caudacuta* is a very rare straggler.

Swift Moth (Homoneura: Hepialidae). A group of five primitive moths, including the ghost moth.

Sword-Fish (Percomorphi: *Xiphias gladius*). The largest bony fish, attaining 15 ft, with its upper jaw developed into a sword-like projection up to 3 ft long; rare autumn visitor to British seas from the south.

Sycamore (Aceraceae: *Acer pseudoplatanus*). A tall deciduous tree with palmate leaves, introduced long ago and now widely planted, freely regenerating and naturalized throughout the British Isles.

Syke. A common term in the North of England, especially Yorkshire, for a marsh or a marshy ditch or stream.

Symbiosis. In the broad sense, any two different organisms living together, but more strictly, two distinct organisms living together to their mutual advantage, e.g. lichens or the sea-anemones that live on shells inhabited by hermit crabs. ⇨ COMMENSAL, PARASITE.

Sympatric. Adjective used of different species or subspecies with overlapping or coincident areas of distribution. ⇨ ALLOPATRIC.

Symphyla (Arthropoda). A small class of small centipede-like myriapods, living in the soil, with no vernacular name.

Symphyta (Insecta: Hymenoptera). The smaller of the two sub-orders of Hymenoptera, with unwaisted adults and saw-shaped ovipositors, consisting of the superfamily Tenthredinoidea with some 430 species in ten families, of which the most important are the Tenthredinidae (saw-flies), Cephidae (stem saw-flies or borers) and Siricidae (wood-wasps).

Synecology. The ecology of communities of animals or plants, as distinct from Autecology.

Syrphid. ⇨ HOVER-FLY.

Systematics. ⇨ TAXONOMY.

T

Tabanid (Brachycera: Tabanidae). A family of two-winged flies, noted for the blood-sucking habits of the females, and with a wide variety of popular names, such as horse flies, gad flies, clegs, stouts, burrel flies, whame flies, breeze flies and dun flies.

Tabby Moth (Pyraloidea: *Aglossa*). Pyralid moths found in barns, stables, warehouses, etc., their larvae feeding on various vegetable refuse; the commonest and most widespread is the Large Tabby Moth *A. pinguinalis*, which is unusually big for a 'micro'.

Tadpole. Young frog, toad or newt in its gill-breathing aquatic larval stage.

Tadpole Fish. ⇨ Lesser FORK-BEARD.

Talus. The fragments of rock which fall from crags to form a scree.

Tamarisk (Tamaricaceae: *Tamarix gallica*). A pink-flowered shrub with feathery evergreen foliage, extensively planted and more or less naturalized, mainly on the south and east coasts of England.

Tangle. ⇨ OARWEED.

Tansy (Compositae: *Chrysanthemum vulgare*). A tall, pungently aromatic plant, formerly grown as a herb, with pinnate leaves and flattish clusters of rayless yellow button-like flowers, widespread and frequent in grassy waste places.

Tapestry Moth or White-tip Clothes Moth (Tinaeoidea: *Trichophaga tapetzella*). A clothes moth found more often in stables and outhouses than inside houses, its larvae preferring coarser materials to the common clothes moth.

Tapeworm (Platyhelminthes: Cestoda). An exclusively endoparasitic class of flatworms, named from their resemblance to dirty white tape, and consisting of an extensive series of separate segments, each fully equipped with organs and ending up full of eggs, the whole animal often extending to several feet. A large number of different tapeworm species are found in the guts of British terrestrial, freshwater and marine vertebrates, one of the best known being *Ligula intestinalis*, which starts life in a small copepod, and passes through various freshwater fish before ending up in ducks and other water birds.

Tardigrada. ⇨ WATER BEAR.

Tare (Leguminosae: *Vicia*). Three smaller vetches, the commonest being Hairy Tare *V. hirsuta*, named from its hairy seed-pods. These are not the biblical tares.

Tarn. A small lake or pond in hill districts of the North of England.

Taxidermy. The preparation and mounting of the skins of animals, especially vertebrates, for display in museums and elsewhere.

Taxis. A directional but non-growth response to a stimulus impelling an organism or cell in a certain direction: *chemotaxis* if the stimulus is chemical, as with the scent of female moths attracting males, *geotaxis* if it is gravity, and *phototaxis* if it is light, as with moths attracted by light. ⇨ TROPISM.

Taxon. Any named taxonomic group or unit, from phylum down to variety or form.

Taxonomy or Systematics. The science of Classification of living things, taxonomy comprising the principles of classification and systematics the actual groups into which animals or plants are classified. ⇨ NOMENCLATURE.

Teal (Anatidae: *Anas crecca*). A surface-feeding duck, the smallest British duck, widespread and common in winter, but breeding mainly in the north and west.

Teart Land. The teart pastures of Somerset are noted for causing acute diarrhoea in sheep pastured on them; research has now shown that this was due to an excess of molybdenum in the soil.

Teasel (Dipsacaceae: *Dipsacus fullonum*). The wild form of the plant, still cultivated in Somerset for the use of its prickly heads in the dressing of broadcloth, is common in bushy places, especially in the south; its lilac flowers are set in compound conical heads.

Tectibranch (Opisthobranchia: Tectibranchiata). Sea slugs, which may have an external shell, e.g. actaeon shell, or a much reduced internal shell, e.g. sea-hare, or no shell at all, e.g. the widespread and locally common *Pelta coronata.*

Tectonic Forces. The forces which contribute to the process of building up the physical features of the earth's crust, i.e. by throwing up mountain chains, as distinct from those which wear down hills and valleys into a uniform plain.

Teleost Fish (Pisces: Neopterygii). The vast majority of the bony fishes, i.e. the whole sub-class Neopterygii less two non-British orders classified as Holostei.

Tellin (Lamellibranchia: Tellinidae). Smooth cockle-like marine bivalves, mostly living in sand and common on the shore, especially *Tellina tenuis* and *Macoma balthica.*

Tench (Cyprinidae: *Tinca tinca*). A sluggish bottom-living freshwater fish of muddy lakes and ponds; widespread and frequent. The Golden Tench is an ornamental form like the goldfish.

Tenchweed. A Norfolk name for broad-leaved pondweed.

Tendril. An extension or development of a stem or leaf, which curls round any available support to enable the plant to climb or clamber; in British plants mostly among the vetches and peas.

Tepal. A term covering both petals and sepals and used especially where flowers do not have distinct petals and sepals, as with many of the monocotyledons, such as some of the Lily Family.

Teratology. The scientific study of malformations and monstrosities of animals and plants, e.g. fasciation.

Terete. Botanical adjective used of smooth stems that are not ridged or grooved.

Tern, or Sea-Swallow (Charadriiformes: Laridae). Medium to small sea and marsh birds, closely allied to the gulls, but slenderer and always with forked tails. Five species of sea tern breed as summer visitors on sand dunes, shingle and rocky islets; the widespread Common Tern *Sterna hirundo* and Little Tern *S. albifrons*, the northern Arctic Tern *S. macrura*, the large and rather local Sandwich Tern *S. sandvicensis* and the very local and mainly Irish Roseate Tern *S. dougallii*. Five other sea terns are rare visitors, of which the Gull-billed Tern *Gelochelidon nilotica* has bred once. Three species of marsh tern occur as passage migrants, but only the Black Tern *Chlidonias niger*, a former breeder in eastern England, is regular and at all frequent.

Ternate. ⇨ LEAF-SHAPES.

Terrapin. A term often applied to freshwater tortoises.

Territory. A mechanism for dispersing animals evenly over the available habitat, whereby individuals, or sometimes groups, defend, mainly against members of their own species, a particular area, which may or may not be related to their actual food supply. The most conspicuous instances of territorial behaviour are provided by song birds, among which the robin is rather exceptional in defending both breeding-season and winter territories, the hen birds also doing so in winter.

Tertiary Period. ⇨ GEOLOGICAL PERIODS.

Testudines or Chelonia. (Reptilia). The order of reptiles containing the tortoises, terrapins and turtles, characterized by their hard shells or carapaces.

Tetrad. A two-kilometre square, used for mapping the distribution of animals and plants.

Tetraploid. ⇨ POLYPLOIDY.

Thallophyte. A collective name for the lower plants other than mosses and liverworts (bryophytes); i.e. the bacteria, algae, lichens and fungi.

Thallus. The vegetative body of a lower plant, not differentiated into such organs as stems and leaves.

Therophyte. A plant which overwinters as a seed, i.e. an annual.

Thicket. Dense scrub.

Thistle. Composites, mostly purple-flowered, with spines on the leaves, stems and flowerheads. Our three commonest and most widespread species are the pale-flowered Creeping Thistle *Cirsium arvense*, the larger, spinier and darker purple Spear Thistle *C. vulgare*, both common in grassy and waste places, and the tall, slender Marsh Thistle *C. palustre* of damp and dry grassland. Other common species are Dwarf or Stemless Thistle *C. acaulon*, and Musk Thistle *Carduus nutans*, both of calcareous grassland, and Welted Thistle *Carduus acanthoides*, common on hedge-banks. Scotch Thistle *Onopordon acanthium* is a tall whitely woolly plant rare in Scotland but not infrequent on light soils in eastern England. Melancholy Thistle *Cirsium heterophyllum*, of damp places in the north, was formerly used to cure melancholia. Milk Thistle *Silybum marianum*, growing in waste places, is named from the milk-white veins in its leaves. Carline Thistle *Carlina vulgaris* with yellow flowers is another chalk grassland specialist. Several casual relatives of the knapweeds have spines around their flower-heads only, and are 'honorary' thistles, e.g. Star Thistle *Centaurea calcitrapa* with purple flowers and St Barnaby's Thistle or Yellow Star Thistle *C. solstitialis* with yellow ones. Watling Street Thistle ⇨ ERYNGO, FIELD. Thistle Weed ⇨ COCKLEBUR.

Thongweed (Phaeophyta: *Himanthalia elongata*). A brown seaweed with straplike fronds anything from 3 to 8 ft long, locally common on rocky shores.

Thorax. The portion of an insect between its head and abdomen, in three segments, each bearing a pair of legs. The last two may bear a pair of wings each.

Thorn. A sharply pointed woody structure on a stem.

Thorn-Apple (Solanaceae: *Datura stramonium*). Poisonous plant with white trumpet-shaped flowers, named from its spiny fruits; a weed varying in quantity from year to year.

Thorn Insect (Homoptera: Membracidae). Two small bugs that frequent thorny plants and mimic the thorns; *Centrotus cornutus*, with thorn-like spines on the thorax is the commoner of our two species.

Thorn Moth (Notodontoidea: Selidosemidae). A group of geometrid moths with rather angled wings, most of which emerge either early or late in the year; the Early Thorn *Selenia bilunaria* is one of the most widespread and common.

Thorn Tree. Usually refers to a hawthorn.

Thorow-Wax (Umbelliferae: *Bupleurum rotundifolium*). An atypical yellow-flowered umbellifer with rounded perfoliate leaves; a decreasing and now rare cornfield weed.

Three-legged Bug (Heteroptera: Emesinae). A group of four highly attenuated, almost mosquito-like assassin bugs; four British species, the commonest being *Empicoris vagabundus*.

Threadworm. ⇨ HORSEHAIR WORM, ROUNDWORM.

Thresher. ⇨ FOX SHARK.

Thrift or Sea Pink (Plumbaginaceae: *Armeria maritima*). A common cushion plant of seashores, also on mountains, with tight heads of small fragrant pink flowers.

Thrips (Insecta: Thysanoptera). An order of small, sap-sucking flies characterized by the fringes of long hairs on the four wings, sometimes known as Thunder Flies. The tiny flies so common among the florets of dandelions and other composites are thrips. Some of the *c.*180 British species are regarded as pests, e.g. the Onion Thrips *Thrips tabaci* and the Pea Thrips *Kakothrips robustus*. Note that 'thrips' is both singular and plural.

Throstle. Old name for the song thrush.

Thrum-eyed Flower. Individual flowers, especially in the Primrose Family, in which the stigma is borne at the end of a short style halfway down the corolla tube, while the anthers on long stamens occupy its mouth; as opposed to Pin-eyed.

Thrumwort. ⇨ STARFRUIT.

Thrush (Passeriformes: *Turdus*). Medium-sized song birds with songs of high quality. Three resident species (blackbird, song and mistle thrushes), one summer visitor (ring ouzel), two regular winter visitors (fieldfare and redwing), and five stragglers. Mistle Thrush *T. viscivorus*, the largest, named for its fondness for mistletoe berries, lays brown-spotted white eggs; Song Thrush *T. philomelos*, formerly also known as mavis or throstle, lays black-spotted blue eggs in a mud-lined nest; both are brown above with black-spotted breast, and are widespread and common. The Rock Thrush *Monticola saxatilis* is a brightly coloured rare straggler from southern Europe. Two thrushes of the genus *Hylocichla* are stragglers from North America.

Thunder. The loud sound made by the return of air into the vacuum created by the sudden overheating of the air molecules by the passage of a Lightning discharge. Most thunderstorms occur in summer, and much more frequently in eastern England and the Midlands than elsewhere.

Thunder Fly. ⇨ THRIPS.

Thyme, Wild (Labiatae: *Thymus*). Low-growing aromatic native plant with pinkish purple flowers, of downs, dunes, heaths and moors, with three micro-species, the most widespread being *T. drucei*. The related Basil Thyme *Acinos arvensis* has blue-violet flowers and grows mainly on chalky soils. Garden Thyme is *T. vulgaris*.

Thysanoptera. ⇨ THRIPS.

Thysanura. Three-pronged ⇨ BRISTLE-TAIL.

Tick (Acari: Metastigmata). An order or sub-order, also known as Ixodoidea, of large mites, highly specialized to blood-sucking. The best known is the common sheep tick, grass tick or castor bean tick *Ixodes ricinus* (⇨ KED) which carries the virus of the cattle and sheep disease louping-ill; it will feed on many mammals and birds, and is often found on dogs which have been in a field where sheep have been kept. The hedgehog tick *I. hexagonus* also infests many small mammals, including dogs and cats, occasionally transferring itself to man; other members of the genus, especially the guillemot tick *I. uriae*, specialize in birds. The yellowish pigeon tick *Argas reflexus* occasionally bites man, and together with the fowl tick *A. persicus* carries fowl relapsing fever.

Tide. Twice-daily movement of the sea, produced by the combined gravitational force of the sun and moon. At full and new moon, sun and moon are pulling together to produce the very high and low tides known as Spring Tides; in between the less high and low tides are known as Neap Tides, also occurring at fortnightly intervals. The highest and lowest spring tides occur around the spring and autumn equinoxes, and these are of great importance for students of marine life, for at their extreme low water animals and plants may be uncovered which are accessible at no other time of year. The average tidal range, or distance between high and low water marks, on our coasts is about 15 ft at spring tides but only 11½ ft at neaps. These averages, however, conceal many local differences; in the Bristol Channel, for instance, the corresponding figures are 42 ft and 21 ft.

Tiercel. The male falcon.

Tiger Beetle (Adephaga: *Cicindela*). Five fiercely predatory beetles, some of them brightly coloured; the best known is the Common or Green Tiger Beetle *C. campestris.*

Tiger Moth (Caradrinoidea: Arctiidae). Four moths with 'woolly bear' caterpillars, distinguished by having dark blotches and streaks on their whitish forewings and red or yellow underwings: the largest and commonest is the Garden Tiger *Arctia caja.* Two handsome but rare or local Hypsidae are also so named: the Jersey Tiger *Euplagia quadripunctaria,* confined to south Devon, and Scarlet Tiger *Panaxia dominula,* the subject of important genetical observations at Cothill, Berkshire.

Till. ⇨ CLAY.

Tilth. The depth of soil affected by tillage or cultivation.

Timberman. A name for certain longhorn beetles with wood-boring larvae.

Timothy or Catstail (Gramineae: *Phleum*). A group of six grasses with sausage-shaped flower-spikes similar to the foxtails, the two commonest being *P. pratense,* often sown in grass leys, and the smaller *P. bertolonii,* of calcareous grassland.

Tinaeoidea (Lepidoptera: Heteroneura). A large superfamily of small moths, consisting of the Sesiidae (clearwings), commonly classed among the Macrolepidoptera, and twenty families of the Microlepidoptera, of which the best known are the Yponomeutidae, which includes the small ermines, the Oecophoridae (house moths) and the Tinaeidae (clothes moths).

Tinder Fungus (Asomycetes: *Fomes fomentarius*). A bracket fungus found on birch in the Highlands.

Tinker. ⇨ Ten-spined STICKLEBACK.

Tit (Passeriformes: Paridae). Small song birds, usually with not very highly developed songs, those belonging to the genera *Parus* and *Aegithalos* often banding together in parties in winter. Eight resident species. Bearded Tit *Panurus biarmicus* is confined to extensive reed-beds in East Anglia. Blue Tit *Parus caeruleus,* the only British bird with substantially blue plumage, is widespread and common; a frequent visitor to bird tables in suburban gardens. Coal Tit *P. ater,* the smallest British tit, is widespread and common, especially among conifers. Crested Tit *P. cristatus* is very local in pinewoods in the Scottish Highlands, mainly in Strathspey. Great Tit *P. major,* the largest British tit, is widespread and common, and another frequent visitor to suburban bird tables. Long-tailed Tit *Ae. caudatus,* noted for its strikingly long tail and domed oval nest, is widespread and frequent. Marsh Tit *P. palustris* is widespread and frequent in England and Wales, mainly in the south, in woodland rather than marshes. Willow Tit *P. montanus* was overlooked as a British bird until 1900 owing to its similarity to the marsh tit; widespread but more thinly scattered than marsh tit, though extending into the Highlands.

Tit Bell, Tit Cone or Tit Cylinder. A receptacle to hold food for tits.

Titlark. ⇨ Meadow PIPIT.

Titmouse. Outmoded term for the tits.

Tittlebat. ⇨ Three-spined STICKLEBACK.

Toad (Salientia: *Bufo*). Frog-like amphibians; two native species, the local Natterjack and the Common Toad *B. bufo*, which is widespread and common. Common toads secrete a poison in the warty glands scattered over their bodies. Three small colonies of the European Midwife Toad *Alytes obstetricans* are naturalized at Bedford.

Toadflax (Scrophulariaceae: *Linaria*). Plant with yellow or purple snap-dragon-like flowers; seven species, the commonest being Yellow Toadflax *L. vulgaris*, widespread in waste and grassy places. The related Ivy-leaved Toadflax or Mother of Thousands *Cymbalaria muralis* has pale purple flowers and grows on walls. Bastard Toadflax (Santalaceae: *Thesium humifusum*) is a local plant of southern calcareous turf with quite different-looking open yellow flowers.

Toadstone. A volcanic rock found in Derbyshire, composed of solidified lava.

Toadstool (Fungi: Basidiomycetes). A general name for umbrella-shaped larger fungi, including the common mushroom; the term is sometimes used to denote only poisonous fungi, but Ramsbottom recommends it should be used in its wider original sense, while restricting the term Mushroom to the edible members of the genus *Psalliota*. The typical umbrella-shape, however, is mainly found in the Agaricaceae or gill fungi, which contains both the edible mushrooms and the mostly highly poisonous genus *Amanita*; all these have a marked stem surmounted by a flat cap, under which are a series of ridged gills bearing the basidiospores.

Tobacco Moth. ⇨ CACAO MOTH.

Tomentose. Botanical adjective meaning covered with white woolly down or tomentum.

Tompot. ⇨ BLENNY.

Tom Thumb. A vernacular name for birdsfoot trefoil.

Tooth Fungus (Basidiomycetes: Hydnaceae). A rather miscellaneous group of fungi, some with tooth-shaped scales, some toadstool-shaped, some edible, e.g. *Hydnum repandum*, *H. rufescens* and *Auriscalpium vulgare*.

Toothpick Plant (Umbelliferae: *Ammi*). Two rare carrot-like weeds, the Greater *A. majus* mainly in carrot and shoddy fields and the Lesser *A. visnaga* mainly on rubbish tips from bird-seed waste.

Toothwort (Orobanchaceae: *Lathraea squamaria*). A low-growing parasite on the roots of hazel and other shrubs and trees, with pinkish flowers and no green leaves; locally frequent, in woods and hedgerows, mainly in England.

Tope (Pleurotremata: *Eugaleus galeus*). A small true shark, related to the blue shark, but not much bigger than a large dogfish; common around our coasts.

Topknot (Heterosomata: Bothidae). Three rather oval, left-sided flatfish, with spiny scales. The Common Topknot *Zeugopterus punctatus* is the most frequent, the Norwegian Topknot *Phrynorhombus norvegicus* having a northern distribution, and the scarce Bloch's Topknot *P. regius* a southern one.

Top Shell (Prosobranchia: Trochidae). Primitive winkle-like marine operculate molluscs with conical or flattened shells; the Common or Painted Top Shell *Calliostoma zizyphinum*, yellow, white or pink with red streaks, is the commonest and most widely distributed. ⇨ SILVER TOMMY.

Tor. Outcrop of granite on the moors of Devon and Cornwall, often eroded into curious shapes.

Torgoch (Salmonidae: *Salvelinus alpinus perisii*). A race of char inhabiting two or three lakes in North Wales.

Tor Grass or Heath False Brome (Gramineae: *Brachypodium pinnatum*). A locally common yellow-green grass of calcareous grassland, often dominant to the point of being a pest; closely related to the slender false brome.

Tormentil (Rosaceae: *Potentilla erecta*). Small trailing yellow-flowered cinquefoil common on heaths and moors.

Torsk or Tusk (Gadidae: *Brosme brosme*). A deep-water marine fish of the Cod Family, found mainly in northern waters.

Tortoise (Reptilia: Testudines). No land tortoises are native to Britain, and any found in the countryside must be escapes from gardens and probably the Greek Tortoise *Testudo graeca*. The European Pond Tortoise *Emys orbicularis*, properly a terrapin, is also easily acclimatized, but cannot breed in our cool summers; occasionally found in ponds away from gardens.

Tortoise Beetle (Phytophaga: Cassinidae). A sub-family of leaf beetles, named from the fact that their bodies are covered with a carapace-like shield.

Tortoise Bug (Heteroptera: *Eurygaster*). Two brown plant-bugs, mainly in the south.

Tortoiseshell Butterfly (Papilionina: Nymphalidae). Two vanessid butterflies: the common Small Tortoiseshell *Aglais urticae*, often seen hibernating in houses, emerges to fly in March and April and has blackish nettle-feeding larvae; and the now very rare Large Tortoiseshell *Nymphalis polychloros*, feeding on elm.

Tortricoidea (Lepidoptera: Heteroneura). A large superfamily of small moths, all classed as Microlepidoptera, consisting of three families, Phaloniidae, Tortricidae and Eucosomidae. Its best known species are the green oak moth (Tortricidae) and the codling moth (Eucosomidae).

Totter Grass. ⇨ QUAKING GRASS.

Tower Shell (Prosobranchia: *Turritella communis*). A marine mesogastropod operculate mollusc, common in muddy gravels offshore, whose distinctive narrowly spiral empty shell is often found on beaches.

Town-Hall Clock. ⇨ MOSCHATEL.

Trap-Door Spider or Purse-Web Spider (Araneae: *Atypus affinis*). The only British spider related to the tropical trap-door spiders, which unlike them does not make a hinged door to the closed silken tube within which it lives. The spider seizes its prey with its exceptionally long jaws while the prey is resting on the aerial portion of the tube, and hauls it inside through a rapidly made slit.

Traveller's Joy (Ranunculaceae: *Clematis vitalba*), also known as Old

Man's Beard from its massed feathery fruits, is a woody climber related to the garden clematises; locally frequent on calcareous soils in the south.

Tree. A tall woody plant with a single, usually stout, stem. Trees may be divided into conifers, whose wood is called softwood and broad-leaved trees, whose wood is known as hardwood. These terms are not completely accurate, for some broad-leaved trees, e.g. willows, have narrow leaves, while some softwood is hard and some hardwood is soft.

Treecreeper (Passeriformes: *Certhia familiaris*). Small brown song bird with curved beak, noted for its habit of creeping up trees in search of food; widespread and frequent.

Tree Snail (Pulmonata: *Balea perversa*). A widespread land snail, mainly frequenting crevices in the bark of trees, also on old walls and rocks.

Tree Wasp (Aculeata: *Vespula sylvestris* and *Vespa norvegica*). Two of the social wasps which hang their papery nests in trees or bushes instead of in holes like the other social wasps.

Trefoil (Leguminosae). Name given to several small yellow-flowered pea-flowers with trefoil leaves, notably Hop Trefoil *Trifolium campestre*, Common Yellow Trefoil *T. dubium* and Birdsfoot Trefoil *Lotus corniculatus*, all of which are widespread and common in grassland. Hop trefoil is named from the hop-like appearance of its dead flower-heads, birdsfoot trefoil from the claw-like appearance of its seed-pods; the latter also has many folk names, such as Eggs and Bacon, and Tom Thumb.

Trematoda. ⇨ FLUKE.

Tremallales. ⇨ JELLY FUNGUS.

Triassic Period. ⇨ GEOLOGICAL PERIODS.

Trichoptera. ⇨ CADDIS-FLY.

Triclad (Turbellaria: Tricladida). A class of free-living flatworms, including the planarians.

Trifid. Botanical adjective meaning divided into three but not split right to the base.

Trifoliate. ⇨ LEAF-SHAPES.

Trigonous. Botanical adjective to describe bluntly three-sided stems, as distinct from Triquetrous.

Triploid. ⇨ POLYPLOIDY.

Triquetrous. Botanical adjective to describe stems that are sharply three-sided, as distinct from Trigonous.

Triungulin Larva. Larva of the oil beetle.

Tropism. A directional growth-response of a plant or sedentary animal to a stimulus; *chemotropism* if the stimulus is chemical, *geotropism* if it is gravity, as with roots growing downwards, and *phototropism* if it is light, as, for instance, a pot plant will grow towards the light or plants be drawn upwards when growing in dense shade. ⇨⇨ TAXIS.

Trough Shell (Lamellibranchia: Mactridae). Small offshore bivalves, with shells like smooth cockles, which are often cast up on the beach.

Trout (Salmonidae: *Salmo trutta*). An important game fish, with numerous forms, including the Sea Trout, Salmon Trout or Sewin, which are anadromous, and the River, Lake or Brown Trout, which live entirely in fresh water. The Gillaroo is an Irish lake trout, and there are innumerable other local names. Sea trout have a similar life history to salmon, also pass through a smolt stage, and are known as finnocks or whitlings when they first return from the sea. Slob trout is a trout living in an estuary, not quite resembling either a river or a sea trout. The North American Rainbow Trout *S. gairdnerii* is established in several rivers and lakes, including the Derbyshire Wye. The American Brook Trout *Salvelinus fontinalis*, a species of char, is also naturalized in several small lakes and tarns.

Truffle (Ascomycetes: Tuberales). More or less round fungi growing underground and so rarely seen. The famous Continental truffles *Tuber melonosporum* and *T. magnatum* do not occur in Britain, but two native species, *T. aestivum* and *Hydnotria tulasnei*, are edible and are still hunted in the south with dogs which detect them by scent in the same way as the true truffles. *T. aestivum* grows mainly in beechwoods on chalk, especially in Wiltshire, where Winterslow used to be the main truffle-hunting centre.

Trumpet Snail. ⇨ RAMSHORN SNAIL.

Truncate. Of a leaf with a sharply truncated base.

Trypanosome (Mastigophora: *Trypanosoma*). Parasitic flagellates common in vertebrates, the most notorious of which, the cause of sleeping sickness in man and nagana in cattle, transmitted by tsetse flies, is fortunately non-British.

Tuber. A swollen portion of an underground root or stem, not lasting more than one year.

Tuberculate. Botanical adjective meaning warty.

Tub-Fish. ⇨ Yellow GURNARD.

Tulip, Wild (Liliaceae: *Tulipa sylvestris*). A yellow flower recognizably akin to the garden tulips, formerly widely naturalized for ornament, but now decreasing.

Tunic. The rather papery brownish covering of a bulb.

Tunicate (Chordata: Urochordata). A sub-phylum of primitive chordates, which show their affinity to the vertebrates by passing through a 'tadpole' stage of development. The adults consist mainly of a stout bag with two openings called siphons. The British species belong mainly to the Class Ascidiacea or sea squirts.

Tunny (Percomorphi: *Thunnus thynnus*). A large mackerel-like fish of the open seas off our coasts, up to 900 lb in weight. Four related species of tunny or bonito are rare visitors.

Turbellaria (Platyhelminthes). A class of free-living flatworms including the triclads, polyclads and planarians.

Turbot (Bothidae: *Scophthalmus maximus*). The largest of the inshore flatfish, diamond-shaped, left-sided, highly esteemed for its flesh and weighing up to 30 lb; mainly off our southern coasts.

Turion. A bud which falls off an aquatic plant into the mud of a pond or river and overwinters there.

Turkey Brown. Angler's name for both dun and spinner of the mayfly *Paraleptophlebia submarginata*.

Turlough. A deep grassy hollow on the limestone formation of western

Ireland, which fills with water during wet weather, especially in winter, because the water wells up through fissures on the rock; a few of the deeper turloughs are permanently flooded.

Turnip Fly. A turnip ⇨ FLEA BEETLE.

Turnip Moth (Heteroneura: *Agrotis segetum*). A noctuid moth whose larva is a pest of turnips and swedes.

Turnstone (Charadriiformes: *Arenaria interpres*). A small wader related to the plovers, frequent on our coasts throughout the year, a few non-breeding birds remaining in summer, especially in the north. Named from its feeding habits.

Turtle (Testudines: Cheloniidae). No marine turtle is native to British waters, but four species, mostly from the Caribbean, are occasionally stranded. The most frequent is the Common Loggerhead *Caretta caretta*; the others are Kemp's Loggerhead or the Atlantic Ridley *C. kempi*, the Hawksbill *Eretmochelys imbricata*; and in the family Dermochelidae the Leathery Turtle or Luth *Dermochelys coriacea*.

Turtlebug (Heteroptera: *Podops inuncta*). A brown plant-bug, common in the south.

Tusk. ⇨ TORSK.

Tusk Shell or Elephant-tooth Shell (Scaphopoda: *Dentalium entalis*). A deep-sea mollusc whose tubular or tusk-shaped shell is often found washed up on our shores, especially in the north.

Tussock Moth (Caradrinoidea: Lymantriidae). A family of eleven moths, with furry caterpillars, including such well known ones as the vapourer, yellow-tail and gipsy moths.

Tutsan (Hypericaceae: *Hypericum androsaemum*). A yellow-flowered undershrub with purplish-black berries, widespread but local in woods and hedge-banks; closely related to the St John's worts, named from the French *toute-saine*.

Twachel. Angler's name for the common earthworm *Lumbricus terrestris*.

Twayblade (Orchidaceae: *Listera ovata*). One of the least conspicuous of our larger orchids, with yellow-green flowers; named from the two large oval leaves at the base of the flower spike; widespread and common in woods and grassy places. Lesser Twayblade *L. cordata*, with reddish flowers, grows on moors in the north and west.

Twitch. ⇨ Common COUCH GRASS.

Twite (Passeriformes: *Carduelis flavirostris*). The moorland counterpart of the linnet, a small brown finch named from its call note; locally common in the north and west, and in coastal areas of the south in winter.

Type Species. In taxonomy, the species chosen as representative of a genus.

Type Specimen. ⇨ HOLOTYPE.

Typographer. ⇨ BARK BEETLE.

Typology. The scientific study of types.

Tystie. The Shetland name of the black guillemot.

U

Umbel. An umbrella-shaped compound flower-head, in which all the flowers are borne on stalks arising like the spokes of an umbrella from a single point.

Umbellifer (Dicotyledones: Umbelliferae). A large family containing such familiar wild and garden plants as carrot, parsley and hemlock, with a highly distinctive arrangement of the flowers in umbels, which in most species have spokes ending in secondary umbels of small five-petalled individual flowers. The leaves are usually pinnate or twice or thrice pinnate, and sometimes mistaken by the uninitiated for ferns. ⇨ UMBEL.

Uncinate. Zoological adjective meaning hooked.

Undercliff. The terrace, or lower cliff, made as the result of a cliff-fall from the top.

Undershrub. A low woody plant, not always bushy.

Ungulate. Herbivorous hoofed mammal belonging to the orders Perissodactyla (horses) and Artiodactyla (pigs, deer, cattle, sheep).

Univalve. ⇨ GASTROPOD.

Urceolate. Botanical adjective used to describe the shape of corollas which are more or less round in outline but markedly contracted at the mouth.

Urchin. An old country name for the hedgehog; ⇨⇨ SEA-URCHIN.

Uredinales. ⇨ RUST.

Urodela. Alternative name for Caudata (Amphibia).

Ustiginales. ⇨ SMUT.

V

Vaagmar or Deal-fish (Pisces: *Trachypterus arcticus*). An elongated northern deep-sea fish, uncommon in British waters.

Vagrant. A migratory bird which has a defective sense of direction or has been blown far off its normal course to arrive in a district where it is very rarely seen. Also applicable to winged insects.

Vale. A broad flat valley, e.g. the Vale of Aylesbury at the foot of the Chilterns in Buckinghamshire and the Vale of Berkeley at the foot of the southern Cotswolds in Gloucestershire.

Valerian (Valerianaceae: *Valeriana* and *Centranthus*). Four plants with loose heads of small red or pink flowers, especially Common Valerian or All-heal *V. officinalis*, widespread and frequent in damp woods and damp and dry grassland; the smaller Marsh Valerian *V. dioica*, confined to marshes; and the well established alien Spur or Red Valerian *C. ruber*, locally common on walls, banks and rocks.

Valve Snail (Prosobranchia: *Valvata*). Three species of freshwater operculate mollusc, the commonest being *V. piscinalis*, in streams and other running water.

Vanessid Butterfly (Papilionina: Nymphalidae). A group of seven brightly coloured and strongly flying species allied to the red admiral and painted lady (*Vanessa*), and including also the tortoiseshells, the peacock, the comma and the Camberwell beauty.

Vapourer Moth (Caradrinoidea: *Orgyia antiqua*). A tussock moth whose variegated furry caterpillars are common on trees and shrubs even in Central London; the male is warm brown in colour; the female is wingless.

Variety. Technically a distinct form of an animal or plant, including a Colour Phase, occurring either by mutation or by the maintenance of a state of balanced polymorphism. Often more loosely used to indicate any variation within a species or subspecies, and in common speech even more loosely used to mean species also.

Vascular Plant. Higher plants which possess a vascular (i.e. containing channels for conducting liquids) system of plant tissue and well-developed roots, stems and leaves, comprising the seed-plants and the vascular cryptogams or Pteridophyta (ferns and their allies).

Vector. A carrier of disease or parasites.

Vegetable Fly or Caterpillar Fungus (Ascomycetes: *Cordyceps*). Fungi growing on the dead bodies of insects, especially the club-shaped *C. militaris* on the larvae and pupae of Lepidoptera.

Vegetable Kingdom. One of the two major divisions of the living world, the other being the Animal Kingdom, divided broadly into the higher and lower plants, the higher being the vascular plants and the lower those which lack a well-defined vascular system, the bryophytes and the thallophytes. Also known as Plant Kingdom.

Vegetation. A plant community or communities.

Velvet-Ant (Aculeata: *Mutilla*). Two species of solitary wasp, whose females are black, wingless and ant-like, parasitic on bumblebees: the larger *M. europaea* and the smaller *M. rufipes*. Similarly ant-like is the female of the allied *Myrmosa melanocephala*, parasitic on solitary wasps of the genus *Crabro*.

Velvet Mite (Prostigmata: *Allothrombium fuliginosum*). An important predator on the woolly aphis pest of apple trees.

Velvet Shell (Prosobranchia: *Velutina velutina*). A marine mesogastropod operculate mollusc, with its oval shell largely covered by a velvety brown overlay, somewhat resembling a cross betwen a limpet and a slug; widespread but uncommon near low tide mark.

Vendace (Salmonidae: *Coregonus vandesius*). A small whitefish confined to Lakes Derwentwater and Bassenthwaite in Cumberland and a few small lochs near Lochmaben, Dumfriesshire.

Venus's Comb. ⇨ SHEPHERD'S NEEDLE.

Venus's Looking Glass or Corn Campanula (Campanulaceae: *Legousia hybrida*). A locally frequent cornfield weed in the south, mainly on limy soils.

Venus Shell (Lamellibranchia: *Venus*). Four species of widespread and mostly common marine bivalves, resembling rounder smooth cockles, the most abundant being the Banded Venus Shell *V. fasciata*.

Verdigris Toadstool (Basidiomycetes: *Stropharia aeruginosa*). A small toadstool with a slimy blue-green cap, common in woods and grassy places.

Vermes. Old-fashioned comprehensive term for worm-like animals.

Vermin. A loaded word used to signify the predators of game, and now becoming old-fashioned.

Vertebrate (Chordata: Vertebrata). The most advanced sub-phylum of the Animal Kingdom, all of whose members have their central nerve-cord enclosed in a bony or cartilaginous spine or backbone and their brain enclosed in a skull: the mammals, birds, reptiles, amphibians and fishes.

Vervain (Verbenaceae: *Verbena officinalis*). Plant with small two-lipped pale lilac flowers, growing in bare places, mainly in the south.

Vetch (Leguminosae: *Vicia*). A group of herbaceous peaflowers, with tendrilled pinnate leaves, often scrambling over other vegetation, all except the local Yellow Vetch *V. lutea* with pink or purple flowers. The commonest are Common Vetch *V. angustifolia*, Bush Vetch *V. sepium* and Tufted Vetch *V. cracca*. ⇨⇨ TARE. Bitter Vetch is a name of two untendrilled peaflowers, the widespread *Lathyrus montanus* and the local *V. orobus*. Other peaflowers named vetch are also untendrilled. Crown Vetch *Coronilla varia* is a garden escape. Horseshoe Vetch *Hippocrepis comosa* of calcareous grassland looks like a birdsfoot trefoil. Kidney Vetch or Lady's Fingers *Anthyllis vulneraria* is common in calcareous grassland and by the sea. Milk Vetch is a name of three lime-loving species of *Astragalus*, especially Wild **Liquorice** *A. glycyphyllos*, and Purple Milk **Vetch** *A. danicus*, and of the two rare Mountain Milk Vetches *Oxytropis*.

Vetchling (Leguminosae: *Lathyrus*). Name of two atypical peas, Grass Vetchling *L. nissolia* with solitary crimson flowers and grass-like apparent leaves, and Yellow Vetchling *L. aphaca* with solitary yellow flowers and broad triangular apparent leaves. Meadow Vetchling, ⇨ MEADOW PEA.

Villous. Botanical adjective meaning shaggily hairy.

Vinegar Fly. ⇨ FRUIT FLY (Drosophilidae).

Violet (Violaceae: *Viola*). Small violet or blue-violet flowers, mostly spring flowering; nine British species, the commonest being Common Dog Violet *V. riviniana*, widespread and often abundant in woods and hedgerows; Sweet Violet *V. odorata* is the only scented one; Hairy Violet *V. hirta* is frequent in limestone turf; Marsh Violet *V. palustris* is frequent on wet moors and heaths. Dame's Violet and Water Violet are unrelated.

Violet Sea Snail (Prosobranchia: *Janthina janthina*). An oceanic mesogastropod operculate mollusc, living on the surface of the Atlantic and occasionally swept ashore by storms, complete with the air-filled mucous sac similar to those of the Portuguese man-o'war and By-the-wind-sailor; more often its empty snail-like shell, partly coloured a beautiful violet, is found on our western beaches.

Viper. ⇨ ADDER.

Virus. Minute self-reproducing protein particle, sometimes crystalline sometimes molecular, filterable through meshes that exclude bacteria parasitic and causing both animal and plant diseases, such as the mosaic diseases of many cultivated plants, psittacosis of birds and foot-and-mouth disease of cattle.

Vivipary. Producing live young. Mammals and some reptiles are viviparous. Viviparous plants have green shoots instead of flowers, e.g. viviparous fescue.

Voe. An arm of the sea in the Shetlands.

Volcanic Rocks. Igneous rocks which have solidified after reaching the surface during a volcanic eruption.

Vole (Rodentia: Muridae). Small rodents with blunter snouts than mice. The four British species are the abundant Field Vole *Microtus agrestis*, often erroneously called the short-tailed field mouse; the closely related, *M. orcadensis*, confined to the Orkneys; the common and widespread Bank Vole *Clethrionomys glareolus*, which prefers less open country than the field vole; and the much larger Water Vole *Arvicola terrestris*, often misleadingly called Water Rat. Like the field mice, most of these also have island races.

Vulture (Falconiformes: Aegypiidae). Very large broad-winged short-tailed birds of prey. Two very rare stragglers, the Egyptian Vulture *Neophron percnopterus* and the Griffon Vulture *Gyps fulvus*.

W

Wader. A group of birds, mainly frequenting shallow water and mostly long-legged and long-billed, formerly known as the Limicolae but now included in the Charadriiformes. Comprises the oystercatcher (Haematopodidae), plovers (Charadriidae), sandpipers and snipes (Scolopacidae), avocet and stilt (Recurvirostridae), phalaropes (Phalaropodidae), stone-curlew (Burhinidae) and pratincole and courser (Glareolidae).

Wagtail (Passeriformes: *Motacilla*). Small, long-tailed song birds, often associated with water, named from their habit of restlessly moving their tails up and down. Four British species: Blue-headed Wagtail, Continental race of the yellow wagtail, scarce passage migrant and has bred; Grey Wagtail *M. cinerea*, common on fast streams, especially in the north and west; Grey-headed Wagtail, Scandinavian race of the yellow wagtail, scarce passage migrant; Pied Wagtail *M. alba*, widespread and common, also known as Dishwasher; Water Wagtail, the pied wagtail; White Wagtail, Continental race of pied wagtail, regular passage migrant and has bred; Yellow Wagtail *M. flava*, summer visitor, mainly to damp grassland, England and Wales; Yellow-headed wagtail *M. citreola*, very rare straggler.

Wainscot Moth (Caradrinoidea: Caradrinidae). A group of noctuid moths, mostly of the genus *Leucania*, named from their pale colouring, faintly streaked like an old-fashioned wainscot, and mainly frequenting fens, marshes and reed-beds.

Wall Butterfly or Wall Brown (Papilionina: *Pararge megaera*). One of the commoner Brown Butterflies, named from its habit of sunning itself on a wall.

Wallcreeper (Passeriformes: *Tichodroma muraria*). Small bird with bright red wing-patches, a rare straggler from southern Europe, where it is adapted to creeping up rocks and cliffs.

Wallflower (Cruciferae: *Cheiranthus cheiri*). A familiar garden flower long established on old walls, cliffs, etc.; flowers usually yellow, fragrant.

Wallow. Place where deer roll in a muddy pool or peat hag.

Wall-Pepper or Biting Stonecrop (Crassulaceae: *Sedum acre*). The commonest of our yellow-flowered stonecrops, named from its biting taste.

Wall-Rue (Aspleniaceae: *Asplenium ruta-muraria*). A small fern related to the spleenworts; widespread and common, especially on walls.

Walnut (Juglandaceae: *Juglans regia*). A deciduous catkin-bearing tree from the Mediterranean, much planted for its nuts, and occasionally bird-sown in hedgerows in the south.

Walrus (Pinnipedia: *Odobenus rosmarus*). A large tusked marine mammal allied to the seals; an increasingly rare visitor to our northern coasts.

Wandering Snail (Pulmonata: *Limnaea pereger*). The commonest British pond snail, widespread in fresh and brackish waters, and very variable.

Warble Fly (Cyclorrhapha: *Hypoderma*). Three species of two-winged flies related to the bluebottles, whose larvae feed inside cattle (*H. lineatum* and *H. bovis*) or deer (*H. diana*), producing a swelling under the skin called a warble. They are serious economic pests as they can ruin hides.

Warbler (Passeriformes: Sylviidae). Small insectivorous song birds, mostly with rather plain brown plumage but songs of moderately high quality. Twelve species breed, all but one being summer visitors; half a dozen more are fairly regular passage migrants, and there are nineteen rare stragglers, as well as four from the quite distinct family (Parulidae) of North American warblers. Aquatic Warbler *Acrocephalus paludicola* and Barred Warbler *Sylvia nisoria* are scarce passage migrants. Blackcap. Chiffchaff. Dartford Warbler *S. undata*, our only resident warbler, confined to heaths in southern England. Garden Warbler *S. borin*, one of the better songsters; frequent in England and Wales, very local elsewhere, in woods and scrub, but only in the larger gardens. Grasshopper Warbler *Locustella naevia*, a shy bird announcing its presence by its reeling song which resembles some Continental grasshoppers, widespread but local, often in marshy places. Icterine Warbler *Hippolais icterina*, scarce passage migrant. Leaf Warbler, the genus *Phylloscopus*. Marsh Warbler *A. palustris* closely resembles reed warbler except in song, very local near water in southern England. Melodious Warbler *H. polyglotta*, scarce passage migrant; has bred. Moustached Warbler *Lusciniola melonopogon*, very rare straggler, but bred near Cambridge 1946. Reed Warbler *A. scirpaceus*, frequent by water, in England and Wales, mainly in the south; especially associated with the reed *Phragmites*. Savi's Warbler *L. luscinioides*, similar to grasshopper warbler, formerly bred in the Fens, and has recently established a colony in Kent. Sedge Warbler *A. schoenobaenus*, widespread and common near water; mimetic. Common and Lesser Whitethroats. Willow Warbler *Phylloscopus trochilus*, one of our commonest breeding birds, distinguishable from chiffchaff mainly by song, formerly miscalled willow wren; widespread and often abundant, not especially associated with willows. Wood Warbler *P. sibilatrix*, formerly miscalled wood wren, widespread but local in woods without much undergrowth; has two quite distinct songs. Yellow-browed Warbler *P. inornatus*, scarce passage migrant.

Warm-blooded. ⇨ HOMOIOTHERMIC.

Warning Coloration. Also known as aposematic coloration, is the evolution by an animal of colours or markings that make it deliberately conspicuous to its predators, often by Mimicry, as an indication that it has built-in defences against them such as poisonous or distasteful flesh or the capacity to strike back in some way, e.g. the common wasp, strikingly barred black and yellow, and the red and black burnet and cinnabar moths, the latter with black and yellow caterpillars, are rejected by both birds and spiders.

Warp. Mud or silt carried in river water and eventually deposited as sediment or alluvium. This is a feature of some estuaries, such as the Humber.

Warren. The burrows of a rabbit colony.

Wart-Cress or Swine's Cress (Cruciferae: *Coronopus*). Two low-growing, white-flowered plants of waste ground in the south: *C. squamatus*, the commoner, and *C. didymus*.

Wasp (Apocrita: Aculeata). Like the bees, the wasps can conveniently be divided into a small number of social wasps, including what most people regard as wasps, and a large number of solitary wasps. The Social Wasps comprise the Hornet and six other species which are the black-and-yellow banded insects, the Common Wasp *V. vulgaris* and the German Wasp *V. germanica* being the most frequent visitors to jam-jars in the house. They make large rounded nests of wood pulp which they chew into a papery substance, either in a hole in the ground or a tree, or like the Tree Wasps in a tree. The social wasps are true Social Insects, and all predators of two-winged flies and other insects, so that apart from their raids in the kitchen they can be ranked as beneficial. The great majority of our wasps, however, are Solitary Wasps, which make nest burrows in the ground or in plant stems against the hatching of their young, with which they do not subsequently have anything to do. Most species specialize in one kind or group of insects or spiders. Those which dig holes in the ground are generally known as Digger Wasps, but this name is sometimes applied generally to all Solitary Wasps. The principal families include the Chrysididae (ruby-tail wasps), Myrmosidae and Mutillidae (velvet-ants), Vespidae (potter and mason wasps, also the social wasps), Sphecidae (sand, wood-boring, digger, and black wasps) and Pompilidae (spider-hunting wasps).

Wasp Orchid. A variety of the bee orchid.

Water Bear (Arthropoda: Tardigrada). A group of tiny aquatic or moisture-loving eight-legged animals, formerly included in the Arachnida, and named from their superficially bear-like appearance. Most species live in such damp terrestrial habitats as damp soil or moss, or the detritus which collects in roof gutters, but a few are truly aquatic in fresh water; *Macrobiotus macronyx* is one of the commonest species.

Water Beetle (Coleoptera: Adephaga). The commonest aquatic beetles are the carnivorous ones belonging to four families of the sub-order Adephaga: the small water beetles (Haliplidae: eighteen species), the screech beetle (Hygrobiidae), the water beetles proper (Dytiscidae; 107 species), including the great diving beetle, and the whirligig beetles. ⇨ SCAVENGER BEETLE.

Water-Boatman (Heteroptera: *Notonecta*). Active, predatory water bugs, propelling themselves by their powerful hindlegs, the 'oars' of the Wherry-men, Backswimmers or Boat-flies, as they have also been called. *N. glauca* is the commonest species. The less widespread Lesser Water-boatman *Plea atomaria* belongs to a different family.

Water Bug (Hemiptera: Heteroptera). Two groups of families, the water bugs proper (Hydrocorisae) and the surface water bugs (Amphibicorisae). The Water Bugs include the corixids (Corixidae), saucer-bugs (Naucoridae), water-boatmen (Notonectidae), lesser water-boatmen (Pleidae) and water-scorpions (Nepidae). The Surface Water Bugs include the pond-skaters (Gerridae), pondweed bug (Mesoveliidae), sphagnum bugs (Hebridae), water-crickets (Veliidae) and water measurers (Hydrometridae).

Watercress (Cruciferae: *Rorippa*). Two closely similar species, *R. nasturtium-aquaticum* and *R. microphyllum*, widely distributed in shallow moving fresh water, together with their hybrid, which is the chief commercial watercress, when it is known as Winter Cress, and must not be confused with the true Wintercress (*Barbarea*). The leaves of Fool's Watercress (Umbelliferae: *Apium nodifiorum*) are similar (and apparently may be eaten with impunity), but its smaller white flowers are quite differently arranged, in umbels.

Water-Cricket (Heteroptera: Veliidae). Five rather stout predatory surface water bugs, *Velia caprai* being the commonest.

Water Dropwort (Umbelliferae: *Oenanthe*). Seven white-flowered poisonous umbellifers of wet or damp places, the best known and one of the most poisonous being the tall stout Hemlock Water Dropwort ***Oe.*** *crocata*, locally frequent in south and west England. ⇨ SULPHURWORT.

Water Flea or Cladoceran (Branchiopoda: Cladocera). An order of tiny primitive freshwater crustaceans, forming the principal constituents of the zooplankton of lakes and ponds; species of *Daphnia* are often abundant

Waterfowl. Aquatic birds, sometimes restricted to ducks, geese and swans.

Water-Gnat. ⇨ WATER MEASURER.

Waterhen. ⇨ MOORHEN.

Water-Lily (Nymphaeaceae). Large aquatic plants with floating leaves, growing mainly in still water, quite unrelated to the lilies. Three species, the two commoner being the White Water-lily *Nymphaea alba* and the Yellow Water-lily or Brandy-bottle *Nuphar lutea* named from the scent of its flowers. The Fringed Water-lily (Menyanthaceae: *Nymphoides peltata*) is smaller than the true water-lilies and found mainly in the south.

Water-Louse, Water Slater or Hog Slater (Isopoda: *Asellus*). Common small freshwater crustaceans, related to the woodlice, frequenting weedy streams and ponds; *A. aquaticus* and *A. meridianus* are the commonest of our six species.

Water Measurer or Water-Gnat (Heteroptera: Hydrometridae). Two slender, elongated predatory surface water bugs; *Hydrometra stagnorum* is the common species, *H. gracilenta* being rare.

Water Milfoil (Haloragaceae: *Myriophyllum*). Three species of submerged waterweed with whorls of feathery pinnate leaves up the stem, and emergent spikes of inconspicuous flowers.

Water Mite (Prostigmata: Hydrachnellae). Well over 200 species of mite live in fresh water, parasitizing as larvae such diverse creatures as beetles, mussels and sponges, with a few species also living in salt water.

Water-Pepper (Polygonaceae: *Polygonum*). Three narrow-spiked relatives of the bistorts, the commonest being the green-flowered Water-pepper *P. hydropiper* of damp places, with acrid-tasting leaves, while the two others, *P. minus* and *P. mite*, are scarce, tasteless and pink-flowered.

Water-Plantain (Alismataceae). Five aquatic monocotyledons (*Alisma, Baldellia, Luronium*), with white or pale pink flowers, the most widespread being *A. plantago-aquatica*.

Water-Scorpion (Heteroptera: *Nepa cinerea*). A widespread and common elongated water bug, superficially resembling a scorpion because of its 'tail', actually a respiratory siphon.

Watershed. The high ground separating the catchment areas of two separate river systems, now often used in America and elsewhere to mean the catchment areas themselves.

Water Singer (Heteroptera: *Micronecta poweri*). A common corixid bug noted for its vocal performances.

Water Slater. ⇨ WATER-LOUSE.

Water Soldier (Alismataceae: *Stratiotes aloides*). A singular free-floating aquatic monocotyledon, which rises to the surface of ponds and ditches to display its white flowers in the middle of a tuft of sharply toothed sword-shaped leaves, and then sinks again; widespread but now very local.

Water Spider (Araneae: *Argyroneta aquatica*). The only spider that lives in the water, constructing an underwater bell which it fills with air bubbles trapped in its body hairs; common in fresh water.

Waterspout. A column of water formed at sea by a whirlwind which carries water upwards to meet a downward projection from low clouds above, occasionally seen in the English Channel and elsewhere, but much commoner on a larger scale in the tropics.

Water Starwort (Callitrichaceae: *Callitriche*). Small-leaved waterweeds, sometimes completely submerged; six British species, the commonest being *C. stagnalis*.

Water Stick-Insect (Heteroptera: *Ranatra linearis*). A remarkably long thin water bug, closely related to the water-scorpion, mainly southern in distribution.

Water Strider. ⇨ POND SKATER.

Water-Thyme. Name given to various monocotyledon waterweeds of the Hydrocharitaceae, especially Canadian Waterweed *Elodea canadensis* and various plants grown by aquarists and sometimes discarded and more or less naturalized.

Water Vapour. Water held invisibly in the air before a change in the temperature forces it to condense and become visible as cloud. Warm air can hold more water vapour than cold air, so that cloud is produced by a fall in temperature.

Water Violet (Primulaceae: *Hottonia palustris*). A floating plant with leafless spikes of pale lilac-white flowers, mainly in eastern England.

Waterweed. General name for aquatic plants growing right in the water, especially those, such as pondweeds, water starwort and water milfoil, which have rather inconspicuous flowers. Canadian Waterweed or Water-thyme (Hydrocharitaceae: *Elodea canadensis*) is an invader from North America, now well established in fresh water.

Waterwort (Elatinaceae: *Elatine*). Two species of diminutive, often submerged, aquatic plants, with tiny pink flowers; very local in peaty waters.

Wave Moth (Notodontoidea: Sterrhidae). Name given to many small geometrid moths, especially of this family.

Waves. Approach the beach across an uninterrupted stretch of sea called the *fetch*; they break on the shore as the *swash*, and return down the slope of the beach as the *backwash.*

Waxwing (Passeriformes: *Bombycilla garrulus*). Starling-sized crested song bird, named from the 'waxy' red and yellow markings on its wings, formed by the fusion of the feather shafts with the tip of the outer vane. An irregular winter visitor to Britain from northern Europe, sometimes in considerable numbers. Formerly called Bohemian Chatterer; known as Bohemian Waxwing in North America.

Wayfaring Tree (Caprifoliaceae: *Viburnum lantana*). Deciduous shrub (not, despite its name, a tree) close to wild guelder; common on chalk and limestone in the south, with berries turning first red and then black.

Weasel (Carnivora: *Mustela nivalis*). Our smallest native carnivore, widespread and common in Great Britain, but absent from Ireland, where the name 'weasel' is applied to the stoat. It resembles a small stoat, but has no black tip to the tail, and does not turn white in winter. Males are about 8½ ins. long and females about 6 ins. In many country districts the larger male weasel is called the whitrick and the smaller female the cane; it was once believed that these were separate species.

Weasel's Snout. ⇨ Lesser SNAPDRAGON.

Weed. A name loosely applied to any plant growing in a place inconvenient to man, but more specifically to unwanted plants growing in cultivated ground. Most weeds, such as groundsel, scarlet pimpernel and various speedwells are annuals, but some very tiresome weeds, such as ground elder and field bindweed, are perennials.

Weever or Sting-fish (Percomorphi: *Trachinus*). The Greater *T. draco* and Lesser Weaver *T. vipera* are small marine fish of sandy shallows, equipped with a poisonous spine, which can inflict a painful wound. The lesser weever is much the commoner.

Weevil (Rhynchophora: Cucurlionidae). A large family of small beetles, readily recognized by their beak-like rostrum and elbowed antennae; the family is the largest in the Animal Kingdom, and even in Britain there are 511 species. The Grain Weevil *Sitophilus granarius* and the Rice Weevil *S. oryzae* are serious pests of stored products. Pea and Bean Weevils are unrelated, belonging to the Family Bruchidae in the Phytophaga; some of them are pests of stored food.

Weld or Dyer's Rocket (Resedaceae: *Reseda luteola*). A tall relative of wild mignonette, with narrow spikes of small greenish-yellow flowers, frequent in disturbed ground.

Wels. ⇨ CAT-FISH.

Welshman's Button or Halford's Welshman's Button (Trichoptera: *Sericostoma personatum*). A widespread caddis-fly of lakes and rivers, whose adults are an important angling fly.

Wendeltrap or Wentletrap (Prosobranchia: *Clathrus clathrus*). A relative of the winkles, with a strikingly beautiful narrowly spiral shell, with marked ridges crossing the spiral whorls at right angles, whence the curious name derived from the Dutch for 'winding staircase'; widespread and common offshore, the shells being often thrown up on the beach.

Wet-Rot Fungus. ⇨ DRY ROT.

Whale (Cetacea). A very distinct order of marine mammals, ranging up to 100 ft in length and furnished with a thick layer of fat (blubber) yielding a valuable oil which has been much sought after for centuries. Though fish-like in form, whales are not fish, but come to the surface regularly to breathe air like any other mammal. They are conveniently divided into the large whalebone or baleen whales (Mystacoceti), which feed on plankton through the sieve formed by the whalebone (baleen) plates that replace the teeth in their upper jaws; and the smaller toothed whales (Odontoceti), which have upper teeth and feed on fish and squids. Some two dozen species are known in British seas, but only about half a dozen smaller ones are at all commonly seen close inshore. Some baleen whales are called Rorquals and some toothed ones Dolphins. The ferocious Killer Whale *Orcinus orca*, the largest of the dolphins, is not infrequent in British seas, and the giant Blue Whale *Balaenoptera musculus*, a rorqual and the largest living animal (80–100 ft) occurs out towards the Atlantic. Caa'ing Whale or Pilot Whale, ⇨ BLACKFISH; Sperm Whale, ⇨ CACHALOT; White Whale, ⇨ BELUGA. ⇨⇨ HUMPBACK.

Whame Fly. A local name for the blood-sucking tabanid flies.

Wheat-Bulb Fly (Muscidae: *Hylemia coarctata*). A two-winged fly whose larvae attack wheat, and to destroy which seed-corn is dressed with the deadly aldrin and dieldrin insecticides, whose indiscriminate use in the countryside in the early 1960s led to widespread deaths of wild birds and mammals and subsequently to a partial ban on the use of these poisons.

Wheatear (Passeriformes; *Oenanthe oenanthe*). Small song bird allied to the thrushes, named from its white rump, and formerly caught in large numbers for the table on the Sussex downs. Summer visitor, one of the earliest to arrive, in mid-March; widespread on migration, but breeding mainly in hill districts of the north and west. Greenland Wheatear is the somewhat larger subspecies breeding in Greenland, not infrequent on migration in Britain. Five other species are stragglers.

Wheel Animalcule. ⇨ ROTIFER.

Whelk (Prosobranchia: Stenoglossa). Marine operculate molluscs with broadly spiral, winkle-like shells, the commonest and most widespread intertidal species being the edible Common Whelk or Buckie *Buccinium undatum*, which may grow to 6 ins. and whose large sponge-like egg-masses are familiar jetsam on the shore; the Common Dog Whelk or Dog Winkle *Nucella lapullus*, which feeds mainly on barnacles, and two other dog whelks *Nassarius*. ⇨⇨ DRILL.

Wherryman. ⇨ WATER-BOATMAN.

Whimbrel (Charadriiformes: *Numenius phaeopus*). A small curlew, frequent as a passage migrant, and breeding very sparsely in northern Scotland. Also known from its principal cry as Titterel or Seven Whistler.

Whin. A Scottish name for gorse. Petty Whin or Needle Furze. (Leguminosae: *Genista anglica*) is a spiny heathland relative of dyer's greenweed.

Whinchat (Passeriformes: *Saxicola rubetra*). Small song bird allied to the thrushes, not nowadays especially associated with whin. Summer visitor, mainly to the hill country of the north and west, but also breeding locally in the south.

Whinstone. A hard dolerite rock, forming the Great Whin Sill which runs across the North of England to the Northumberland coast. ⇨ SILL.

Whirligig Beetle (Adephaga: Gyrinidae). The small water beetles that gyrate round one another on the surface of still or slow-moving fresh water; twelve species.

Whirlwind. A pillar of rotating air, with low atmospheric pressure at its centre, and sometimes associated with thunderstorms, comparatively rare in Britain on a scale which causes damage to trees and buildings in its path, but more frequent in hot weather as the minor phenomenon known as a 'dust devil', which catches up dust and debris and whirls it around.

White Admiral (Papilionina: *Limenitis camilla*). A handsome white-barred vanessid butterfly, once rare and still local in southern woodlands.

Whitebait. General name for young herrings or other tiny young fish.

White Beam (Rosaceae: *Sorbus aria*). Handsome tree or large shrub, with leaves silvery white beneath; found mainly on chalk and limestone.

White Butterflies (Papilionina: Pieridae). Six predominantly white butterflies, of which three are common and three rare. Large White *Pieris brassicae* and Small White *P. rapae*, collectively known as Cabbage Whites, feed on garden brassicas and are regarded as pests; both are migratory, the Large White especially so. Green-veined White *P. napi* is common in the countryside but quite innocuous. The very local Wood White, *Leptidea sinapis* is our sole representative of a South American group of the Pieridae. Marbled White *Melanargia galathea*, though black and white is actually related to the Brown Butterflies; locally frequent in calcareous grassland. ⇨ ORANGE-TIP BUTTERFLY.

White Fish. A general term for marine fish with white flesh, especially the Cod Family and flat-fish, as opposed to herring, mackerel, salmon, etc., which have brown or pink flesh; also called demersal, or bottom-living, fish, as opposed to the pelagic species, which swim near the surface. The White Fish Authority, however, in fact covers all marine fish except herring, salmon and sea trout, and includes within its purview both kinds of shellfish.

Whitefish (Salmonidae: *Coregonus*). Small herring-like relatives of the trout, mostly landlocked in various lakes in North Wales, the Lake District, Scotland and Ireland: gwyniad, pollan, powan, schelly and vendace, but also including the anadromous houting.

White-Fly (Homoptera: Aleurodidae). A family of small bugs whose wings and bodies are coated with a mealy white powder and so resemble minute moths or midges. They exude honey-dew, and include a number of economic pests, including the Cabbage White-fly *Aleurodes brassicae* and the Greenhouse White-fly *Trialeurodes vaporariorum*.

White-front. Birdwatchers' slang for white-fronted goose.

White Grub. Farmers' name for the larvae of cockchafers.

White Horses. The foaming white crests of waves at sea, whipped up by a strong wind.

White Moor. The Exmoor term for moorland under grass, especially blue moor-grass or sedge, as distinct from the heather moor.

White Nun. Wildfowler's name for the drake smew.

Whitethroat (Passeriformes: *Sylvia*). Summer visitor warblers with whitish throats, frequenting woods and scrub. Common Whitethroat *S. communis*, widespread and common. Lesser Whitethroat *S. curruca*, frequent in England and Wales, mainly in the south.

White Weed (Hydrozoa: Sertularia). A species of sea-fir, growing offshore and often cast up on the beach after storms, which, according to Barrett and Yonge, is sold commercially for decorative purposes.

White Worm (Oligochaeta: Enchytraeidae). Small white or pinkish bristled terrestrial oligochaete worms, also known as pot worms, enchytraeid worms or gentles, common in damp and wet soil.

Whiting (Gadidae: *Gadus merlangus*). A common shallow-water fish of British seas, mainly the North Sea, with an important fishery. Couch's Whiting or Poutassou *Gadus poutassou* is an uncommon deep-water fish.

Whitlow-Grass (Cruciferae: *Erophila verna*). Tiny early-flowering white-flowered annual, formerly used to cure whitlows. Name also used for three white- and one yellow-flowered crucifers of the genus *Draba*, all rare or uncommon.

Whitrick. A country name for the male weasel, which is larger than the female. ⇨ CANE.

Whorl. A ring of leaves or flowers around a stem.

Whorl Snail (Pulmonata: *Truncatellina* and *Vertigo*). A group of eleven tiny land snails, often frequenting moist places, the commonest being *V. pygmaea*.

Whortleberry. Southern name for the bilberry.

Wigeon (Anatidae: *Anas penelope*). A surface-feeding duck, widespread and common as a winter visitor, breeding mainly in the north.

Wildfowl. Birds sought after by the wildfowler, usually comprising the ducks, geese, swans and waders.

Wildfowl Trust. The national society for the study and conservation of wildfowl (ducks, geese and swans), founded in 1946 as the Severn Wildfowl Trust; maintains collections of wildfowl at Slimbridge, Gloucestershire, where there is also a laboratory, and Peakirk, Northants; publishes an *Annual Report*.

Wildlife. A general term covering all living things, both flora and fauna.

Wild Liquorice. ⇨ Milk VETCH.

Wild Service Tree (Rosaceae: *Sorbus torminalis*). Small deciduous tree with jagged leaves, related to white beam and rowan. Rather local in England and Wales. ⇨⇨ SERVICE TREE.

Will o'the Wisp. ⇨ MARSH GAS.

Willow (Salicaceae: *Salix*). Medium to tall deciduous catkin-bearing trees, growing in damp places, often by rivers, the larger ones often pollarded. Eleven native species are trees or large shrubs, the commonest being the Crack Willow *S. fragilis*, the White Willow *S. alba*, with silvery foliage, the Bay Willow *S. pentandra*, the Osier, and the Pussy Willows or Sallows.

Seven more species are low or creeping shrubs or undershrubs, all but the Creeping Willow *S. repens* of wet heaths and dune slacks being rare or local mountain plants, from the Lake District northwards. The Cricket Bat and Golden Willows are varieties of the white willow, but the widely planted Weeping Willow is a separate species, *S. babylonica.*

Willow Fly (Plecoptera: *Leuctra geniculata*). A common autumn-emerging stone-fly, especially on the chalk streams of the south.

Willow-Herb (Onagraceae: *Epilobium*). Pink-flowered plants, mostly of damp or shady places, named from the willow-like shape of their leaves; seed-pods long and thin. A dozen species, of which the largest and commonest is the Great Willow-herb or Codlins-and-Cream *E. hirsutum.* ⇨ ROSEBAY.

Wind. A moving current of air, also known as a breeze when it is light and a gale when it is strong, technically over about 40 m.p.h. The strength or velocity of the wind is measured on the Beaufort Scale, ⇨ p. 27.

Windflower. Old-fashioned name for the wood anemone.

Windhover. ⇨ KESTREL.

Window Fly (Brachycera: Scenopinidae). A small family of two-winged flies with three species, named from their habit of congregating in large numbers on windows: *Scenopinus fenestralis,* sometimes erroneously called the Carpet Fly (it does not harm carpets) is the commonest species.

Windward. The direction from which the wind is blowing.

Wintercress (Cruciferae: *Barbarea vulgaris*). The commonest of four yellow-flowered crucifers found on waysides and river banks; not to be confused with the winter cress of commerce, which is a hybrid form of watercress.

Winter Fungus or Velvet-stemmed Agaric (Basidiomycetes: *Flammulina velutipes*). A common yellowish toadstool, growing on tree stumps and branches right through the winter.

Wintergreen (Pyrolaceae: *Pyrola*). Three mainly woodland plants with white flowers rather like lily of the valley but not fragrant; the commonest is Common Wintergreen *P. minor*; Yavering Bells and St Olaf's Candlestick are closely related. The unrelated Chickweed Wintergreen (Primulaceae: *Trientalis europaea*) is also a woodland plant with white flowers; mainly in the north.

Winter Heliotrope (Compositae: *Petasites fragrans*). A relative of the butterburs, whose fragrant flowers appear with the leaves during the winter; a widespread and increasing garden escape, especially on road verges.

Winter Moth (Notodontoidea: *Opheroptera brumata*). A dull-coloured geometrid moth, allied to the carpets and pugs, whose green looper caterpillars are common on many deciduous trees and shrubs in spring; the males fly mostly between November and January, but the females are wingless.

Winkle or Periwinkle (Prosobranchia: *Littorina*). Mesogastropod operculate molluscs, with squatly spiral shells, the commonest group of gastropods on rocky shores between the tides. The common and widespread species include the Common or Edible Winkle or Periwinkle *L. littorea,* Rough Periwinkle *L. saxatilis,* Flat Periwinkle *L. littoralis* and Small

Periwinkle *L. neritoides*; most are rather variable in colour, including red, black, orange, yellow, green, purple, brown and streaked. Two species of Freshwater Winkle or River Snail *Viviparus viviparus* and *V. fasciatus*, found in slow-moving streams in England, have large snail-like shells up to 1½ ins. long. Dog Winkle, ⇨ Dog WHELK. Sting Winkle, ⇨ DRILL,

Wireworm. The larva of certain click beetles. The name is also incorrectly applied to millipedes.

Witch (Pleuronectidae: *Glyptocephalus cynoglossus*). A smallish, right-sided, deep-water flatfish, shaped like a sole, locally common off our western coasts; also known as Pole Dab or Pole Flounder.

Witches' Broom. A cluster gall due to deformed leaf or twig structures, and making a birds-nest-like mass of twigs in a tree or shrub. The cause may be either a fungus, as with *Exoascus cerasi* on wild cherry, or *E. betulinus* on birch, or an eriophyid mite. ⇨ GALL.

Witches' Butter (Ascomycetes: *Tremella mesenterica*). A common gelatinous, orange-yellow fungus growing on dead wood, supposedly resembling butter or the convolutions of the brain.

Withy. ⇨ OSIER.

Wold. Open chalk or limestone hills with few trees; chalk in Yorkshire and Lincolnshire, and limestone in the Cotswolds and East Midlands.

Wolf (Carnivora: *Canis lupus*). Extinct in England since the reign of Henry VIII (1485–1509), in Scotland since 1743 and Ireland since the 1760's. The last wolf on the mainland of Great Britain was killed in Morayshire in 1743.

Wolf-Fish. ⇨ CAT-FISH.

Wolfsbane. ⇨ MONKSHOOD.

Wolf Spider (Araneae: Lycosidae and Pisauridae). Two families of spiders, related to the celebrated tarantula of southern Europe and named from the habit of some species of hunting down their prey on foot instead of lying in wait for it in webs. Some species even carry their eggs about with them.

Woodbine. ⇨ HONEYSUCKLE.

Wood-boring Beetle (Polyphaga: Clavicornia). A group of families of clavicorn beetles whose larvae feed on wood, and including the powder-post beetles and furniture beetles. The larvae of some other groups of beetles, such as the longhorns and ambrosia beetles also bore in wood.

Wood-boring Wasp (Aculeata: *Trypoxylon*). Three slender solitary wasps, which make their nests in the stems of shrubs and other plants, provisioning them with spiders.

Woodchat. ⇨ SHRIKE.

Woodcock (Charadriiformes: *Scolopax rusticola*). A medium-sized wader, widespread, frequenting woodland and heathland, with a curious crepuscular display flight known as 'roding'. Commoner in winter when many migrants arrive from northern Europe.

Wood Hedgehog (Basidiomycetes: *Hydnum repandum*). An edible, buff-coloured toadstool, growing, often in rings, in broad-leaved woods, and named from the short teeth or spines under the cap.

Woodland. The two main types of natural woodland are the deciduous forest, mainly of oak with varying admixtures of ash, to be found all over the country, and the fragments of coniferous forest in the Scottish Highlands, dominated by Scots pine. The British Isles were once largely covered by woods, of which only isolated wholly natural fragments remain, e.g. Wistman's Wood on Dartmoor. They were gradually destroyed largely to make way for grassland for grazing animals, with no national effort at replacement until the establishment of the Forestry Commission in 1919. Oak-ash forest is the climatic climax over much of the British Isles, but on the shallow calcareous soils of southern England beech is dominant. Virtually all deciduous woodland today is semi-natural, i.e. natural areas modified by man or plantations by man of native trees which regenerate naturally. The conifer plantations of the Forestry Commission carry a very depauperate ground vegetation.

Woodlouse (pl. Woodlice) (Isopoda: Oniscoidea). A sub-order of mainly terrestrial small crustaceans, also known as slaters; mainly living in damp and shady places, under stones, logs, etc. Among the commoner of the thirty-six British species are *Porcellio scaber*, *Oniscus asellus* and *Philoscia muscorum*. The well-known Pill Woodlouse or Pill Bug *Armadillidium vulgare*, which can roll into a ball, is said to have been used as a pill by medieval doctors. ⇨ SEA SLATER.

Woodpecker (Piciformes: Picidae). Birds which specialize in feeding by searching tree trunks and branches for grubs and other insects and nest in tree holes. Three British breeding species and one straggler. Barred or Lesser Spotted Woodpecker *Dendrocopos minor*, the smallest British species, is widespread but local in England and Wales, mainly in the south. Green Woodpecker *Picus viridis*, our largest species, has green plumage with a red crown; widespread and common in England and Wales, mainly in the south, and spreading slowly into southern Scotland. Pied or Great Spotted Woodpecker *D. major* is widespread and locally common throughout Great Britain; the commonest in the suburbs of large towns. Black Woodpecker *Dryocopus martius* breeds in western Europe and has been reported from Britain more than eighty times, but is not yet officially accepted as a member of the British List. ⇨ WRYNECK.

Woodpigeon (Columbiformes: *Columba palumbus*). The largest native British pigeon or dove, formerly known as Ring Dove, and in Scotland as Cushat or Cushie-doo. Widespread and often abundant, it is the only serious agricultural pest among British birds, especially harmful to green crops. Nest a flat platform of twigs with two white eggs.

Woodruff, Sweet (Rubiaceae: *Galium odoratum*). A close relative of the bedstraws, named from its woodland habitat and its new-mown-hay scent when dried, which once made it a favourite stuffing for bedding.

Wood-Rush (Juncaceae: *Luzula*). Eight species of monocotyledons, closely related to the rushes, characterized by the long white hairs on the leaves and stems, generally growing in grassy or wooded situations. The commoner species are Field Wood-rush *L. campestris*, also known as Sweep's Brush or Good Friday Grass; Heath Wood-rush *L. multiflora*, the early flowering Hairy Wood-rush *L. pilosa*, and the tall woodland Great Wood-rush *L. sylvatica*.

Wood-Sage (Labiatae: *Teucrium scorodonia*). A common labiate of heathy places, related to the germanders, whose yellow-green flowers only have a lower lip.

Woodsia (Athyriaceae: *Woodsia*). Two tiny ferns, with fronds only ½–3 ins. long, rare and decreasing on rocks in mountains in the north: *W. ilvensis* and *W. alpina*.

Wood Wasp or Horn-tail (Symphyta: Siricidae). A small group of large superficially wasp-like Hymenoptera, more closely related to the saw-flies, with wood-boring larvae. The Giant Wood Wasp or Greater Horn-tail *Urocelus gigas* is yellow, has a long ovipositor, measures 2 ins. across and is often mistaken for a hornet; it frequents coniferous woods. There are also three smaller species (*Sirex*).

Wood Woolly-Foot (Basidiomycetes: *Collybia peronata*). A common yellowish woodland toadstool, related to spindle-shank, and named from the woolly hairs at the base of its stem.

Wool Alien or Shoddy Alien. Plants whose seeds have become entangled with sheep's wool in Australia and other countries exporting wool to Britain, and which germinate and grow in fields where wool waste (shoddy) is used as manure and in mill yards and railway sidings where the shoddy is loaded and unloaded. A list of 529 species, including 187 grasses, was compiled in 1961.

Wool-Carder Bee (Aculeata: *Anthidium manicatum*). A carpenter bee named from its habit of stripping the woolly hairs or down from the stems of plants to line its nest cells.

Woolly Bear. Hairy or furry larvae, especially caterpillars of the tiger and ermine moths; most often seen are those of the garden tiger and white and buff ermine.

Worker. The infertile females of social ants, bees and wasps, which do all the work of the colonies.

World Wildlife Fund. Founded in 1961 to raise funds for the conservation of wildlife all over the world, with headquarters in Switzerland and national appeals in various countries. President of the British National Appeal is H.R.H. Prince Philip.

Worm. A vernacular name for any elongated, cylindrical, legless animal, formerly used even for snakes, and still applying to a legless lizard, the slow-worm, though now 'worm' by itself most often refers to earthworms. Zoologists in their early days lumped all worm-like animals together as Vermes, but today they are classified into several quite distinct phyla, of which the most important are the Platyhelminthes (flatworms), the Nemertina (ribbon worms), the Aschelminthes (including the roundworms), and the Annelida (segmented worms, including earthworms). Smaller worm-like phyla include the Acanthocephala, Chaetognatha, Echiuroidea, Phoronida, Pogonophora and Sipunculoidea, while even the chordate enteropneusts are known as acorn worms.

Worm Cast. Some earthworms pass their food of decaying vegetable matter through their bodies, evacuating it on the surface as 'worm casts' consisting of soil-like organic refuse. Lugworms take in sand or mud, absorb the organic matter and reject the remainder as sandy casts on the intertidal sand or mud.

Wort. Old term for Herb (1).

Woundwort (Labiatae: *Stachys*). Five or six species with reddish- or pinkish-purple flowers, named from their leaves being formerly used to dress wounds; the commonest are Hedge Woundwort *S. sylvatica* and Marsh Woundwort *S. palustris*.

Wrack. (1) or Tide Wrack. Fragments of seaweed cast up on the shore by the tide. (2) or Sea Wrack (Phaeophyta: Fucaceae). Large to medium-sized brown seaweeds, often abundant on rocky shores, including Bladder Wrack *Fucus vesiculosus* and Egg or Knotted Wrack *Ascophyllum nodosum*, both with conspicuous air-bladders, Channel Wrack *Pelvetia canaliculata* growing sometimes even above high water mark, Flat Wrack *F. spiralis* and Saw Wrack *F. serratus*. When cast ashore they are used as manure in western Scotland and western Ireland, and sometimes also burnt for their ash like oarweeds.

Wrasse (Percomorphi: Labridae). A family of brilliantly coloured but relatively inedible marine fish. Seven species, the commonest being the Ballan Wrasse *Labrus bergylta*, Striped or Cuckoo Wrasse *L. mixtus* and Baillon's Wrasse *Crenilabrus melops*, the last also known as Corkwing, Gilt-head and Sea Partridge. ⇨⇨ GOLD-SINNY, ROCK COOK.

Wreck Fish. The Stone Bass, ⇨ PERCH, SEA.

Wren (Passeriformes: *Troglodytes troglodytes*). The second smallest British song bird, noted for its loud, vigorous song, and habits of cocking up its tail and the making of several domed nests by the cock, only one being eventually selected by the hen; widespread and common. The St Kilda Wren is a subspecies confined to the islands of the St Kilda group. Fire-crested Wren, ⇨ FIRECREST. Golden-crested Wren, ⇨ GOLDCREST. Willow Wren, ⇨ Willow WARBLER. Wood Wren, ⇨ Wood WARBLER.

Wryneck (Piciformes: *Jynx torquilla*). A small brown woodpecker, named from its ability to turn its head right round to look backwards. Summer visitor, formerly widespread in England and Wales, but now rare as a breeder and confined to south-east England, especially Kent; more frequent as a passage migrant, especially in autumn.

Wyke. A common term for a coastal creek in the north of England, especially Yorkshire.

X

Xanthism or Xanthochroism. A colour variation of animals in which the normal colourings are replaced by or suffused with yellow.

Xeromorphic. Adapted to withstand dry conditions.

Xerophyte. A plant adapted to growing in dry conditions, e.g. by reduction of transpiration, water storage, deep root systems, etc.

Xerosere. The natural succession of vegetation on dry soils.

Y

Yaffle. ⇨ Green WOODPECKER.

Yarrow (Compositae: *Achillea millefolium*). An abundant aromatic grassland plant, with flat umbel-like heads of small white or pink daisy-like flowers and finely cut leaf-segments.

Yavering Bells or Serrated Wintergreen (Pyrolaceae: *Orthilia secunda*). A small green-flowered relative of the wintergreens, now mainly in Highland pinewoods.

Yeast (Ascomycetes: Saccharomycetes). Minute, mostly one-celled fungi, which turn sugar into carbon dioxide and water if they are well supplied with oxygen and into carbon dioxide and alcohol if they are starved of oxygen. The latter process is known as fermentation, and is used in bread baking, brewing, and the production of wine and industrial alcohol.

Yellow Archangel or Yellow Dead-nettle (Labiatae: *Galeobdolon luteum*). A woodland plant with creamy-yellow flowers, related to the dead-nettles. The white dead-nettle is also called White Archangel.

Yellow-Cress (Cruciferae: *Rorippa*). Three yellow-flowered relatives of watercress: Creeping Yellow-cress *R. sylvestris* and Marsh Yellow-cress *R. islandica*, of bare ground by fresh water, and the taller Greater Yellow-cress *R. amphibia*, among waterside vegetation.

Yellow-eyed Grass (Iridaceae: *Sisyrinchium californicum*). Very rare yellow-flowered relative of blue-eyed grass, confined to one meadow in south-east Ireland.

Yellow Flag (Iridaceae: *Iris pseudacorus*). Our commonest native iris, with yellow flowers, widespread in marshes and by fresh water.

Yellowhammer or Yellow Bunting (Passeriformes: *Emberiza citrinella*). The commonest British bunting, widespread in hedgerows and bushy places; noted for its song, said to suggest the phrase, 'a little bit of bread and no cheese'.

Yellowlegs (Charadriiformes: *Tringa*). North American waders with yellow legs, the Greater Yellowlegs *T. melanoleuca* and the Lesser Yellowlegs *T. flavipes*, both rare stragglers to Europe on migration.

Yellow Rattle (Scrophulariaceae: *Rhinanthus*). Hemiparasite with two-lipped yellow flowers, named from the way its ripe seeds rattle in the inflated calyx; widespread and common in grassy places. ⇨ RED RATTLE.

Yellow Sally (Plecoptera: *Isoperla grammatica*). A widespread small yellow stone-fly, mainly in the lowlands, common on some chalk and limestone streams, emerging April–August. The Small Yellow Sally *Chloroperla torrentium* is its upland counterpart.

Yellowshank. Outmoded term for the American waders now known by their American name Yellowlegs.

Yellow-Stainer. ⇨ Yellow-staining MUSHROOM.

Yellow-Tail Moth (Caradrinoidea: *Euproctis similis*). A common tussock moth, with pure white wings and a tuft of yellow hairs at the tip of the abdomen, whose brightly coloured red, black and white caterpillar is known as the palmer worm.

Yellow Underwing Moth (Caradrinoidea: Caradrinidae). A rather heterogeneous assemblage of noctuids whose hind wings are bright yellow with a black border, the most familiar being the Large Yellow Underwing *Noctua pronuba*, which readily flies when disturbed from its daytime hiding place.

Yellow-Wort or Yellow Centaury (Gentianaceae: *Blackstonia perfoliata*). A yellow-flowered relative of the centauries, locally frequent in calcareous grassland or on dunes, except in Scotland. The related Slender Yellow-wort *Cicendia filiformis* is scarce on damp heaths in the south and west.

Yew (Taxaceae: *Taxus baccata*). A native evergreen coniferous tree, forming woods in a few places on the southern chalk, and frequent elsewhere, especially in calcareous soils; also widely planted. Its fruit is a red berry, technically known as an aril. The yew is highly poisonous to cattle. The Irish Yew of cultivation derives from two wild fastigiate plants found in Co. Fermanagh about 1767.

Yorkshire Fog (Gramineae: *Holcus lanatus*). A common softly downy grass, closely related to soft grass.

Young Bird. A young bird is a *Nestling*, *Chick* or *Pullus* while still in the downy stage, a *Juvenile* after its first true feathers have grown, and an *Immature* after its first moult.

Z

Zeugloptera (Arthropoda: Insecta). A small order of moth-like insects, containing fourteen species in one family, sometimes classed as a sub-order of the Lepidoptera.
Zoea. The first stage of shrimps and prawns after hatching from the egg.
Zoogeography. The study of the geographical distribution of animals.
Zooid or Polyp. The attached stage of the marine animals formerly included in the Coelenterata, often looking remarkably like a flower, even to the extent of being attached to a common stalk with other zooids.
Zoological Society of London. The national society for the study of zoology and animal physiology, founded in 1826; maintains the London Zoo in Regent's Park, and a zoological park at Whipsnade, Bedfordshire; publishes *Transactions* and the *Journal of Zoology* (formerly *Proceedings*).
Zoology. The scientific study of animals.
Zoophyte. An outmoded term for the colonial marine animals which resemble plants, the sea-firs and sea-mats, now divided respectively into the quite distinct phyla Cnidaria (Hydrozoa) and Bryozoa or Polyzoa.
Zooplankton. Animals in the plankton.
Zygodactyle (Of birds) with two toes pointing forwards and two backwards, as in the woodpeckers.
Zygomorphic. Bilaterally symmetrical, but only in one plane.

INDEX OF SPECIES

Index of Species

Index of Species

Index of Species

Index of Species

Index of Species

Index of Species

Index of Species

Index of Species

Index of Species

Index of Species

Index of Species

Index of Species

Index of Species

Index of Species

Index of Species

Index of Species

Index of Species

Index of Species

Index of Species

Index of Species